Processing Foods to Design Structures for Optimal Functionality

Processing Foods to Design Structures for Optimal Functionality

Editors

Alejandra Acevedo-Fani

Harjinder Singh

MDPI • Basel • Beijing • Wuhan • Barcelona • Belgrade • Manchester • Tokyo • Cluj • Tianjin

Editors
Alejandra Acevedo-Fani
Riddet Institute
Massey University
Palmerston North
New Zealand

Harjinder Singh
Riddet Institute
Massey University
Palmerston North
New Zealand

Editorial Office
MDPI
St. Alban-Anlage 66
4052 Basel, Switzerland

This is a reprint of articles from the Special Issue published online in the open access journal *Foods* (ISSN 2304-8158) (available at: www.mdpi.com/journal/foods/special_issues/Design_Structures).

For citation purposes, cite each article independently as indicated on the article page online and as indicated below:

LastName, A.A.; LastName, B.B.; LastName, C.C. Article Title. *Journal Name* **Year**, *Volume Number*, Page Range.

ISBN 978-3-0365-4390-1 (Hbk)
ISBN 978-3-0365-4389-5 (PDF)

Contents

About the Editors . . . . **vii**

Preface to "Processing Foods to Design Structures for Optimal Functionality" . . . . **ix**

Tianxiao Wang, Lovedeep Kaur, Yasufumi Furuhata, Hiroaki Aoyama and Jaspreet Singh
3D Printing of Textured Soft Hybrid Meat Analogues
Reprinted from: *Foods* **2022**, *11*, 478, doi:10.3390/foods11030478 . . . . **1**

Keying Yang, Ruoting Xu, Xiyu Xu and Qing Guo
Role of Flaxseed Gum and Whey Protein Microparticles in Formulating Low-Fat Model
Mayonnaises
Reprinted from: *Foods* **2022**, *11*, 282, doi:10.3390/foods11030282 . . . . **19**

Qi Wang, Yang Zhu, Zhichao Ji and Jianshe Chen
Lubrication and Sensory Properties of Emulsion Systems and Effects of Droplet Size
Distribution
Reprinted from: *Foods* **2021**, *10*, 3024, doi:10.3390/foods10123024 . . . . **35**

Francesca Louise Garcia, Sihan Ma, Anant Dave and Alejandra Acevedo-Fani
Structural and Physicochemical Characteristics of Oil Bodies from Hemp Seeds (*Cannabis sativa*
L.)
Reprinted from: *Foods* **2021**, *10*, 2930, doi:10.3390/foods10122930 . . . . **53**

**Xiao Chen, Senghak Chhun, Jiqian Xiang, Pipat Tangjaidee, Yaoyao Peng and Siew Young
Quek**
Microencapsulation of *Cyclocarya paliurus* (Batal.) Iljinskaja Extracts: A Promising Technique to
Protect Phenolic Compounds and Antioxidant Capacities
Reprinted from: *Foods* **2021**, *10*, 2910, doi:10.3390/foods10122910 . . . . **69**

Ran Feng, Søren K. Lillevang and Lilia Ahrné
Effect of Water Temperature and Time during Heating on Mass Loss and Rheology of Cheese
Curds
Reprinted from: *Foods* **2021**, *10*, 2881, doi:10.3390/foods10112881 . . . . **87**

Marbie Alpos, Sze Ying Leong, Veronica Liesaputra, Candace E. Martin and Indrawati Oey
Understanding In Vivo Mastication Behaviour and In Vitro Starch and Protein Digestibility of
Pulsed Electric Field-Treated Black Beans after Cooking
Reprinted from: *Foods* **2021**, *10*, 2540, doi:10.3390/foods10112540 . . . . **101**

**Cristina Gabriela Burca-Busaga, Noelia Betoret, Lucía Seguí, Jorge García-Hernández,
Manuel Hernández and Cristina Barrera**
[-25]Antioxidants Bioaccessibility and *Lactobacillus salivarius* (CECT 4063) Survival Following
the In Vitro Digestion of Vacuum Impregnated Apple Slices: Effect of the Drying Technique, the
Addition of Trehalose, and High-Pressure Homogenization
Reprinted from: *Foods* **2021**, *10*, 2155, doi:10.3390/foods10092155 . . . . **129**

Zheng Pan, Aiqian Ye, Siqi Li, Anant Dave, Karl Fraser and Harjinder Singh
Dynamic In Vitro Gastric Digestion of Sheep Milk: Influence of Homogenization and Heat
Treatment
Reprinted from: *Foods* **2021**, *10*, 1938, doi:10.3390/foods10081938 . . . . **145**

Louise Krebs, Amélie Bérubé, Jean Iung, Alice Marciniak, Sylvie L. Turgeon and Guillaume Brisson
Impact of Ultra-High-Pressure Homogenization of Buttermilk for the Production of Yogurt
Reprinted from: *Foods* **2021**, *10*, 1757, doi:10.3390/foods10081757 . . . . **163**

Aislinn M. Richardson, Andrey A. Tyuftin, Kieran N. Kilcawley, Eimear Gallagher, Maurice G. O'Sullivan and Joseph P. Kerry
The Application of Pureed Butter Beans and a Combination of Inulin and Rebaudioside A for the Replacement of Fat and Sucrose in Sponge Cake: Sensory and Physicochemical Analysis
Reprinted from: *Foods* **2021**, *10*, 254, doi:10.3390/foods10020254 . . . . **179**

About the Editors

Alejandra Acevedo-Fani

Dr Alejandra Acevedo-Fani is a Research Officer at the Riddet Institute, Massey University, New Zealand. She is also a Marie-Curie Fellow at the International Iberian Nanotechnology Laboratory in Portugal jointly with her role at the Riddet Institute. Her research interests include: relationships between the food structure, digestion and release of nutrients and bioactive compounds; encapsulation of active food ingredients; plant-based foods strategies to develop future food systems with better nutrition value.

Harjinder Singh

Harjinder Singh is Distinguished Professor, and the Director of the Riddet Institute (Centre of Research Excellence), Massey University, New Zealand. His research expertise includes food proteins, food colloids, and food structures and digestion interface. He has published over 400 research papers in international journals. He is a Fellow of the Royal Society of New Zealand and a Fellow of the International Academy of Food Science and Technology, and has received several international awards, including the New Zealand Prime Minister's Science Prize.

Preface to "Processing Foods to Design Structures for Optimal Functionality"

Foods are designed by nature through complex assembly of macromolecules that interact to form hierarchical structures. These structures are transformed by processing. Traditional processing of foods is designed to preserve, destroy, and re-create food structures to develop desirable food products and functional food ingredients. For example, whole foods, such as meat, fish, nuts, fruits and vegetables, undergo minimal processing in order to preserve the natural structures. On the other hand, the controlled destruction of natural structures results in a variety of ingredients, including starches, flours, oils, and protein isolates. These ingredients are then reassembled and transformed into various fabricated foods. Interactions of food components and creation of an appropriate structural characteristics are achieved through the processing operations, such as thermal treatment, homogenisation, shear, as well as microbiological or biochemical transformations. In recent years, these approaches have led to the concept of "food structure design" as a new way of guiding food product development, particularly in relation to developing foods with reduced levels of fat, salt, and sugar. Also efforts have been made in restructuring plant-based materials to develop meat and dairy analogues with acceptable product functionality.

In this Special Issue, we have selected high quality papers that deal with understanding of the impact of food processing and formulation on interactions and structures of food components and how these structures/interactions influnce the functionality and acceptability of food products.

Alejandra Acevedo-Fani and Harjinder Singh
Editors

foods

Article

3D Printing of Textured Soft Hybrid Meat Analogues

Tianxiao Wang [1,2], Lovedeep Kaur [1,2,*], Yasufumi Furuhata [3], Hiroaki Aoyama [3] and Jaspreet Singh [1,2,*]

[1] School of Food and Advanced Technology, Massey University, Palmerston North 4442, New Zealand
[2] Riddet Institute, Palmerston North 4442, New Zealand
[3] Ajinomoto Co., Inc., Suzuki-cho 3-1, Kawasaki-ku, Kawasaki-shi 210-0801, Japan
*   Correspondence: l.kaur@massey.ac.nz (L.K.); j.x.singh@massey.ac.nz (J.S.)

Abstract: Meat analogue is a food product mainly made of plant proteins. It is considered to be a sustainable food and has gained a lot of interest in recent years. Hybrid meat is a next generation meat analogue prepared by the co-processing of both plant and animal protein ingredients at different ratios and is considered to be nutritionally superior to the currently available plant-only meat analogues. Three-dimensional (3D) printing technology is becoming increasingly popular in food processing. Three-dimensional food printing involves the modification of food structures, which leads to the creation of soft food. Currently, there is no available research on 3D printing of meat analogues. This study was carried out to create plant and animal protein-based formulations for 3D printing of hybrid meat analogues with soft textures. Pea protein isolate (PPI) and chicken mince were selected as the main plant protein and meat sources, respectively, for 3D printing tests. Then, rheology and forward extrusion tests were carried out on these selected samples to obtain a basic understanding of their potential printability. Afterwards, extrusion-based 3D printing was conducted to print a 3D chicken nugget shape. The addition of 20% chicken mince paste to PPI based paste achieved better printability and fibre structure.

Keywords: food 3D printing; hybrid meat analogues; rheological properties; pea protein isolate; chicken

Citation: Wang, T.; Kaur, L.; Furuhata, Y.; Aoyama, H.; Singh, J. 3D Printing of Textured Soft Hybrid Meat Analogues. *Foods* **2022**, *11*, 478. https://doi.org/10.3390/foods11030478

Academic Editor: Pierre Sylvain Mirade

Received: 27 November 2021
Accepted: 28 January 2022
Published: 6 February 2022

Publisher's Note: MDPI stays neutral with regard to jurisdictional claims in published maps and institutional affiliations.

1. Introduction

Meat-based products are popular among consumers due to their unique taste, texture and nutritional values [1]. With the development of technology, meat is becoming more accessible to humans' daily diets. According to predictions, the demand of meat will continue to grow in the future [2]. However, growing meat consumption is associated with increasing environmental concerns, including large land use and green gas emission [3–5]. To ensure sustainability, alternative diets or food sources have been suggested to decrease the average individual meat consumption [6]. Meat analogue is a type of food considered as a replacer that mimics characteristics of meat.

Traditional meat analogues have been created and developed for many centuries. These meat analogues were generally made by vegetables or plants rich in protein, to deal with low meat accessibility or due to religious reasons in different parts of the world [7–9]. Due to the limitations of the traditional processing technique, traditional meat analogues could not properly simulate the sensory characteristics and texture of meat. Therefore, scientists have started researching new methods to improve the quality of meat analogues [9,10].

Currently, the most common technology producing meat analogue is high-moisture extrusion. It can be used on various plant protein sources and produce different kinds of meat analogues [9,11,12]. The percentages of moisture in high-moisture extrusion normally vary from 40 to 80% [13]. The extruded products tend to show higher similarity to meat, compared with conventional meat analogues. Other novel technologies such as shearing, and spinning have also been developed to further imitate the meat-like fibres and microstructure [14–16]. However, these methods have not been widely promoted on an industrial scale. Aside from its meat-like structure, the nutritional value and other physical sensations of meat analogue also need to be improved.

Consumers' preference plays an important role in the commercialisation of meat analogues. Currently, the challenges of developing meat analogues mainly include lower nutrition quality of plant proteins, lack of meaty sensations and high price [17,18]. To improve the nutritional and sensorial characteristics of meat analogues, one option is to add low value animal proteins to formulate a hybrid meat analogue. Animal proteins are generally known to be nutritionally superior to plant proteins.

Additionally, it is worthwhile to develop some meat analogues, which are suitable for the elderly. There are increasing numbers of elderly that need to be fed. However, the decreasing tooth strength with aging limits the elderly's' food choices [19].

Three-dimensional printing is a novel technology, which could be introduced into food manufacturing to modify food structure and texture [20]. In addition, it creates desirable food shapes and improves nutrient selection [21]. Because of these, it has been used to produce some health care food products for the elderly. Aged people would be benefited because they have more options, instead of simply consuming conventional pureed food. Scientists have already shown their interest in producing meat analogues from plant proteins through 3D printing. Some printed plant-based meat products with different shapes have been reported [22–24]. However, there is no available literature on the evaluation of properties of printed meat analogue. The printing of hybrid meat analogue has also not yet been mentioned by currently available research.

Conventionally, soy is the most common material to produce meat analogues. However, pea has fewer food allergy issues and GMO concerns than soy [25]. The interest in research on pea protein has increased in recent years, especially for simulating chicken products [26,27]. Hence, pea protein and chicken were selected as the main materials to produce printed soft hybrid meats.

The main objectives of this study were to: (1) Develop a printable formulation mainly made from pea protein isolate and chicken mince paste. (2) Optimize the formulation to printing process based on rheology and extrusion tests.

2. Materials and Methods

2.1. Materials

Pea protein isolate (PPI) (containing 80% protein) was purchased from Davis Food Ingredients (Palmerston North, New Zealand). Commercially available chicken mince (typically 95% meat and 5% fat) was purchased from a local market (Palmerston North, New Zealand). It was then blended with a Moulinex Food blender (Masterchef 650, Groupe Seb New Zealand, Auckland, New Zealand) and finely minced by a Silverson L4RT High shear mixer (Advanced Packaging System Limited, Auckland, New Zealand) as much as possible into a paste-like texture.

Other materials included pre-gelatinized maize starch (Hi-Maize 1043; Ingredion ANZ Pty Ltd., Auckland, New Zealand), beef fat (Premium 100% pure beef dripping, Farmland foods, purchased from a local market, Palmerston North, New Zealand) and soy lecithin (Hawkins Watts Ltd., Auckland, New Zealand).

2.2. Sample Preparation

The formulations of PPI based pastes and PPI-chicken pastes (Table 1) were finalized based on preliminary extrusion trials. Soy lecithin was mixed and dispersed in water (69%) with the help of a MicroMix stick blender (Robot Coupe, 220W, Robot-Coupe Australia Pty Ltd., Auckland, New Zealand). Then, water containing soy lecithin, PPI (24%), maize starch (3.6%) and beef fat (2.4%) were added and mixed in a Moulinex food blender (Masterchef 650). All the ingredients were blended for 2 min to prepare a paste. The blended paste was transferred into metal beakers. The metal beaker containing the paste was placed in a pot containing boiling water and heated on a hotplate (MR 3001, Heidolph, Schwabach, Germany) for 10 min. The samples were further mixed during heating, using a Silverson L4RT high shear mixer (Advanced Packaging System Ltd., New Zealand) at 4000 rpm (556× *g*). The cooked paste was naturally cooled down to room temperature and used for further experimentation. Fresh pastes were prepared before each experiment.

Table 1. Pea protein isolate (PPI) based paste and PPI-chicken paste (PCP) formulations.

Ingredients	Percentage (% *w/w*, Wet Basis)					
	P Control [1]	PS [1]	PF [1]	PSF [1]	20CHK [1]	50CHK [1]
PPI	30	26.4	27.6	24	19.2	12
Starch	0	3.6	0	3.6	3.6	3.6
Fat	0	0	2.4	2.4	2.4	2.4
Chicken paste	0	0	0	0	19.2	48
Soy lecithin	1	1	1	1	1	1
Water	69	69	69	69	54.6 [2]	33 [2]

[1] P control represents pea protein isolate paste, PS represents PPI paste with starch, PF represents PPI paste with fat, PSF represents PPI paste with both starch and fat. 20CHK represents 20% chicken added into PSF paste; 50CHK represents 50% chicken added into PSF pastes. [2] The amount of chicken paste and water are based on moisture content analysis of the raw chicken pastes. The moisture of the chicken paste was 75.30 ± 0.23% measured by hot air oven method. The total moisture of these PCP samples was kept as 69%, which is similar to PPI pastes.

In PPI-chicken pastes (PCP) samples, the amount of added starch, fat, soy lecithin and total moisture was same as the PPI paste. The total dry matter in the chicken paste was aimed to replace PPI powder by 20 and 50%. When PPI paste was prepared, it was mixed with a certain amount of raw chicken paste, as shown in Table 1, using a Moulinex food blender (Masterchef 650) for 2 min.

2.3. Rheological Properties

Rheological properties of PPI and PCP pastes were studied according to the methodology described by Wang et al. [28] with slight modifications, using a dynamic rheometer (AR-G2, TA Instruments, New Castle, DE, USA). Temperature and frequency sweep experiments were performed on raw and cooked pastes, respectively. Steady shear viscosity tests were performed on cooked PPI pastes. A 40 mm parallel steel plate geometry was chosen to test of all samples, and a gap of 2 mm was set between two plates. The strain was set as 0.4%, which ensured all samples were in their linear viscoelastic region. The measurements on each selected sample were conducted in triplicate. Data were collected and analysed by TA software (TA Universal Analysis Version 4.5A, TA Instruments, New Castle, DE, USA).

2.3.1. Temperature Sweeps

A temperature sweep test aims to find how the viscoelastic properties of experimental samples change with heating and cooling. Four different formulations were prepared as shown in Table 1: PPI control, PPI + starch (PS), PPI + fat (PF) and PPI + starch + fat (PSF). The preparation method was the same as given in Section 2.2. For chicken paste added samples, raw chicken pastes were blended with uncooked PSF paste in the proportions of 20 and 50% (Table 1).

Pastes were loaded on the rheometer plate. The temperature was set at 25 °C at the beginning of the test and heated until 95 °C at the rate of 4 °C/min. After holding for 30 s at 95 °C, samples were cooled down from 95 to 25 °C at the rate of 4 °C/min. At the end of the test, cooled samples were held for 30 s at 25 °C. Storage modulus (G'), Loss modulus (G'') and *tan δ* at were recorded. A little amount of mineral oil (Bio-Rad Laboratories, Rosedale, New Zealand) was applied to the sample edges to minimize the moisture loss.

2.3.2. Frequency Sweeps

P control, PS, PF and PSF pastes were prepared as described in Section 2.2 and cooked in a boiling water for 10 min. PSF pastes with 20 and 50% raw chicken paste (20CHK and 50CHK) were prepared in the same manner as described in Section 2.2. Chicken paste was not cooked because denaturation would decrease the flow ability of samples. Viscoelastic parameters such as G', G'' and *tan δ* were determined at 25 ± 0.1 °C, with angular frequency increasing from 0.1 to 100 rad/s. Ten points were recorded within each decade.

2.3.3. Shear Flow Behaviour Tests

Samples used in shear flow behaviour tests were the same as mentioned in the frequency sweep tests. Tests were carried out at 25 ± 0.1 °C, while the shear rate was ramped from 0.1 to 100 s^{-1}. Shear-viscosity curves were obtained after testing, with 10 points shown within each decade.

2.4. Forward Extrusion Testing of PPI and Chicken Pastes

Forward extrusion tests were conducted with a textural analyser (TA.XT.plus, Stable Micro Systems, Godalming, UK) with a 50 kg load cell, using the method described by Kim et al. [29] and Zhu et al. [30] with minor modification. A device consisting of a syringe and piston was set up on the texture analyser (Supplementary Materials Figure S1). Four PPI-based pastes, 20CHK, 50CHK and chicken pastes were prepared as described in Section 2.2. They were carefully scooped to fill into a 60 mL polypropylene syringe with a spatula. The syringe was placed vertically on a heavy-duty platform (HDP) with a hole in the centre. The test was conducted by a single compression with a 61 mm cylindrical probe. Compression force-time curve was obtained, and the maximum compression force was defined as the extrusion hardness of each food paste.

The compression speed for the forward extrusion test was calculated based on a range of equations. The relationship between compression distance and paste length is shown in Equation (1).

$$\frac{\pi d_f^2}{4} \times \text{compression distance} = \frac{\pi d_n^2}{4} \times \text{paste length} \tag{1}$$

where d_f is the diameter of the filament, which also equals to the syringe diameter in this study; d_n is diameter of the nozzle. Paste length refers to the length that paste extruded out from the syringe and nozzles.

If the extrusion time is controlled, Equation (1) can be modified to Equation (2), which shows the relationship between compression speed and extrusion speed. If extrusion speed is set, then the compression speed could be calculated by Equation (3).

$$\frac{\pi d_f^2}{4} \times \text{Speed}_c = \frac{\pi d_n^2}{4} \times \text{Speed}_e \tag{2}$$

$$\text{Speed}_c = \text{Speed}_e \times \left(\frac{d_n}{d_f}\right)^2 \tag{3}$$

where Speed_c is the compression speed; Speed_e is the speed at which paste extruded out from the nozzle.

The diameter of syringe used in this study was 28.5 mm. The diameters of two selected nozzles were 1.54 and 2.16 mm. Thus, the compression speed was set as 0.04 mm/s for the 1.54 mm nozzle and 0.09 mm/s for the 2.16 mm nozzle. This referred to the extrusion speed of 15 mm/s for each nozzle size. The compression distance was 5 mm and the average compression force measured and calculated from triplicated observations.

2.5. 3D Printing Process

The printer used in this study was a newly assembled LVE 3D printer (Supplementary Materials Figure S2). It is a combination of a plastic filament 3D printer frame and an extruder unit. The frame of the printer belonged to Creality Ender-3 (Creality 3D, Shenzhen, China). However, the extruder unit was created based on the design from Pusch et al. [31].

In this study, all 3D models for experiments were downloaded from online sources, which are open to public access through 3D builder (Version 18.0.1931.0; Microsoft Co.,

Redmond, WA, USA). Then, they were loaded and sliced by Repetier Host (Version 2.1.4; Hot-World GmbH and Co. KG., Willich, Germany).

PSF, 20CHK and 50CHK were selected for 3D printing based on their rheological properties. The pastes were prepared as described in Section 2.2 and filled into syringes. A small nugget shape sample (Figure 1) was printed using a large volume extrusion (LVE) 3D printer. The effect of nozzle size (1.54 and 2.16 mm) on the printability (ability to form layers) was observed along with the appearance of the printed samples. Printing was carried out at ambient temperature, with a printing speed of 15 mm/s and 100% infill density. Printed samples were cooked in sealed polypropylene bags in a boiling water bath. The structures of both raw and cooked samples were visually evaluated. Fibre formation was particularly noticed.

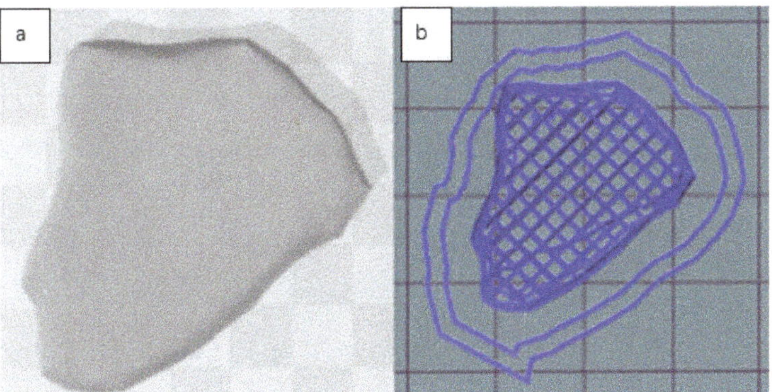

Figure 1. A 3D model of a chicken nugget shape downloaded through 3D builder (**a**) and sliced by Repetier Host (**b**).

2.6. Statistical Analysis

The data presented in the results and discussions are the mean values of triplicated measurements. In forward extrusion tests, standard deviation (SD) is also presented. One-way analysis of variation (ANOVA) and Tukey's pairwise comparisons were conducted by Minitab (version 18.1, Minitab Inc., State College, PA, USA) to analyse the significance of the data. Statistical significance was defined by a *p* value lower than 0.05.

3. Results and Discussion

3.1. Rheology

3.1.1. Temperature Sweeps

PPI Pastes

In temperature sweep tests, the comparison of storage modulus (G'), loss modulus (data not presented) and loss factor (*tan δ*) of four PPI pastes are shown in Figure 2. Both G' and G'' of all samples decreased during the heating process. They dropped down rapidly before 55 °C and then slowly decreased after that. A change in rheological properties in protein-based food is normally associated with protein denaturation. According to Shand et al. [32], the denaturation temperature of non-globulin and globulin fractions in lab-prepared PPI were 67 and 85 °C, respectively. However, there are no obvious changes on moduli at either temperature in Figure 2a. It was reported by Aryee et al. [33] that processing methods vary the characteristics of PPI products. The decreasing trend in moduli during heating is generally similar to the study by Moreno et al. [34], in which PPI with a greater denaturation degree showed a continuous reduction in G' and G'' during heating. This was explained to be the result of the destruction of polar interactions, mainly leading to decreased moduli. Hence, the PPI used in this study is assumed to be already denatured to some extent due to processing. As demonstrated by Jiang et al. [35], denatured protein shows a higher extrudability since most native proteins have poor shape-holding capacity. This could be the reason that PPI paste can be manually extruded in the trials.

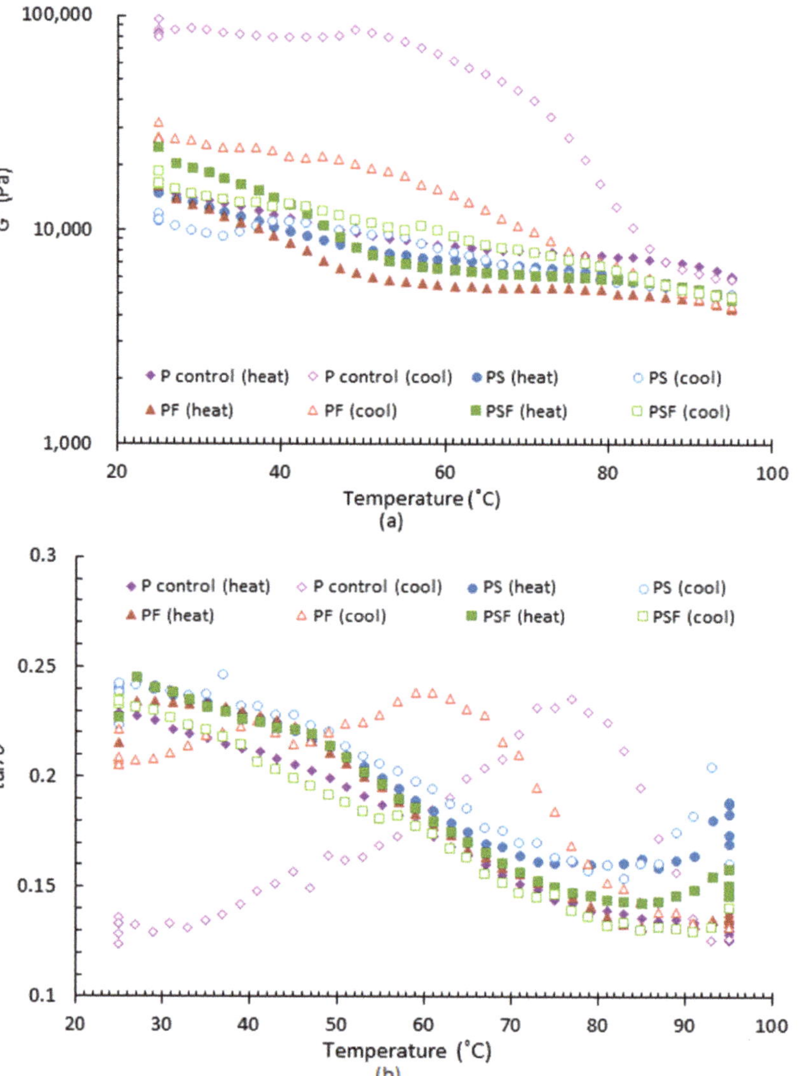

Figure 2. Viscoelastic properties of four PPI pastes during heating (25 to 95 °C) and cooling (95 to 25 °C) at rate of 4 °C/min. (**a**) Storage modulus (*G′*) and (**b**) *tan δ*. In the figure, P control represents PPI paste; PS represents PPI Paste containing starch; PF represents PPI paste containing fat; PSF represents PPI paste containing both starch and fat. Result shown was the mean value of triplicated tests.

During cooling, *G′* and *G″* of the four pastes increased to different extents. A similar phenomenon was reported by Sun and Arntfield [36], who investigated the rheological properties of salt-extracted PPI at different temperatures. In their research, *G′* of all PPI samples increased steeply from 95 to 25 °C. The increasing curves of *G′* differ from this study. This could be caused by the different heating and cooling rates and sample compositions. As reported by Oyinloye and Yoon. [37], a slower cooling rate led to a faster increase in G' and G''. Moreover, different formulations influenced rheological properties. The storage and loss moduli for the samples with starch added (PS and PSF) returned to a slightly lower level than at the beginning of heating cycle. This finding suggested that starch limited the moduli change during heating and cooling. Therefore, cooling of PS and PSF can be considered as a roughly reverse process to heating. However, G' and G'' of P control and PF were higher than the initial values before heating. As can be seen, viscoelastic moduli of P control and PF rose from approximately 90 to 50 °C, while G'' of these two samples increased dramatically from 90 to 70 °C.

Phase changes during heating and cooling did not exist, since *tan δ* of all these samples was lower than 1 at all temperatures (Figure 2b). *Tan δ* of all these samples ranged between 0.1 and 0.25, which means samples were always predominantly elastic [38]. Before the heating process, the initial *tan δ* of all paste samples was between 0.2 and 0.25. During heating, all pastes showed, generally, a decreasing tendency in *tan δ*, indicating that the gel strength is reinforced. The reason was associated with the protein–protein interactions generated from temperature change, which contributed to a more elastic gel network [36]. *Tan δ* of P control and PF was lower than 0.15 at 95 °C. For PS and PSF however, *tan δ* rose slightly when the temperature was higher than 85 °C. During cooling, *tan δ* of PS increased dramatically at the beginning, dropping down afterwards and rising again when the temperature was below 83 °C. As for the PSF sample, *tan δ* steadily increased. The changes in *tan δ* of P control and PF during cooling were more complicated. *Tan δ* of PF increased drastically from 95 to 59 °C and decreased smoothly afterwards. PS, PF and PSF samples did not change hugely during the entire heating and cooling process. *Tan δ* of P control rose rapidly and dropped down after 77 °C to nearly 0.13, which is much lower than the initial *tan δ*. This indicates that the cooling procedure increased the gel firmness of the P control sample. In general, results show that starch added samples were suitable to be heated and cooled before printing. This is because they demonstrate a minor reduction in viscoelastic moduli and a similar *tan δ*, which potentially create a proper paste fluidity for printing [28]. More studies on plant protein-based food are needed to explore the interactions among protein, starch, and fat. A deeper understanding of component interactions helps to design formulations with ideal rheological properties for printing.

PPI-Chicken Paste

For chicken added samples, the change of viscoelasticity is shown in Figure 3. During both heating (50 °C onwards) and cooling, chicken and chicken paste added samples showed an increased G' over that of PPI paste alone. G' of chicken and chicken added samples increased steeply when the temperature was between 60 and 80 °C (Figure 3a). A similar rheology investigation was reported by Rabeler and Feyissa [39], in which the G' of chicken breast showed a rapid increase from 60 to 80 °C. Minor differences in G' values reported in this study could be explained by the different protein contents of the meat samples. As reviewed by Lesiów and Xiong [40], the denaturation of majority of proteins, including myosin and myofibrils in chicken meat occurred between 35 and 40 °C. The denatured proteins started to aggregate and form gels with the continuously increasing temperature. Tornberg [41] indicated that denaturation by heating may have caused meat fibre contraction and connective tissue solubilization. Such changes formed a denser network that enhanced G' of meat during heating.

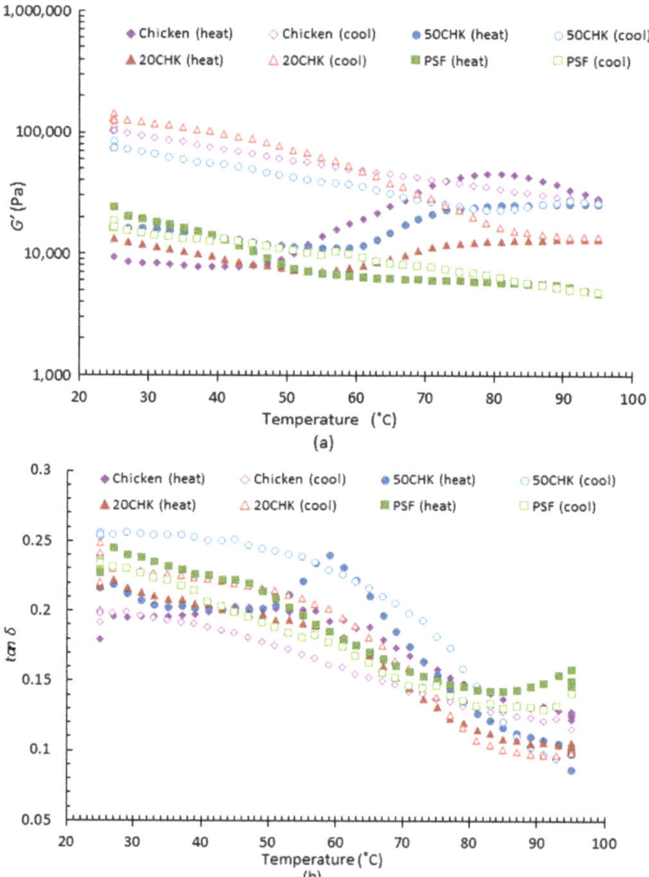

Figure 3. Viscoelastic properties of PPI and chicken mixture samples during heating (25 to 95 °C) and cooling (95 to 25 °C). (**a**) Storage modulus (*G′*) and (**b**) *tan δ*. In the figure, PSF represents PPI paste containing both starch and fat; 20CHK represents 20% chicken added into PSF paste; 50CHK represents 50% chicken added into PSF pastes. The result shown was the mean value of triplicated tests.

The variation of *G″* in the four samples showed a similar trend to *G′* in the entire process (data not shown). However, *G″* of chicken and 50CHK reduced slightly from 75 °C onwards during heating. *G″* of 20CHK had a slight increase during heating after 50 °C. The increase in *G′* and *G″* during heating declined with the increasing amount of chicken paste, indicating that meat protein has more sensitivity to temperature than PPI at a high temperature. Both *G′* and *G″* of chicken and chicken-added samples increased steeply after the heating and cooling process.

Compared with PSF and chicken, the extent of increase in *tan δ* after the entire process for the 20CHK and 50CHK samples were greater. This might demonstrate that interactions between plant and animal proteins simulate the change in *tan δ* during temperature change. Rheological properties of emulsion gel systems containing pea protein and animal protein have been studied by Graca, Raymundo and de Sousa [42]. Their investigation showed that *tan δ* of a food sample containing pea protein and collagen protein in a ratio of 50:50 increased to over 1 from 40 °C during heating. It represented a breakdown of the original emulsion structure. The deformation was maintained until 80 °C when a new gel structure formed (*tan δ* < 1). This trend did not appear in samples containing either a higher amount of pea protein or purely collagen protein. It could explain why only 50CHK exhibits an increasing *tan δ* between 50 and 60 °C in this study (Figure 3b).

3.1.2. Frequency Sweeps

During frequency sweeps, G' and G'' of four PPI pastes both progressively increased with the growing angular frequency (Figure 4). In addition, G' was always higher than G'', indicating that these pastes show a weak gel behaviour [43]. For both G' and G'', PS paste showed the highest value than other pastes during the whole frequency sweep test among all plant protein only samples (Figure 4a; data for G'' not presented). Interestingly, cooked P control and PF pastes did not exhibit a high value for viscoelastic moduli after the heating and cooling cycle, similar to the temperature sweeps (Figure 2). This may be because cooked samples were cooled at a different rate than in the conditions of the temperature sweep. Higher G' and G'' values could be caused by adding starch, which potentially decrease the fluidity [28,44]. In contrast, the addition of fat could enhance fluidity, which was agreed by Lille et al. [45]. This phenomenon may benefit the printing process.

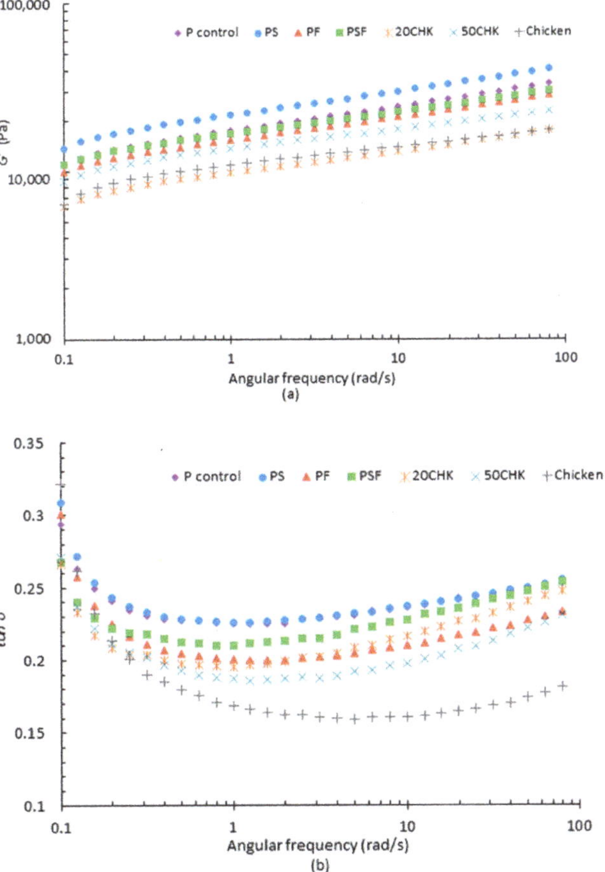

Figure 4. Viscoelastic properties of different PPI pastes (**a**) Storage modulus (G') and (**b**) *tan δ*. In the figure, P control represents PPI paste; PS represents PPI Paste containing starch; PF represents PPI paste containing fat; PSF represents PPI paste containing both starch and fat; 20CHK represents 20% chicken added into PSF paste; 50CHK represents 50% chicken added into PSF paste. Result shown was the mean value of triplicated tests.

Similar to PPI pastes, G' of 20CHK, 50CHK and chicken was higher than G'', and both increased progressively during the frequency sweep. It shows that chicken added samples also behave similar to a weak gel. G' and G'' of cooked PSF pastes were reduced by the addition of raw chicken paste, with 20% chicken paste showing lower values than 50% during the experiment. Raw chicken showed a higher G' than 20CHK, but a lower value than 50CHK (Figure 4a). G'' of 20CHK was lower than chicken when the angular frequency was lower than 0.6 rad/s but surpassed it afterwards (Figure 4b). This demonstrates that the 20CHK sample had the highest fluidity among all three samples, which demonstrates a better flowability [28].

All seven pastes showed a decreasing tendency on *tan δ* below 1 rad/s (Figure 4b). In this stage, *tan δ* of chicken paste showed the sharpest reduction. When the angular frequency is over 1 rad/s, *tan δ* of PPI and chicken added pastes increased slowly. For chicken paste, the *tan δ* value remained stable between 1 and 10 rad/s and rose slightly to a higher angular frequency. Savadkoohi et al. [46] explained that this change was associated with the damage to the gel network. In their research, similar plateaus of *tan δ* of chicken-based samples in a higher angular frequency range (1 to 100 rad/s) was observed. PF expressed the lowest *tan δ* among all PPI pastes after the frequency sweep. *Tan δ* of PSF showed minimal change before and after the whole process.

3.1.3. Shear Flow Behaviour

In this study, all samples were pseudoplastic materials, showing a shear-thinning behaviour (Figure 5). This indicates that all samples are suitable to extrude, which is in agreement with the findings of Lipton [47]. The reduced viscosity caused by applied shear force enables food gels to be extruded from nozzles [35]. Among all PPI pastes, PS showed higher viscosity than the three other pastes (Figure 5). This indicates that starch increases the viscosity, which could be explained on the basis that cooked starch absorbs water and forms an intensive gel structure [48,49]. However, adding both starch and fat did not negatively influence the flow behaviour. This might be because the addition of fat reduced the viscosity, as PSF and PF samples both showed low initial viscosity among all the PPI pastes. A similar finding was reported by Lille et al. [45], namely that a food formulation with semi-skimmed milk powder was less viscous than with skimmed milk powder. In addition, the viscosity of the paste was decreased when the raw chicken paste was added.

At a shear rate higher than $10\ \mathrm{s}^{-1}$, PPI pastes (except the PS sample) become less viscous than chicken-added pastes and chicken. The most drastic reduction existed in the shear-viscosity curve of the P control paste, showing that P control paste is easy to deform when a shear force is added. A slight infinite shear viscosity plateau was shown in PF when the shear rate was above $40\ \mathrm{s}^{-1}$ (Figure 5).

As phase change was not involved during rheological testing in this study, shear-flow behaviour can be an important parameter to evaluate the sample's extrudability. According to Hölzl et al. [50], the viscosity of bio-ink for extrusion-based printing can range from 3×10^{-2} to 6×10^{4} Pa s. In addition, a high zero-shear viscosity was also considered as a suitable property for extrusion type printing [51]. Nevertheless, Wang et al. [28] claimed that food material showed a poor printing performance if the zero-shear viscosity was too high. In their research, a food gel with zero-shear viscosity (viscosity at shear rate $0.1\ \mathrm{s}^{-1}$) around 30,000 Pa s was extrudable but not capable of expressing a proper printing appearance. In this study, the zero-shear viscosity of PS and P control pastes are both close to 30,000 Pa s (25,723 and 22,427 Pa s, respectively), which potentially means these two pastes are not suitable for a smooth 3D printing process.

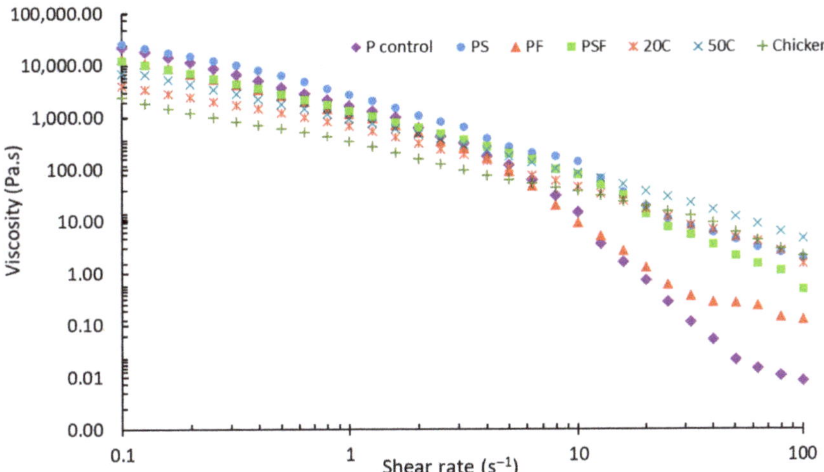

Figure 5. Apparent viscosity of samples at shear rate between 0.1 and 100 s^{-1}. Comparison of shear viscosity curve of PPI pastes, chicken and PPI-chicken pastes. In the figure, P control represents PPI paste; PS represents PPI paste containing starch; PF represents PPI paste containing fat; PSF represents PPI paste containing both starch and fat; 20CHK represents 20% chicken added into PSF paste; 50CHK represents 50% chicken added into PSF paste. The result shown was the mean value of the triplicated tests.

3.2. Forward Extrusion Test

The extrusion force of various samples was measured by a forward extrusion test. As shown in Table 2, paste samples (except PS) showed a lower extrusion force from a 2.16 mm nozzle than a 1.54 mm nozzle. It was also shown by Zhu et al. [30] that tomato puree exhibited lower extrusion stress through a bigger (1.2 mm) nozzle than a smaller (0.8 mm) one. These findings suggest that a bigger nozzle is easier for extrusion. Raw chicken paste was not able to be extruded from a 1.54 mm nozzle. The reason for this is that the presence of big particles constrains the flow, which leads to big variations in different tests. It suggests that mincing at 556× g is not enough to break down some big muscle particles in chicken paste. Mincing at a higher rotation rate or filtering out big muscle particles may be helpful to produce smooth flows. However, it would possibly lead to a low shape-forming capacity during extrusion (as preliminary trials). Hence, chicken paste was not suitable to be directly used for 3D printing in this study. However, it can be added into PPI based pastes since 20CHK and 50CHK samples were able to extrude from both nozzle sizes. The P control paste exhibited the lowest extrusion force, demonstrating that it was easiest to extrude. The reason might be that the viscosity of the P control paste decreased dramatically at a high shear rate (Figure 5). PF showed the highest extrusion force of all the samples (Table 2).

A correlation between extrusion force (or extrusion stress) and rheological properties was shown by Zhu et al. [30]. They pointed out that the extrusion stress of food samples expressed a linear correlation with flow stress. A higher flow stress contributes to higher extrusion stress. However, there was no correlation between extrusion stress and viscoelastic properties. The reason was assumed to be that viscoelastic properties refer to the characteristics of a sample in a non-deforming stage, while flow stress and extrusion stress are both parameters related to deformation. Such correlation was suitable for various water-based food pastes, but not for oil-based food pastes. In addition, extrusion force was related to the materials' printability, which is defined as the capacity of deposited materials to support their own weight [21]. Kim et al. [29] showed that the printability of hydrocolloid samples was positively correlated with the extrusion force. According to their finding, hydrocolloids with a higher methylcellulose concentration exhibited an increasing extrusion force, which simultaneously led to a lower deformation rate of a printed cylinder shape. Nevertheless, this correlation was not suitable for food samples with multiple ingredients. This is because interactions among ingredients creates a complex food matrix, which deserves further investigations. Currently, there is no available research that systematically assesses the relationship between extrusion force and printing performance. It is still necessary to correlate extrusion force and printing experiments, especially for food with numerous ingredients and complex structures.

Table 2. The extrusion force of tested materials with 1.54 and 2.16 mm nozzle sizes.

| Samples [3] | Extrusion Force (N) [1,2] | |
| | Nozzle Size | |
	1.54 mm	2.16 mm
P control	57.74 ± 1.86 [a]	49.91 ± 1.98 [a]
PS	73.47 ± 4.30 [b]	83.64 ± 2.18 [d]
PF	141.10 ± 9.43 [e]	98.80 ± 2.13 [e]
PSF	87.88 ± 3.63 [c]	62.13 ± 2.85 [b]
20CHK	106.31 ± 3.06 [d]	83.22 ± 2.50 [d]
50CHK	84.27 ± 0.85 [b,c]	74.07 ± 0.54 [c]
Chicken	N/A	67.73 ± 1.78 [b]

[1] Results are shown as means ± SD (n = 3). [2] According to Tukey's pairwise comparison, different letters in each column show a significant difference ($p < 0.05$). [3] P control represents PPI paste; PS represents PPI Paste containing starch; PF represents PPI paste containing fat; PSF represents PPI paste containing both starch and fat; 20CHK represents 20% chicken added into PSF paste; 50CHK represents 50% chicken added into PSF past.

3.3. Printing Performance

3.3.1. Appearance of Printed Meat Analogues in Different Formulations and Nozzle Sizes

Printed samples with a small chicken nugget shape are shown in Figure 6. In general, both PSF and chicken paste added samples formed more desirable shapes through a 1.54 mm nozzle, compared with a 2.16 mm. Since PSF, 20CHK and 50CHK all showed lower extrusion force with a 2.16 mm nozzle size, it may indicate that a lower extrusion force results in lower printability. This agrees with the finding of Kim et al. [29]. During printing through a 2.16 mm nozzle, the poor printing performance could be related to the shear rate and viscosity. The relationship between shear rate and viscosity in extrusion printing can be shown by Equations (4)–(6) [52]:

$$\dot{\gamma} = \frac{4Q}{\pi r^3} \tag{4}$$

$$Q = \pi r^2 \times \text{Speed}_e \tag{5}$$

$$\dot{\gamma} = \frac{4 \times \text{Speed}_e}{r} = \frac{8 \times \text{Speed}_e}{d} \tag{6}$$

where $\dot{\gamma}$ is shear rate; Q is the volumetric flow rate, referred to extrusion rate in this study; r is the radius of the nozzle; Speed_e is the extrusion speed; d is the diameter of the nozzle.

Therefore, increasing the nozzle diameter with a controlled printing speed leads to a lower shear rate, resulting in a higher sample viscosity (Figure 5). It is assumed that higher viscosity causes poor extrusion behaviour. Although there is no available research that quantifies the relationship between shear viscosity and printability, similar investigations were reported by Wang et al. [28] showing that fish surimi with a high viscosity would reduce the extrusion smoothness. In this study, the viscous extrusion flow from a 2.16 mm nozzle contributed to an inconsistent deposition line and exhibited a less desirable appearance. Similar findings were reported by Yang et al. [53]. They reported that a bigger nozzle tended to result in poorer printing quality. Wang et al. [28] also tested printing performances through different nozzle diameters. In contrast to this study, however, they found that a printed sample from a smaller nozzle demonstrated poorer printing performance than from a bigger nozzle. The reason might be that printing through a small nozzle demanded a higher pressure, which caused an extremely low viscosity and a low shape-building capacity.

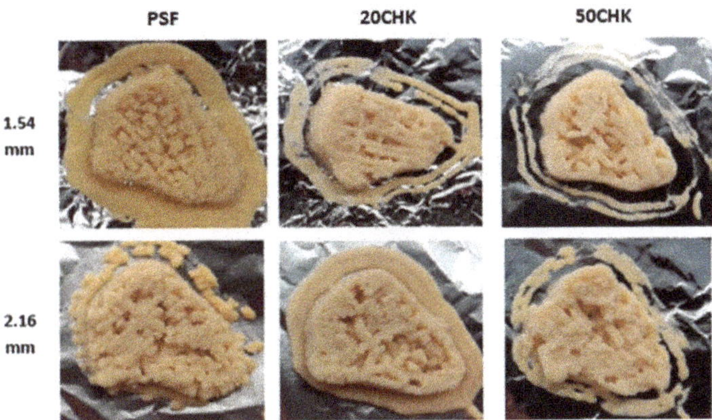

Figure 6. Printed meat analogues using 1.54 and 2.16 mm nozzles (15 mm/s, 100% infill). PSF represents PPI paste containing both starch and fat; 20CHK represents 20% chicken added into PSF paste; 50CHK represents 50% chicken added into PSF paste.

The printing smoothness declined with the increasing amount of chicken paste. Printed 50CHK samples through both nozzle sizes show many printing defects. This is associated with the hypothesis that a difference in smoothness of PSF paste and chicken paste may lead to a heterogeneous mixture and non-continuous flow during extrusion from nozzles. For meat paste samples, Dick et al. [54] recommended the use of nozzle sizes bigger than 2 mm to enable extrusion of some components containing big particles. A similar finding was shown in Section 3.2, namely that big particles in chicken blocked the 1.54 mm nozzle and stopped the extrusion process. Although 50CHK and 20CHK showed extrudability through a 1.54 mm nozzle, varied flow resistance of a chicken portion and a plant protein portion could cause a non-smooth extrusion. PSF paste was able to deposit bottom layers stably. Although printing defects also appeared on the surface layer, the general nugget shape based on the 3D model was formed. The reason could be that PSF paste has a more uniform structure than 50CHK and 20CHK samples.

3.3.2. Appearance and Macrostructure of Printed Meat Analogues after Cooking

The comparison of three cooked samples is presented in Figure 7. Chicken paste added samples showed a more acceptable colour after boiling in water. The shapes of the printed samples were slightly damaged after cooking. Although heat-sealed bags were used to prevent damage of the shape, the edges of the printed samples were not protected perfectly. Lipton et al. [55] tried cooking printed meat in a controlled vapour oven. The overall shape of printed turkey meat was protected from being damaged by the package. Nonetheless, cooking in a controlled vapour oven caused inward shrinkage of the meat, which made the shape bow upwards. To improve the printability of meat analogues, shape-maintenance after cooking is a challenge that needs to be overcome.

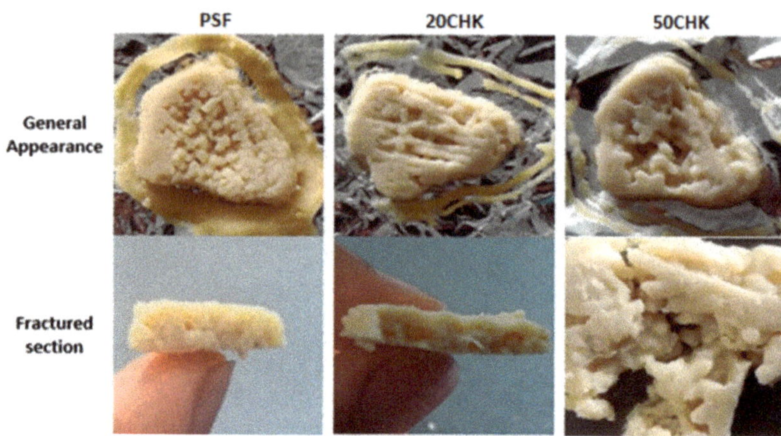

Figure 7. Printed meat analogues using 1.54 mm nozzle after cooking. PSF represents PPI paste containing both starch and fat; 20CHK represents 20% chicken added into PSF paste; 50CHK represents 50% chicken added into PSF paste. Printed samples were cooked in boiling water for 10 min.

Fibrous structure formation is a pivotal characteristic to assess the quality of meat analogue. As found from fractured sections, PSF sample was insufficient to form a fibrous structure without adding chicken (Figure 7). According to previous studies, the fibrous structure of PPI was formed in the thermal extrusion system above 120 °C [25]. In a shearing processing system, the same conditions of temperature were required [56]. To develop a printed product from a PPI based formulation without adding meat, the post-printing cooking method should be changed. This is because the temperature of the boiling water bath (100 °C) was inadequate to build fibrous structures. Pan-frying, microwaving and baking were suggested, since these methods enable cooking samples at a higher temperature. Moreover, the addition of some ingredients such as wheat gluten would help fibre formation [56]. Fibres were found in 20CHK and 50CHK, indicating that meat fibres were generally provided by chicken paste. More fibres were found in 20CHK than the 50CHK sample because of its stable printed shape. This is associated with the proper extrusion and deposition performance of the 20CHK sample. Thus, 20CHK paste is considered as the optimal material for printing in this study.

4. Conclusions

This study investigates the printing of 3D nugget shapes by using plant-only and plant-meat-based formulations. PPI combined with maize starch, beef fat and water expressed suitable rheological properties for extrusion-based 3D printing. Both PPI paste and PPI chicken paste showed a weak gel behaviour according to rheology tests. The addition of maize starch increased the viscosity of food paste, while it minimized the moduli change during temperature sweep. The combination of chicken and PPI-based paste reduced the viscosity change with the increasing shear rate. The addition of raw chicken paste to cooked PPI–based paste was recommended, since it provided a suitable flow behaviour. Forward extrusion tests helped understand the general extrusion difficulties of different samples. However, the correlation between the extrusion force and other characteristics was not identified in this study.

Printing through a 1.54 mm nozzle showed better 3D shape forming capacity than a 2.16 mm nozzle. It was explained that a bigger nozzle size led to a higher shear viscosity, resulting in poor printability. The PPI-paste sample without chicken showed a more desirable appearance, while the addition of chicken paste into PPI paste created a fibre-like structure. Considering both printing performance and fibre formation, PPI paste with 20% chicken was selected as an optimal formulation.

The printing speed in this study was set as 15 mm/s, which is slower than many non-food materials for 3D printing. To scale up and reach the industrial requirements, it is necessary to develop the methods that allow printing of plant protein-based materials at a high speed.

Supplementary Materials: The following are available online at https://www.mdpi.com/article/10.3390/foods11030478/s1, Figure S1. Design of forward extrusion test: A syringe and piston device attached to a texture analyser. (1) The framework of textural analyser (TA.XT.plus, Stable Micro Systems, UK). (2) A 61 mm cylindrical probe. (3) A piston. (4) A syringe. (5) HDP/90 platform with a hole in the centre. (6) A container to collect the paste extruded out from syringe.; Figure S2. Photograph of LVE 3D printer used in this study. (1) Framework of Ender-3 3D printer. (2) Extruder unit. (3) Gears. (4) Motor. (5) Nozzle holder. (6) Platform. (7) Conveyor belts, controlling the movement to x, y, z axis directions. (8) Operation menu. (9) USB connection to computer.

Author Contributions: Conceptualization, L.K. and J.S.; Data curation, T.W.; Formal analysis, T.W. and L.K.; Funding acquisition, L.K. and J.S.; Investigation, T.W.; Methodology, T.W., L.K., Y.F., H.A. and J.S.; Project administration, L.K., Y.F., H.A. and J.S.; Resources, J.S.; Software, T.W.; Supervision, L.K., Y.F., H.A. and J.S.; Validation, J.S., L.K. and T.W.; Visualization, L.K., T.W. and J.S.; Writing—original draft, T.W.; Writing—review & editing, T.W., L.K., Y.F., H.A. and J.S. All authors have read and agreed to the published version of the manuscript.

Funding: The authors would like to acknowledge Massey University (New Zealand), Riddet Institute (NZ) CoRE and Ajinomoto Co. Inc. (Japan) Innovation Alliance Program, for financial support.

Institutional Review Board Statement: Not applicable.

Informed Consent Statement: Not applicable.

Data Availability Statement: We exclude this statement.

Acknowledgments: The authors would like to acknowledge Massey University (NZ), the Riddet Institute (NZ) CoRE, and the Ajinomoto Co. Inc. (Japan) Innovation Alliance Program for financial support. The authors would also like to thank the Foods journal for waiving the Article Processing Charges for this article.

Conflicts of Interest: The authors J. Singh and L. Kaur have received partial funding for the work from Ajinomoto Co. Inc. (Japan). The author T. Wang was the Masters student in the project. The funding sponsors had no role in the collection, analyses, or interpretation of data; in the writing of the manuscript's original draft; and funding sponsors had no influence on the correctness and objectivity of the conclusions of the manuscript.

References

1. Ortega, M.P.; Soto, R. Meat consumption and health. In *Food and Beverage Consumption and Health*; Nova Publishers: New York, NY, USA, 2013.
2. FAO. Animal Production and Health Division: Meat & Meat Products. Available online: https://www.fao.org/ag/againfo/themes/en/meat/home.html (accessed on 17 November 2021).
3. Ilea, R.C. Intensive livestock farming: Global trends, increased environmental concerns, and ethical solutions. *J. Agric. Environ. Ethics* **2009**, *22*, 153–167. [CrossRef]
4. Stehfest, E.; Bouwman, L.; van Vuuren, D.P.; den Elzen, M.G.J.; Eickhout, B.; Kabat, P. Climate benefits of changing diet. *Clim. Chang.* **2009**, *95*, 83–102. [CrossRef]
5. Sabaté, J.; Soret, S. Sustainability of plant-based diets: Back to the future. *Am. J. Clin. Nutr.* **2014**, *100* (Suppl. 1), 476S–482S. [CrossRef] [PubMed]
6. Aiking, H. Future protein supply. *Trends Food Sci. Technol.* **2011**, *22*, 112–120. [CrossRef]
7. Tibbott, S. Food science and technology. In *Handbook of Food and Beverage Fermentation*; Marcel Dekker: New York, NY, USA, 2004.
8. Riaz, M.N. *Soy Applications in Food*; Taylor and Francis: Boca Raton, FL, USA, 2006.
9. Joshi, V.; Kumar, S. Meat analogues: Plant based alternatives to meat products—A review. *Int. J. Food Ferment. Technol.* **2015**, *5*, 107. [CrossRef]
10. Nieuwland, M.; Geerdink, P.; Brier, P.; van den Eijnden, P.; Henket, J.T.M.M.; Langelaan, M.L.P.; Stroeks, N.; van Deventer, H.C.; Martin, A.H. Reprint of "Food-Grade Electrospinning of Proteins". *Innov. Food Sci. Emerg. Technol.* **2014**, *24*, 138–144. [CrossRef]
11. Yao, G.; Liu, K.S.; Hsieh, F. A new method for characterizing fiber formation in meat analogs during high-moisture extrusion. *J. Food Sci.* **2006**, *69*, 303–307. [CrossRef]

12. Liu, K.; Hsieh, F.-H. Protein–Protein interactions during high-moisture extrusion for fibrous meat analogues and comparison of protein solubility methods using different solvent systems. *J. Agric. Food Chem.* **2008**, *56*, 2681–2687. [CrossRef]
13. Beniwal, A.S.; Singh, J.; Kaur, L.; Hardacre, A.; Singh, H. Meat analogs: Protein restructuring during thermomechanical processing. *Compr. Rev. Food Sci. Food Saf.* **2021**, *20*, 1221–1249. [CrossRef]
14. Chen, D.; Jones, O.G.; Campanella, O.H. Plant protein-based fibers: Fabrication, characterization, and potential food applications. *Crit. Rev. Food Sci. Nutr.* **2021**, 1–25. [CrossRef]
15. Manski, J.M.; van der Goot, A.J.; Boom, R.M. Advances in structure formation of anisotropic protein-rich foods through novel processing concepts. *Trends Food Sci. Technol.* **2007**, *18*, 546–557. [CrossRef]
16. Cornet, S.H.V.; Snel, S.J.E.; Schreuders, F.K.G.; van der Sman, R.G.M.; Beyrer, M.; van der Goot, A.J. Thermo-mechanical processing of plant proteins using shear cell and high-moisture extrusion cooking. *Crit. Rev. Food Sci. Nutr.* **2021**, 1–18. [CrossRef] [PubMed]
17. Ministry for Primary Industries of New Zealand. *The Evolution of Plant Protein–Assessing Consumer Response*; Ministry for Primary Industries of New Zealand: Wellington, New Zealeand, 2018.
18. Kyriakopoulou, K.; Dekkers, B.; van der Goot, A.J. Plant-based meat analogues. In *Sustainable Meat Production and Processing*; Elsevier: Amsterdam, The Netherlands, 2019; pp. 103–126. [CrossRef]
19. Cichero, J.A.Y. Adjustment of food textural properties for elderly patients. *J. Texture Stud.* **2016**, *47*, 277–283. [CrossRef]
20. Kouzani, A.Z.; Adams, S.; Whyte, J.D.; Oliver, R.; Hemsley, B.; Palmer, S.; Balandin, S. 3D printing of food for people with swallowing difficulties. In Proceedings of the DesTech 2016: International Conference on Design and Technology, Geelong, Australia, 5–8 December 2016; KnE Engineering: Dubai, United Arab Emirates, 2017; Volume 2017, pp. 23–29. [CrossRef]
21. Godoi, F.C.; Prakash, S.; Bhandari, B.R. 3d printing technologies applied for food design: Status and prospects. *J. Food Eng.* **2016**, *179*, 44–54. [CrossRef]
22. Hitti, N. Novameat Develops 3D-Printed Vegan Steak from Plant-Based Proteins. Available online: https://www.dezeen.com/2018/11/30/novameat-3d-printed-meat-free-steak/ (accessed on 30 November 2018).
23. Plant-Based Meat l NOVAMEAT l Barcelona. Available online: https://www.novameat.com/ (accessed on 17 November 2021).
24. SavorEat. Available online: https://savoreat.com/ (accessed on 17 November 2021).
25. Osen, R.; Toelstede, S.; Wild, F.; Eisner, P.; Schweiggert-Weisz, U. High moisture extrusion cooking of pea protein isolates: Raw Material characteristics, extruder responses, and texture properties. *J. Food Eng.* **2014**, *127*, 67–74. [CrossRef]
26. Plant Based Nuggets l Raised & Rooted. Available online: https://www.raisedandrooted.com/products/plant-based-nuggets/ (accessed on 17 November 2021).
27. Sunfed Bull Free Beef, Boar Free Bacon & Chicken Free Chicken. Available online: https://sunfed.world/ (accessed on 17 November 2021).
28. Wang, L.; Zhang, M.; Bhandari, B.; Yang, C. Investigation on fish surimi gel as promising food material for 3D printing. *J. Food Eng.* **2018**, *220*, 101–108. [CrossRef]
29. Kim, H.W.; Bae, H.; Park, H.J. Classification of the printability of selected food for 3D printing: Development of an assessment method using hydrocolloids as reference material. *J. Food Eng.* **2017**, *215*, 23–32. [CrossRef]
30. Zhu, S.; Stieger, M.A.; van der Goot, A.J.; Schutyser, M.A.I. Extrusion-based 3D printing of food pastes: Correlating rheological properties with printing behaviour. *Innov. Food Sci. Emerg. Technol.* **2019**, *58*, 102214. [CrossRef]
31. Pusch, K.; Hinton, T.J.; Feinberg, A.W. Large volume syringe pump extruder for desktop 3D printers. *HardwareX* **2018**, *3*, 49–61. [CrossRef]
32. Shand, P.J.; Ya, H.; Pietrasik, Z.; Wanasundara, P.K.J.P.D. Physicochemical and textural properties of heat-induced pea protein isolate gels. *Food Chem.* **2007**, *102*, 1119–1130. [CrossRef]
33. Aryee, A.N.A.; Agyei, D.; Udenigwe, C.C. Impact of processing on the chemistry and functionality of food proteins. In *Proteins in Food Processing*; Elsevier: Amsterdam, The Netherlands, 2018; pp. 27–45. [CrossRef]
34. Moreno, H.M.; Domínguez-Timón, F.; Díaz, M.T.; Pedrosa, M.M.; Borderías, A.J.; Tovar, C.A. Evaluation of gels made with different commercial pea protein isolate: Rheological, structural and functional properties. *Food Hydrocoll.* **2020**, *99*, 105375. [CrossRef]
35. Jiang, H.; Zheng, L.; Zou, Y.; Tong, Z.; Han, S.; Wang, S. 3D food printing: Main components selection by considering rheological properties. *Crit. Rev. Food Sci. Nutr.* **2019**, *59*, 2335–2347. [CrossRef] [PubMed]
36. Sun, X.D.; Arntfield, S.D. Gelation properties of salt-extracted pea protein isolate induced by heat treatment: Effect of heating and cooling rate. *Food Chem.* **2011**, *124*, 1011–1016. [CrossRef]
37. Oyinloye, T.M.; Yoon, W.B. Stability of 3D printing using a mixture of pea protein and alginate: Precision and application of additive layer manufacturing simulation approach for stress distribution. *J. Food Eng.* **2021**, *288*, 110127. [CrossRef]
38. Uruakpa, F.O.; Arntfield, S.D. Impact of urea on the microstructure of commercial canola protein–carrageenan network: A research note. *Int. J. Biol. Macromol.* **2006**, *38*, 115–119. [CrossRef]
39. Rabeler, F.; Feyissa, A.H. Kinetic modeling of texture and color changes during thermal treatment of chicken breast meat. *Food Bioprocess Technol.* **2018**, *11*, 1495–1504. [CrossRef]
40. Lesiów, T.; Xiong, Y.L. Mechanism of rheological changes in poultry myofibrillar proteins during gelation. *Avian Poult. Biol. Rev.* **2001**, *12*, 137–149. [CrossRef]
41. Tornberg, E. Effects of heat on meat proteins–Implications on structure and quality of meat products. *Meat Sci.* **2005**, *70*, 493–508. [CrossRef]

42. Graça, C.; Raymundo, A.; de Sousa, I. Rheology changes in oil-in-water emulsions stabilized by a complex system of animal and vegetable proteins induced by thermal processing. *LWT* **2016**, *74*, 263–270. [CrossRef]
43. Lopes da Silva, J.A.; Rao, M.A. Rheological behavior of food gels. In *Rheology of Fluid and Semisolid Foods*; Barbosa-Cánovas, G.V., Ed.; Food Engineering Series; Springer US: Boston, MA, USA, 2007; pp. 339–401. [CrossRef]
44. Sim, S.Y.J.; Moraru, C.I. High-pressure processing of pea protein–starch mixed systems: Effect of starch on structure formation. *J. Food Process. Eng.* **2020**, *43*, 13352. [CrossRef]
45. Lille, M.; Nurmela, A.; Nordlund, E.; Metsä-Kortelainen, S.; Sozer, N. Applicability of protein and fiber-rich food materials in extrusion-based 3D printing. *J. Food Eng.* **2018**, *220*, 20–27. [CrossRef]
46. Savadkoohi, S.; Shamsi, K.; Hoogenkamp, H.; Javadi, A.; Farahnaky, A. Mechanical and gelling properties of comminuted sausages containing chicken MDM. *J. Food Eng.* **2013**, *117*, 255–262. [CrossRef]
47. Lipton, J.I. Printable food: The technology and its application in human health. *Curr. Opin. Biotechnol.* **2017**, *44*, 198–201. [CrossRef]
48. Schirmer, M.; Jekle, M.; Becker, T. Starch gelatinization and its complexity for analysis. *Starch* **2015**, *67*, 30–41. [CrossRef]
49. Liu, Z.; Zhang, M.; Bhandari, B.; Yang, C. Impact of rheological properties of mashed potatoes on 3D printing. *J. Food Eng.* **2018**, *220*, 76–82. [CrossRef]
50. Hölzl, K.; Lin, S.; Tytgat, L.; Van Vlierberghe, S.; Gu, L.; Ovsianikov, A. Bioink properties before, during and after 3D bioprinting. *Biofabrication* **2016**, *8*, 032002. [CrossRef]
51. Faes, M.; Valkenaers, H.; Vogeler, F.; Vleugels, J.; Ferraris, E. Extrusion-based 3D printing of ceramic components. *Procedia CIRP* **2015**, *28*, 76–81. [CrossRef]
52. Kern, C.; Weiss, J.; Hinrichs, J. Additive layer manufacturing of semi-hard model cheese: Effect of calcium levels on thermorheological properties and shear behavior. *J. Food Eng.* **2018**, *235*, 89–97. [CrossRef]
53. Yang, F.; Zhang, M.; Prakash, S.; Liu, Y. Physical properties of 3D printed baking dough as affected by different compositions. *Innov. Food Sci. Emerg. Technol.* **2018**, *49*, 202–210. [CrossRef]
54. Dick, A.; Bhandari, B.; Prakash, S. 3D printing of meat. *Meat Sci.* **2019**, *153*, 35–44. [CrossRef]
55. Lipton, J.; Arnold, D.; Nigl, F.; Lopez, N.; Cohen, D.; Norén, N.; Lipson, H. *Multi-Material Food Printing with Complex Internal Structure Suitable for Conventional Post-Processing*; University of Texas Library: Austin, TX, USA, 2010. [CrossRef]
56. Schreuders, F.K.G.; Dekkers, B.L.; Bodnár, I.; Erni, P.; Boom, R.M.; van der Goot, A.J. Comparing structuring potential of pea and soy protein with gluten for meat analogue preparation. *J. Food Eng.* **2019**, *261*, 32–39. [CrossRef]

MDPI

Article

Role of Flaxseed Gum and Whey Protein Microparticles in Formulating Low-Fat Model Mayonnaises

Keying Yang [†], Ruoting Xu [†] , Xiyu Xu and Qing Guo *

College of Food Science and Nutritional Engineering, China Agricultural University, Beijing 100083, China; s20203061042@cau.edu.cn (K.Y.); xuruoting1996@163.com (R.X.); xuxiyu2021@cau.edu.cn (X.X.)
* Correspondence: qing.guo1115@gmail.com
† These authors contributed equally to this work.

Abstract: Flaxseed gum (FG) and whey protein microparticles (WPMs) were used to substitute fats in model mayonnaises. WPMs were prepared by grinding the heat-set whey protein gel containing 10 mM $CaCl_2$ into small particles (10–20 μm). Then, 3×4 low-fat model mayonnaises were prepared by varying FG (0.3, 0.6, 0.9 wt%) and WPM (0, 8, 16, 24 wt%) concentrations. The effect of the addition of FG and WPMs on rheology, instrumental texture and sensory texture and their correlations were investigated. The results showed that all samples exhibited shear thinning behavior and 'weak gel' properties. Although both FG and WPMs enhanced rheological (e.g., viscosity and storage modulus) and textural properties (e.g., hardness, consistency, adhesiveness, cohesiveness) and kinetic stability, this enhancement was dominated by FG. FG and WPMs affected bulk properties through different mechanisms, (i.e., active filler and entangled polysaccharide networks). Panellists evaluated sensory texture in three stages: extra-oral, intra-oral and after-feel. Likewise, FG dominated sensory texture of model mayonnaises. With increasing FG concentration, sensory scores for creaminess and mouth-coating increased, whereas those of firmness, fluidity and spreadability decreased. Creaminess had a linear negative correlation with firmness, fluidity and spreadability ($R^2 > 0.985$), while it had a linear positive correlation with mouth-coating ($R^2 > 0.97$). A linear positive correlation ($R^2 > 0.975$) was established between creaminess and viscosity at different shear rates/instrumental texture parameters. This study highlights the synergistic role of FG and WPMs in developing low-fat mayonnaises.

Keywords: low-fat mayonnaise; whey protein microparticles; flaxseed gum; texture; rheology

Citation: Yang, K.; Xu, R.; Xu, X.; Guo, Q. Role of Flaxseed Gum and Whey Protein Microparticles in Formulating Low-Fat Model Mayonnaises. *Foods* **2022**, *11*, 282. https://doi.org/10.3390/foods11030282

Academic Editor: Manuel Castillo Zambudio

Received: 24 November 2021
Accepted: 7 January 2022
Published: 21 January 2022

Publisher's Note: MDPI stays neutral with regard to jurisdictional claims in published maps and institutional affiliations.

1. Introduction

Dietary fats significantly contribute to the total energy and intake of lipophilic nutrients and bioactives. However, their excessive intake is an important factor leading to obesity and overweight [1], which has been linked to an increased rate of chronic diseases such as type-2 diabetes, cardiovascular diseases and certain cancers. The reduction in intake of fats is a generally recommended, non-invasive prevention strategy to combat obesity and overweight. However, food cravings and easily accessible, palatable lipid-based foods have made it hard to control their intake levels in certain groups of people.

Mayonnaise is one of the most popular sauces or condiments in the world. It is a mixture of egg yolk, vinegar, oil and spices and typically consists of 70–80% fat [2]. To meet an increasing consumer demand for healthier food products, developing low-fat mayonnaise without sacrificing food texture has emerged as an important task for researchers [2–6]. From the perspective of material science, traditional mayonnaise is a conventional oil-in-water (O/W) emulsion or a high internal phase O/W emulsion (>74% fat) [7]. At high oil concentrations, oil droplets are tightly packed together and become distorted from the spherical shape. The close packing of the droplets allows them to interact with one another and to form the gel-like structures, which imparts mayonnaise its main texture and mouthfeel, e.g., creaminess, thickness and smoothness [8–14]. Oil volume fraction plays a critical role in determining the sensory properties of liquid or semi-solid foods [15]. Oil droplets can effectively reduce friction because of the formation of a fat film following a plate-out mechanism [16]. Reduced friction is highly related to oil-related sensory properties and lubrication properties [14,17,18].

Fat substitution is a traditional approach for fat reduction and provides enhanced sensory properties and higher acceptability of the reduced-fat food products. In the development of these products, researchers employ biopolymers and their complexes to substitute fats in the food. Fat replacers can be categorized into two groups: (1) enhancing lubrication properties; (2) controlling bulk rheology. There have been many efforts to understand the lubrication properties of protein- or polysaccharide-based particles. Tribology research on the food emulsion system has revealed that the bio-based microspheres (e.g., whey protein microparticles) can effectively reduce the friction coefficient between the contacting surfaces via the ball-bearing mechanism, depending on their mechanical properties, surface properties, size and the volume fraction [19–22]. Furthermore, microparticles have successfully simulated sensory attributes of some emulsion-based foods. For example, mayonnaises substituted with the low-methoxyl pectin-based fat mimetics have the similar texture scores as those of full-fat mayonnaise [23]. Whey protein microparticles contribute to creaminess perception of the O/W emulsion due to their lubrication properties [20]. Furthermore, the addition of some biopolymers (e.g., xanthan, gellan and pectin) can significantly influence lubrication properties due to their adsorption onto the surfaces (i.e., forming film) and viscosity effect, subsequently improving lubrication-related sensory properties [24–28]. On the other hand, the bulk properties (e.g., viscosity) are the most important factor, determining many sensory properties of foods, such as firmness, adhesiveness, thickness and creaminess [29]. For example, the extensional viscosity is correlated to a slimy, sticky and mouth-coating perception in xanthan gum solutions; the instrumental viscosity of bovine milk is positively correlated with the sensory viscosity; the thickness perception is correlated to the shear viscosity of thick solutions prepared by xanthan and dextran [30–32]. In practice, biopolymers with high functionalities (e.g., whey proteins, pectin and chia seed mucilage) have been successfully used to replace fats through enhancing bulk properties in different liquid or semi-solid food systems, such as mayonnaises and yogurts [33–37]. However, there is still a lack of a profound understanding of the relationship between bulk rheology at different deformations and sensory properties, which is critical to the development of a low-fat mayonnaise with high sensory acceptability. In particular, the respective role of bio-based microspheres and thickening agents in the bulk rheology and sensory properties of mayonnaises warrants further study.

In this study, we reformulated model mayonnaises using traditional ingredients (e.g., dry egg yolk, corn oil, vinegar, sucrose, salt) and substituted fats using whey protein microparticles (WPMs) of 10–20 μm and flaxseed gum (FG) (i.e., food reformulation), aiming to understand the underlying mechanisms of the effect of the addition of FG and WPMs on bulk properties (i.e., rheology and instrumental texture). Moreover, we investigated the sensory texture of model mayonnaises, including creaminess and creaminess-related textural attributes, aiming to clarify how the addition of FG and WPMs changed the creaminess.

2. Materials and Methods

2.1. Materials

Whey protein isolate (WPI) (Bipro, >91% protein) was purchased from Davisco Foods International (Eden Prairie, MN, USA). $CaCl_2$ (≥97%) and NaCl (≥99%) were purchased from Sigma-Aldrich Merck KGaA (Darmstadt, Germany). FG was purchased from Linseed Biologic Technologies Co., Ltd. (Xinjiang, China) and used without further purification. Nile Red and Fast Green were obtained from Solarbio (Beijing, China). Sucrose, salt, corn oil, dried egg yolk, whole milk, whipped cream, jam, mayonnaise, vinaigrette and vinegar were purchased from a local supermarket (Beijing, China). Double-distilled water was used to prepare all solutions and mayonnaises. All other reagents were obtained from Sinopharm Chemical Reagent Co., Ltd. (Shanghai, China) and were of analytical grade.

2.2. Preparation of Whey Protein Microparticles (WPMs)

A 10 wt% WPI solution was prepared by dissolving 100 g WPI into 900 g water. WPI solutions were stored at 4 °C overnight to allow complete hydration. Then, 10, 15, 20, 25, 35 or 50 mM $CaCl_2$ was added into the WPI solutions. Whey protein gels containing different $CaCl_2$ concentrations were formed by heating the WPI solutions at 90 °C for 30 min and cooling at room temperature. WPMs were prepared by (1) crushing whey protein gels to smaller pieces; (2) adding water into gel pieces with a ratio of 1:9; (3) homogenizing gel pieces at 9000, 12,000, 15,000, 18,000 or 21,000 rpm for 4 min using an Ultra Turrax (S18N-19G, IKA, KG, Staufen, Germany) at room temperature.

2.3. Scanning Electron Microscopy (SEM)

Whey protein gels were vacuum freeze-dried at −83 °C for 48 h. The freeze-dried gels were sputtered with a platinum conductive layer. The microstructure of different samples was captured using a SEM (UHR FE-SEM SU8200 Series, Hitachi, Kyoto, Japan) at an accelerating voltage of 3 kV.

2.4. Particle Size Measurement

The particle size distribution of WPMs was measured using a Particle Size Analyzer (LS 13 320, Beckman Coulter, Indianapolis, IN, USA). The refractive index of water and WPI was 1.33 and 1.48, respectively. The particle size was reported as the volume-weighted mean diameter ($D_{4,3}$).

2.5. Mechanical Properties of Whey Protein Gels

The mechanical properties of the gels containing different $CaCl_2$ concentrations were determined using a texture analyzer (TA.XT2, Stable Micro System Ltd., Haslemere, Surrey, UK). Cylinder samples (20 mm in height and 25 mm in diameter) were compressed between two flat plates up to 30% strain with the test speed of 1 mm s^{-1} and trigger force of 0.1 N. The force at 30% strain was defined as gel hardness. Young's moduli were calculated from the linear regime of the force-strain curves (<5%). The penetration test was performed to a strain of 80% with a probe (6 mm in diameter). The test speed was 1 mm s^{-1}, with a trigger force of 0.08 N. The fracture force and strain of the gels were recorded.

2.6. Swelling Test

To mimic the physicochemical environment of model mayonnaises, the swelling medium was prepared by dissolving 5 wt% sugar and 0.5 wt% NaCl into water, and the pH was adjusted to 4. Cylinder gels (25 mm in diameter and 10 mm in height) were weighed (w_0) and then immersed in the swelling medium for 0, 3, 8, 16, 24, 30 and 54 h. After static soaking, the samples were weighed again (W_n). The swelling ratio was calculated by the following equation:

$$\text{Swelling ratio } (\%) = \frac{w_n - w_0}{w_0} \times 100\% \tag{1}$$

2.7. Preparation of Model Mayonnaises

The ingredients of the model mayonnaises are listed in Table 1. FG and sugar were added into hot water and stirred at 80 °C for 30 min using a magnetic mixer (C-MAG HS 7, IKA, KG, Staufen, Germany) (mixture 1). WPMs, salt and dried egg yolk were added into water, followed by a homogenization at 10,000 rpm for 20 s using an Ultra Turrax (S18N-19G, IKA, KG, Staufen, Germany). Then, corn oil was added and homogenized at 10,000 rpm for 20 s (mixture 2). Finally, model mayonnaises were obtained by blending vinegar, mixture 1 and mixture 2.

Table 1. Ingredients of model mayonnaises.

Number of Sample	FG wt%	WPM wt%	Sucrose wt%	Salt wt%	Vinegar wt%	Corn Oil wt%	Dry Egg Yolk wt%
0.3–0	0.3	0	5	0.5	4	15	4
0.3–8	0.3	8	5	0.5	4	15	4
0.3–16	0.3	16	5	0.5	4	15	4
0.3–24	0.3	24	5	0.5	4	15	4
0.6–0	0.6	0	5	0.5	4	15	4
0.6–8	0.6	8	5	0.5	4	15	4
0.6–16	0.6	16	5	0.5	4	15	4
0.6–24	0.6	24	5	0.5	4	15	4
0.9–0	0.9	0	5	0.5	4	15	4
0.9–8	0.9	8	5	0.5	4	15	4
0.9–16	0.9	16	5	0.5	4	15	4
0.9–24	0.9	24	5	0.5	4	15	4

2.8. Rheological Measurement

The rheological properties of model mayonnaises were investigated using a stress-controlled rheometer (DHR-2, TA Instruments, New Castle, DE, USA) with a serrated parallel plate geometry (40 mm in diameter) and 800 µm gap. A shear rate range of 0.001–1000 s^{-1} at a constant temperature of 25 °C was used to determine the steady shear flow behavior of the samples. Dynamic oscillatory shear testing was performed by means of a strain amplitude sweep (1 Hz, 0.1–50% strain) and a frequency sweep in the linear viscoelastic region (1% strain, 10–0.1 rad/s). The ability of the samples to resist the deformation under the applied strain was recorded as the storage modulus (G'), loss modulus (G'') and phase angle (δ). All experiments were carried out at least in triplicate.

2.9. Instrumental Texture Measurement

The instrumental texture of model mayonnaises was measured by a texture analyzer (TA Instruments, New Castle, DE, USA) with a cylinder probe (36 mm in diameter). Each sample was filled to a 50 mm depth in a beaker with dimensions of 40 mm in diameter and 60 mm in height. The samples were then penetrated to a depth of 30 mm at a speed of 1.5 mm s^{-1} with a trigger force of 3 g. Textural parameters of the samples (i.e., peak force, positive area, negative peak force, negative area) were calculated. All experiments were carried out at least in triplicate.

2.10. Confocal Laser Scanning Microscopy

Microstructure of model mayonnaises was observed using a confocal laser scanning microscope (LSM710, Zeiss, Germany). The oil and protein phases of the samples were stained by Nile Red solution (0.1 wt% in acetone) and Fast Green solution (0.2 wt%), respectively. A small drop of the sample was placed on a concave confocal microscope slide, mixed with 10 mL Nile Red and 10 mL Fast Green, stained for 30 min and then covered with a cover slip. A 488 nm argon laser and a 633 nm He–Ne laser were used to excite the oil and protein phases, respectively. Emission spectra above 505 nm for the oil phase and above 650 nm for the protein phase were collected. It should be noted that model mayonnaises with a high concentration of FG and WPMs were hard to stain. Therefore, high-quality images were acquired for the samples containing 0.3 wt% FG or 0.6 wt% FG + 0 wt% WPMs in this study.

2.11. Stability Test

Fresh model mayonnaises were stored at 25 °C in a glass tube and monitored by a digital camera over 27 days (1, 7, 18, 27 d) to test their storage stability. Model mayonnaises were put into 5 mL test tubes that were sealed with plastic caps and then centrifuged at

4000 rpm for 8 min and at 10,000 rpm for 10 min using a high-speed centrifuge (TG16-WS, Cence, Changsha, China) to test their centrifugation stability.

2.12. Sensory Evaluation

Panellists aged between 18 and 30 years old were recruited and selected based on strict dental criteria as well as their basic texture, taste and flavor detection. Sensory tests were carried out in compliance with the human ethics committee at China Agricultural University.

Training was executed in 4 sessions of 1 h each with 5 common commercial semi-solid foods (i.e., whole milk, whipped cream, jam, mayonnaise, vinaigrette). An attribute list containing preliminary attributes (extra-oral, intra-oral and after-feel) with precise definitions and rating instructions was provided to panellists during the attribute generation by the panel (Table 2). Each panelist was offered a plastic spoon and 20 g of each commercial semi-solid food. Panelists were instructed to rate extra-oral attributes before ingestion of food and intra-oral attributes with the tongue and palate. After-feel attributes were rated after swallowing. The attribute generation and definition were completed after 2 sessions, followed by one more session to reach consensus between panellists on rating the attributes. In the last training session, panellists were asked to rate 5 commercial samples and discuss the results.

Table 2. Textural attributes and definitions.

Stages	Attributes	Definitions
Texture (extra-oral)	Firmness	Degree of resistance when stirring with a spoon
	Fluidity	Degree of the sample flowing on the tongue
Texture (intra-oral)	Spreadability	Degree of spreading the sample into a thin film by moving the tongue up and down against the palate
	Graininess	Extent of perception of particles when the tongue slides along the palate
	Mouth-coating	Degree of coating in the mouth after swallowing
After-feel	Creaminess	Degree to which the sample leaves a soft, velvety, fatty feeling after swallowing. Creaminess is perceived in the whole mouth.

A total of 12 model mayonnaise samples, labeled with random three-digit numbers, were evaluated in individual sensory booths at 25 °C under normal light conditions during 3 sessions of 1 h each. All model mayonnaise samples were prepared 12 h before sensory testing and served in a randomized order to panellists. The panellists rated the mayonnaise samples using a 100 mm continuous line anchored with 'very little' at 0% and 'very much' at 100% of the line scale, on which participants marked an 'X'. Scores were determined by the distance (in mm) from the left starting point of the line to the 'X'. All panellists followed the same evaluation procedure. First, they were instructed to assess textural (extra-oral) attributes, then to eat the samples once to evaluate textural (intra-oral) attributes. Finally, after-feel attributes were evaluated after swallowing. Panellists rinsed and cleaned their mouth with water before evaluating the next sample.

2.13. Statistical Analysis

All experiments were performed at least in triplicate. The results were expressed as means with standard deviations and were analyzed by SPSS version 25 (SPSS Inc., Chicago, IL, USA). A one-way analysis of variance (ANOVA) and Fisher least significant difference test at $p < 0.05$ were used to determine significant differences between means. A Pearson correlation test was performed to explore the correlation between creaminess and other sensory texture attributes/rheological parameters/instrumental texture parameters at $p < 0.05$ or 0.01. Data were normalized for all correlation tests.

3. Results and Discussion

3.1. Whey Protein Microparticle Properties

In this study, the properties of whey protein microparticles were represented by those of the heat-set whey protein gels (Figure S1). Figure 1 shows SEM images of the heat-set whey protein gels containing different $CaCl_2$ concentrations. A relatively uniform film-like network structure was observed at low $CaCl_2$ concentration (Figure 1a,b). As $CaCl_2$ concentration increased to 15 mM, denser and larger protein aggregates were formed with significant phase separation, which assembled into the 3-dimensional (3D) gel network (Figure 1c,d). With further addition of Ca^{2+} (50 mM), the protein molecules aggregated into the spherical microgels of several microns, which formed a particulate gel network. This evolution of the gel structure was mainly ascribed to electrostatic shielding by Ca^{2+}, which could lower repulsion between protein molecules and promote their interactions [38,39].

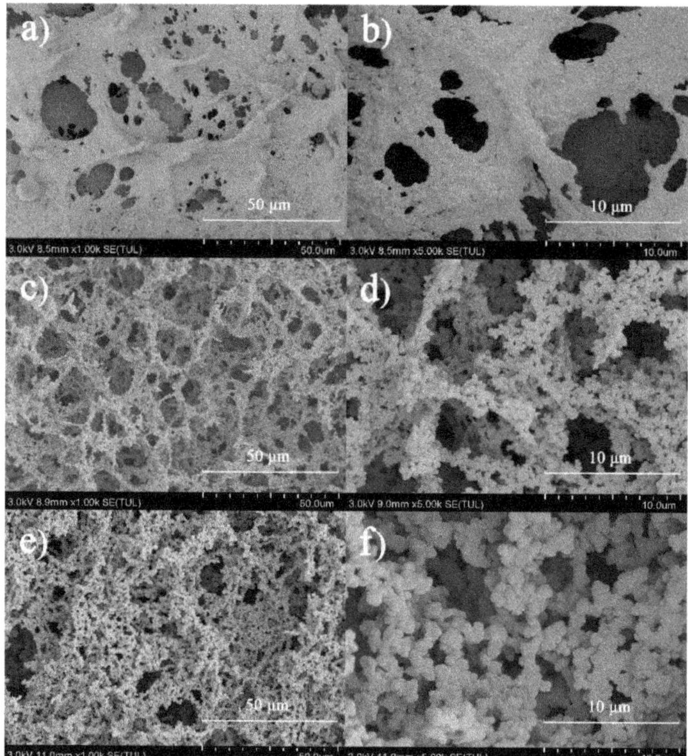

Figure 1. Scanning electron micrographs of the heat-set whey protein gels containing different $CaCl_2$ concentrations with different magnifications. (**a**) 10 mM $CaCl_2$, 1000×; (**b**) 10 mM $CaCl_2$, 5000×; (**c**) 15 mM $CaCl_2$, 1000×; (**d**) 15 mM $CaCl_2$, 5000×; (**e**) 50 mM $CaCl_2$, 1000×; (**f**) 50 mM $CaCl_2$, 5000×.

The effect of $CaCl_2$ concentration on the large deformation properties of the whey protein gels is illustrated in Figure 2a. Hardness and Young's modulus reached the highest value at 15 mM. With further addition of $CaCl_2$, hardness gradually decreased; Young's modulus did not change significantly until $CaCl_2$ concentration reached 25 mM. By contrast, the fracture stress and strain peaked at 15–20 mM, significantly decreased at 25 mM, and plateaued with further addition of $CaCl_2$ (Figure 2b). In theory, the magnitude of interactions between protein molecules per unit volume of the gel increased with $CaCl_2$ concentration [39,40]. However, both large deformation and fracture properties did not correlate with $CaCl_2$ concentration, indicating the important role of the gel structure in determining mechanical properties. As shown in Figure 1, a large volume of pores was generated at higher $CaCl_2$ concentrations (>15 mM), which caused friction during the large deformation and subsequently reduced gel hardness [41,42]. The strength of the gel network itself plays a critical role in determining fracture mechanics, which has a highly positive correlation with $CaCl_2$ concentration [39,43]. This could mediate the adverse effect of pores and cracks on fracture properties that were formed in the gels containing high $CaCl_2$ concentrations [44].

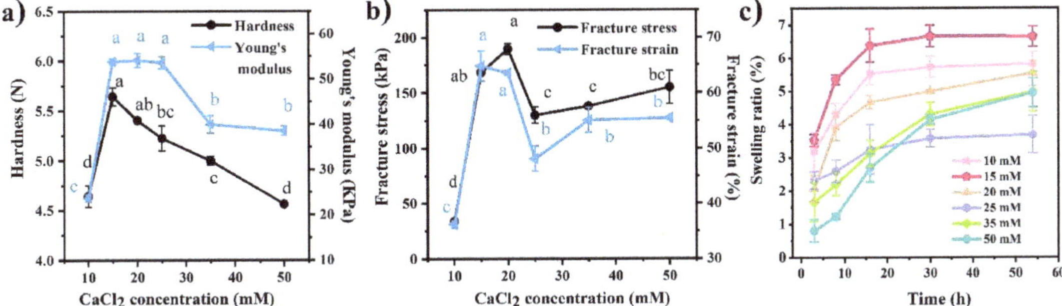

Figure 2. (**a**) Hardness and Young's modulus, (**b**) fracture stress and strain and (**c**) swelling ratios of the heat-set whey protein gels containing different $CaCl_2$ concentrations. Different letters in the same curve indicate significant difference at $p < 0.05$.

The swelling ability of the heat-set whey protein gels is shown in Figure 2. All gels swelled rapidly during the first 8 h of soaking, and then their swelling slowed down. In general, the swelling ratio of the gels containing higher $CaCl_2$ concentrations (25, 35, 50 mM) was larger than that of those containing lower $CaCl_2$ concentrations (10, 15, 20 mM). However, the swelling ratio of all gels was lower than 7% over 54 h of soaking, which could be ascribed to the acidic pH near to the isoelectric point of WPI and high ionic strength of the environmental solution [45,46].

WPMs were prepared by mechanically breaking the whey protein gels. The $D_{4,3}$ of WPMs containing 10 mM $CaCl_2$ decreased from 40 to 10 μm with increasing homogenization speed from 9000 to 21,000 rpm because of increased energy input (Table S1). By contrast, the $D_{4,3}$ of WPMs containing 15 mM $CaCl_2$ became much larger at the same speed because of higher gel strength (Table S1). Therefore, the soft whey protein gel containing 10 mM $CaCl_2$ was selected to produce WPMs of <20 μm in this study.

3.2. Rheology and Instrumental Texture of Model Mayonnaises

A series of 3×4 fresh model mayonnaises were prepared by varying FG weight/WPM weight ratios (Figure 3). With increasing FG and WPM concentration, model mayonnaises visually became more viscous; the addition of FG affected the viscosity of model mayonnaises more significantly.

The viscosity vs. shear rate of model mayonnaises is shown in Figure 4a. Shear thinning occurred in all samples. The thickening effect of WPMs was not significant at low concentration (8 wt%), suggesting that the interparticle interactions and the particle-continuous phase interactions were relatively weak. By contrast, the addition of FG could effectively enhance flow resistance at a relatively low concentration (0.6 wt%), which was ascribed to entangled polysaccharide chains and their interactions with other molecules [47,48]. In this regard, FG played a more important role in enhancing the viscosity of model mayonnaises.

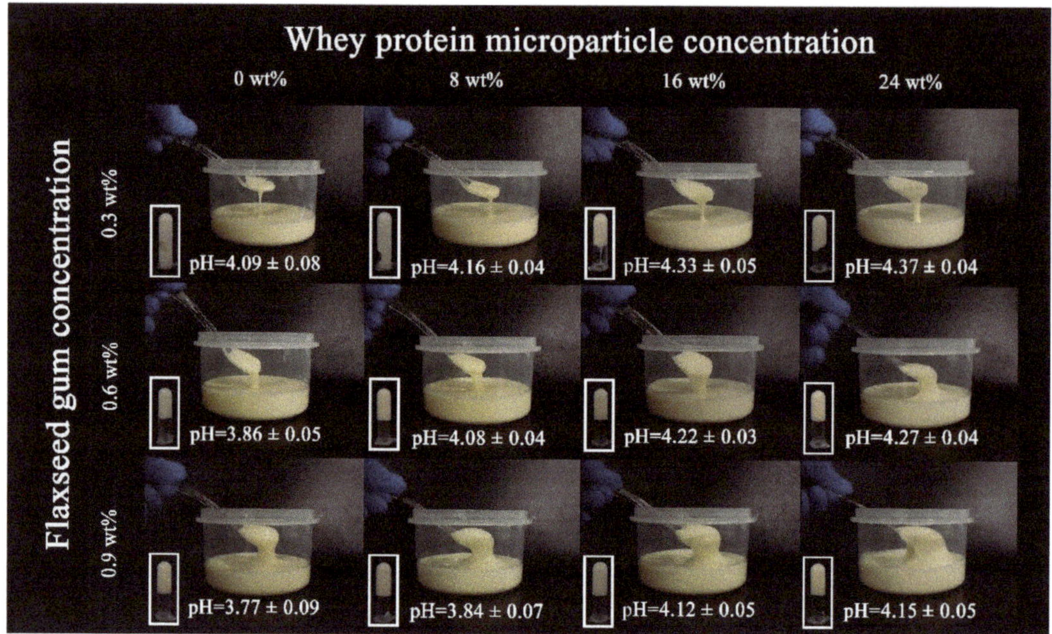

Figure 3. Photographs of model mayonnaises containing different concentrations of the flaxseed gum and whey protein microparticles.

Model mayonnaises had a large linear viscoelastic region up to 10% strain. Although the addition of both FG and WPMs caused a higher yield stress, the former had a relatively larger effect. Both yield stress and strain increased with the addition of FG, whereas the yield stress and strain increased and decreased with the addition of WPMs, respectively, indicating that FG and WPMs influenced the structure of mayonnaises through different mechanisms [49]. WPMs probably acted as an active filler to strengthen the structural network of the samples. However, the slip between WPMs and the continuous phase might occur under the high strain, leading to reduced yield strain [50]. On the other hand, entangled linear FG macromolecules could bear a larger strain due to the formation of a 3D network [51].

Oscillatory shear testing was conducted at the linear viscoelastic region. During the frequency sweep, the storage moduli of all samples were higher than the loss moduli, indicating that solid-like properties dominated in the model mayonnaises. The frequency sweep was conducted from the high to low frequency. At the high frequency, the samples had lower storage and loss moduli, suggesting that the structure of the samples was partially disrupted [52]. With the decrease of the frequency, the storage and loss moduli gradually increased, indicating that this structural disruption was reversible. Moreover, the addition of both FG and WPMs led to a lower phase angle (i.e., a higher elasticity) (Figure S2), suggesting that the structure of the mayonnaise was enhanced by the addition of FG and WPMs.

The instrumental texture of model mayonnaises was measured using the back extrusion test. The peak force, positive area, negative peak force and negative area represent hardness, consistency, adhesiveness and cohesiveness, respectively. As shown in Figure 5, these instrumental textural parameters were enhanced by the addition of FG and WPMs. FG was capable of enhancing textural properties more effectively, which was similar to its effect on rheological properties. A low concentration of WPMs had a limited effect on textural properties. When WPM concentration reached 16 wt%, the enhanced effect became evident probably because of the interparticle interactions and the particle-continuous phase interactions, which was consistent with the results of rheological properties.

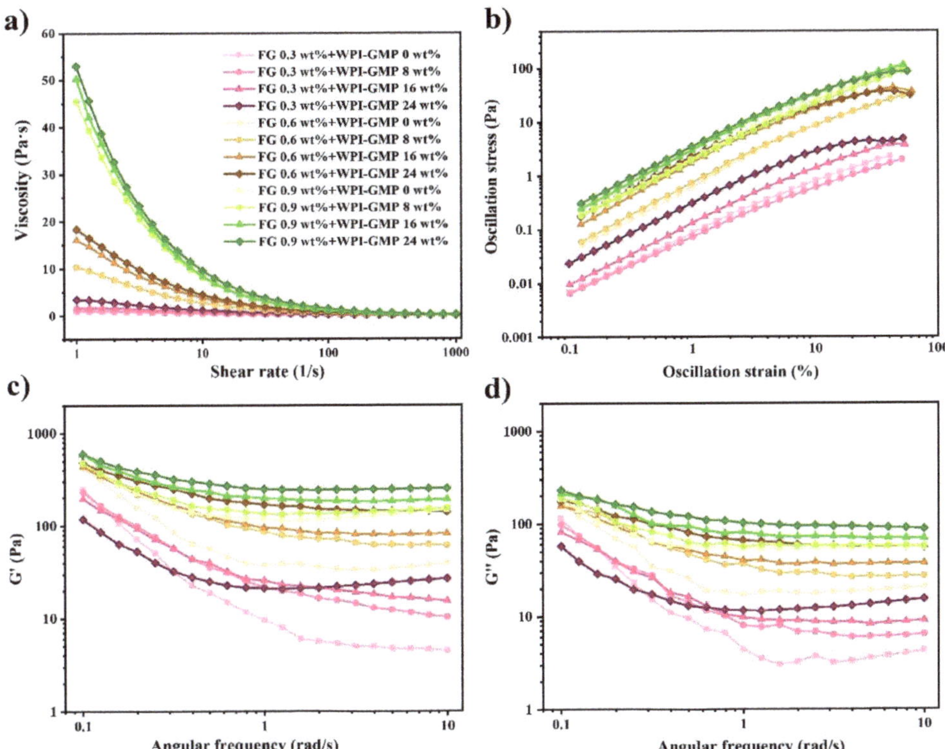

Figure 4. (**a**) Viscosity vs. shear rate flow curves of model mayonnaises, (**b**) strain amplitude sweep of model mayonnaises in the oscillatory shear test, (**c**,**d**) frequency sweep of model mayonnaises in the oscillatory shear test.

The microstructure of model mayonnaises is shown in Figure 6. The oil droplet size of model mayonnaises was of several microns to ~20 microns. With the addition of WPMs, the oil droplet size decreased because of the emulsifying role of WPMs in emulsion formation, i.e., WPMs adsorbed on the oil-water interface (Figure 6a–d). At the high concentration of WPMs (16 and 24 wt%), the aggregation of WPMs occurred with the formation of the network (Figure 6d), which explained that WPMs could significantly enhance rheological and textual properties at the high concentration. Moreover, oil droplets were deformed into a non-spherical shape, although the oil volume fraction (15%) was much lower than 74%, suggesting that FG and WPMs could affect structural and rheological/mechanical properties through a mechanism similar to the compact packing of oil droplets.

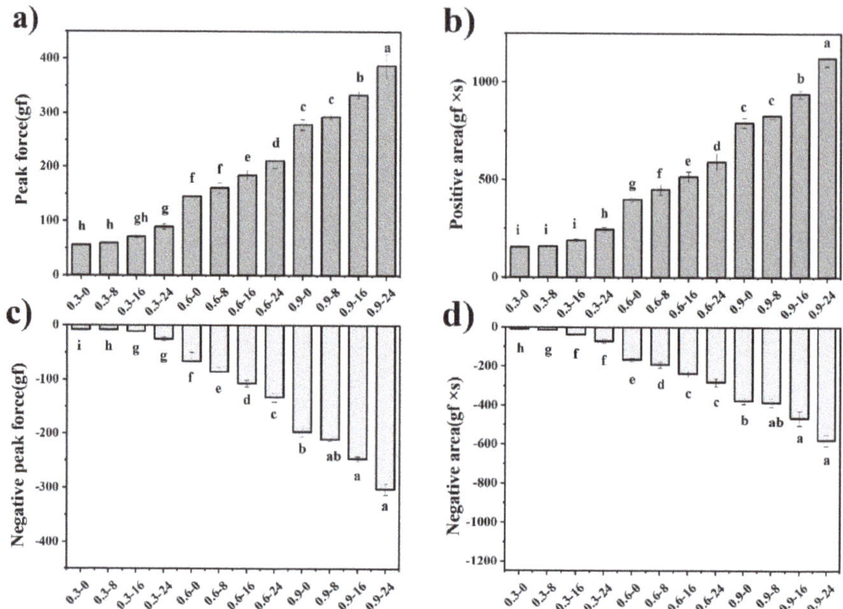

Figure 5. Instrumental texture parameters of model mayonnaises: (**a**) peak force-hardness; (**b**) positive area-consistency; (**c**) negative peak force-adhesiveness; (**d**) negative area-cohesiveness. Different letters above the bars indicate significant difference at $p < 0.05$.

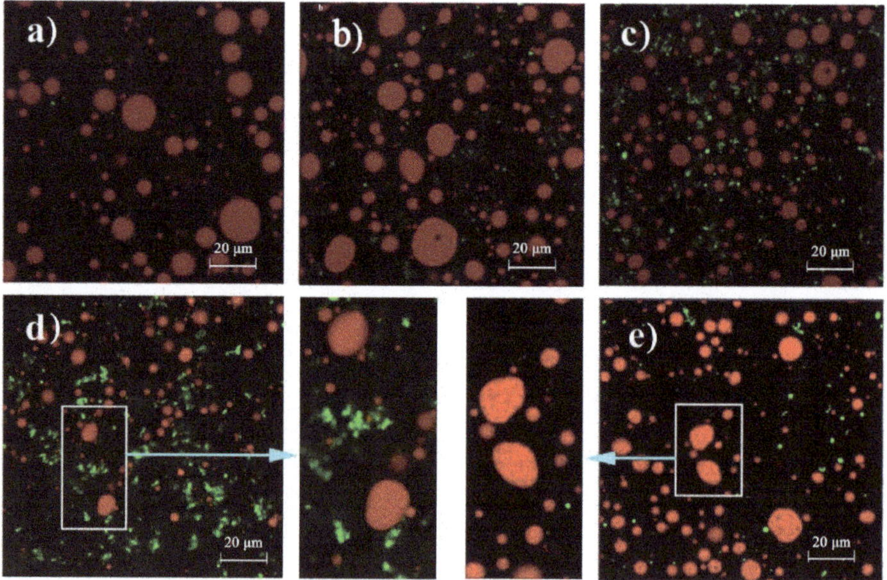

Figure 6. Confocal laser scanning microscopy images of model mayonnaises containing different concentrations of the flaxseed gum (FG) and whey protein microparticles (WPMs). (**a**) 0.3 wt% FG + 0 wt% WPMs, (**b**) 0.3 wt% FG + 8 wt% WPMs, (**c**) 0.3 wt% FG + 16 wt% WPMs, (**d**) 0.3 wt% FG + 24 wt% WPMs, (**e**) 0.6 wt% FG + 0 wt% WPMs.

The centrifugation stability of model mayonnaises is illustrated in Figure 7a. After centrifugation at 4000 rpm, phase separation occurred in model mayonnaises containing 0.3 wt% FG; the samples containing 0.6 and 0.9 wt% FG remained stable. With increasing centrifugation speed to 8000 rpm, severe phase separation occurred in the 0.3 wt% FG samples. Slight phase separation occurred in the 0.6 wt% FG samples. With the addition of 0.9 wt% FG, the mayonnaises were stable. The stability of model mayonnaises during storage is shown in Figure 7b. Creaming occurred in the samples containing 0.3 wt% FG + 0 wt% or 8 wt% WPMs after 27 days of storage, whereas the samples containing 0.3 wt% + 16 or 24 wt% WPMs were stable. All other samples were relatively stable during storage. The role of FG and WPMs in stabilizing model mayonnaises could be ascribed to the increase of bulk viscosity [53,54]. Moreover, the reduction of oil droplet size induced by the addition of WPMs also contributed to the stability of model mayonnaises.

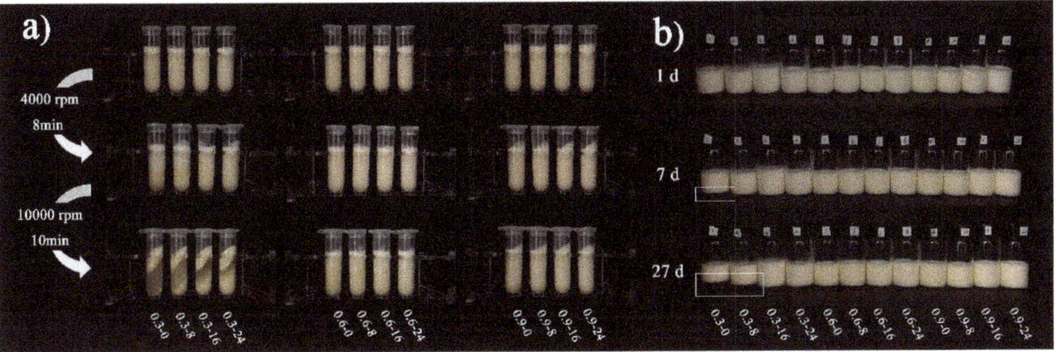

Figure 7. Kinetic stability of model mayonnaises: (**a**) centrifugation stability; (**b**) storage stability.

3.3. Sensory Texture of Model Mayonnaises

Sensory evaluation of mayonnaise texture was conducted in three stages: (1) extra-oral; (2) intra-oral; (3) after-feel (Figure 8). Firmness perceived by stirring the samples increased significantly with FG concentration, whereas the effect of the addition of WPMs on firmness was not significant (Figure 8a). Fluidity in the mouth significantly decreased with FG concentration (Figure 8b). Interestingly, panellists could perceive the fluidity difference between 0.3 wt% FG samples containing different WPM concentrations. With increasing FG concentration, the fluidity difference generated by the addition of WPMs was neglected by panellists because of the much larger contribution of FG to fluidity. Spreadability significantly decreased with FG concentration (Figure 8c). The addition of WPMs also significantly reduced spreadability, regardless of FG concentration, suggesting that a direct relationship existed between spreadability and viscosity [55]. When model mayonnaises were spread into a thin film, panellists could hardly perceive graininess for all samples (Figure 8d), suggesting that the size and mechanical strength of WPMs were below the perceived threshold [56,57]. FG was the key factor generating perceived mouth-coating (Figure 8e). WPMs had a limited effect on perceived mouth-coating of model mayonnaises, although this effect was more significant at a low FG concentration (0.3 wt%).

Creaminess increased greatly with FG concentration (Figure 8f). Although there was no difference in statistics, creaminess increased with the addition of WPMs, which was in agreement with previous studies. For example, Liu et al. reported that microparticulated whey protein contributed to fat-related sensations (e.g., creaminess and fattiness) through enhancing lubrication properties [20]. Moreover, the addition of microparticulated whey protein significantly enhanced creaminess of low-fat yogurts, the effect of which depended on their particle size [58]. Creaminess represents an overall sensory texture, which is a combination of different textural perceptions [59,60]. As shown in Table 3, creaminess had a linear negative correlation with firmness, fluidity and spreadability, while it had a linear positive correlation with mouth-coating. This suggests that thickness perception (i.e., lower fluidity and spreadability) and sticky mouth-coating could be the key factors giving rise to creaminess. Meanwhile, this strongly supports that creaminess is a combination of complex textural attributes.

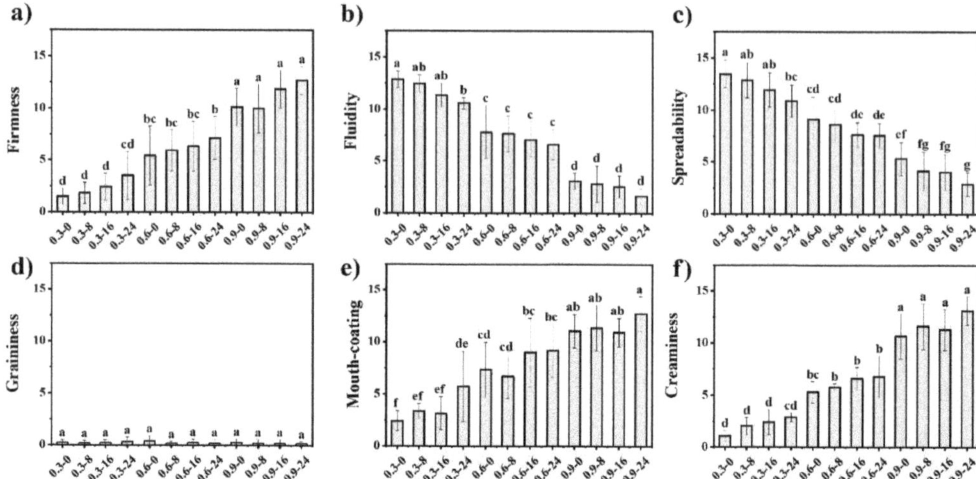

Figure 8. Sensory texture of low-fat model mayonnaises: (**a**) firmness, (**b**) fluidity, (**c**) spreadability, (**d**) graininess, (**e**) mouth-coating, (**f**) creaminess. Different letters above the bars indicate significant difference at $p < 0.05$.

Table 3. Correlation between creaminess and extra- and intra-oral textural attributes.

	Extra-Oral		Intra-Oral	
	Firmness	**Fluidity**	**Spreadability**	**Mouth-Coating**
Creaminess	−0.986 **	−0.994 **	−0.992 **	0.970 **

** indicates significant difference at $p < 0.01$.

3.4. Correlation between Creaminess and Rheology/Instrumental Texture

Viscosities of the 12 samples at different shear rates were used to analyze their correlation with creaminess (Table 4). The results showed that creaminess had a linear positive correlation with the viscosity measured at different shear rates. Although the change in viscosity in the model mayonnaise systems was driven by different mechanisms (e.g., active fillers and entangled linear macromolecules), viscosity that represents the bulk properties could reflect the sensory creaminess of mayonnaises to a relatively high extent. This was in agreement with a previous study on protein beverages, with a key finding that the primary effect on viscosity-related sensory properties (e.g., creaminess, consistency and mouth-coating) was because of increased viscosity for a wide range of viscosity levels [61]. Strong correlation between viscosity and creaminess was also found in other emulsion-based foods, e.g., low-fat yogurts and custards [49,58]. Textural parameters measured by the texture analyzer using different methods (e.g., TPA and back extrusion) are widely used to predict the sensory properties of semisolid and solid foods [62,63]. There was a linear positive correlation between creaminess and four instrumental texture parameters (i.e., hardness, consistency, cohesiveness and adhesiveness), suggesting that instrumental texture measured at large deformations could reflect the creaminess of mayonnaises to some extent [24]. Recently, Conti-Silva et al. investigated the correlations between instrumental texture and oral viscosity using eight different food samples, i.e., water, condensed milk, strawberry yogurt, honey, creamy dairy dessert, UHT cream, petit suisse strawberry flavor, and dulce de leche [64]. The authors found that oral viscosity was positively correlated with positive areas from the curves obtained in the back extrusion test.

Table 4. Correlation between creaminess and instrumental texture parameters/viscosities at different shear rates.

Shear Rate (s^{-1})	R^2	Instrumental Texture	R^2
1	0.977 **	Hardness	0.990 **
10	0.992 **	Consistency	0.988 **
100	0.992 **	Adhesiveness	0.985 **
1000	0.986 **	Cohesiveness	0.984 **

** indicates significant difference at $p < 0.01$.

4. Conclusions

The addition of FG and WPMs enhanced the rheological and textural properties of model mayonnaises; this effect was largely attributed to the addition of FG, which probably formed an entangled 3D network. WPMs worked as active fillers and influenced rheological/textural properties through the interparticle interactions and the particle-continuous phase interactions at a relatively high concentration. We found that apparent viscosities measured at different shear rates and instrumental texture parameters measured by the back extrusion test had a linear positive correlation with creaminess. Moreover, creaminess was highly correlated to extra- and intra-oral textural properties, suggesting that creaminess in the context of texture is a reflection of overall textural perceptions. In particular, the thick perception (low fluidity and spreadability) and sticky mouth-coating were important factors giving rise to creaminess. In future studies, the synergistic effect of FG and WPMs on the tribology of low-fat mayonnaises and its correlation with textural perception will be investigated.

Supplementary Materials: The following supporting information can be downloaded at: https://www.mdpi.com/article/10.3390/foods11030282/s1, Figure S1: Photographs of the heat-set whey protein gels containing different $CaCl_2$ concentrations; Figure S2: Phase angle of low-fat mayonnaises in the oscillatory shear test with the frequency sweep; Table S1: Average particle size (D4,3) of whey protein microparticles after homogenization at different speeds.

Author Contributions: Conceptualization, Q.G. and R.X.; methodology, Q.G. and R.X.; investigation, R.X. and X.X.; formal analysis, R.X., X.X. and K.Y.; writing—original draft preparation, K.Y. and R.X.; writing—review and editing, Q.G.; funding acquisition, Q.G. All authors have read and agreed to the published version of the manuscript.

Funding: This research was funded by the National Natural Science Foundation of China (Grant No. 31901684).

Institutional Review Board Statement: The study was conducted in accordance with the Declaration of Helsinki, and approved by Ethics Committee of China Agricultural University (protocol code CAUHR-2020019 and 2020-11-27).

Informed Consent Statement: Informed consent was obtained from all subjects involved in the study.

Data Availability Statement: Not applicable.

Conflicts of Interest: The authors declare no conflict of interest.

References

1. Bray, G.; Popkin, B. Dietary fat intake does affect obesity! *Am. J. Clin. Nutr.* **1998**, *68*, 1157–1173. [CrossRef]
2. Ma, Z.; Boye, J.I. Advances in the design and production of reduced-fat and reduced-cholesterol salad dressing and mayonnaise: A review. *Food Bioprocess Technol.* **2013**, *6*, 648–670. [CrossRef]
3. Guo, Q.; Ye, A.; Bellissimo, N.; Singh, H.; Rousseau, D. Modulating fat digestion through food structure design. *Prog. Lipid Res.* **2017**, *68*, 109–118. [CrossRef]

4. Guo, Q. Understanding the oral processing of solid foods: Insights from food structure. *Compr. Rev. Food Sci. Food Saf.* **2021**, *20*, 2941–2967. [CrossRef]

5. Di Monaco, R.; Miele, N.A.; Cabisidan, E.K.; Cavella, S. Strategies to reduce sugars in food. *Curr. Opin. Food Sci.* **2018**, *19*, 92–97. [CrossRef]

6. Nastaj, M.; Sołowiej, B.; Terpiłowski, K.; Mleko, S. Effect of erythritol on physicochemical properties of reformulated high protein meringues obtained from whey protein isolate. *Int. Dairy J.* **2020**, *105*, 104672. [CrossRef]

7. Gao, H.; Ma, L.; Cheng, C.; Liu, J.; Liang, R.; Zou, L.; Liu, W.; McClements, D.J. Review of recent advances in the preparation, properties, and applications of high internal phase emulsions. *Trends Food Sci. Technol.* **2021**, *112*, 36–49. [CrossRef]

8. Di Mattia, C.; Balestra, F.; Sacchetti, G.; Neri, L.; Mastrocola, D.; Pittia, P. Physical and structural properties of extra-virgin olive oil based mayonnaise. *LWT-Food Sci. Technol.* **2015**, *62*, 764–770. [CrossRef]

9. Herald, T.J.; Abugoush, M.; Aramouni, F. Physical and sensory properties of egg yolk and egg yolk substitutes in a model mayonnaise system. *J. Texture Stud.* **2009**, *40*, 692–709. [CrossRef]

10. Flamminii, F.; Di Mattia, C.D.; Sacchetti, G.; Neri, L.; Mastrocola, D.; Pittia, P. Physical and sensory properties of mayonnaise enriched with encapsulated olive leaf phenolic extracts. *Foods* **2020**, *9*, 997. [CrossRef] [PubMed]

11. Olsson, V.; Håkansson, A.; Purhagen, J.; Wendin, K. The effect of emulsion intensity on selected sensory and instrumental texture properties of full-fat mayonnaise. *Foods* **2018**, *7*, 9. [CrossRef] [PubMed]

12. Kadian, D.; Kumar, A.; Badgujar, P.C.; Sehrawat, R. Effect of homogenization and microfluidization on physicochemical and rheological properties of mayonnaise. *J. Food Process Eng.* **2021**, *44*, e13661. [CrossRef]

13. Yesiltas, B.; García-Moreno, P.J.; Sørensen, A.-D.M.; Soria Caindec, A.M.; Hyldig, G.; Anankanbil, S.; Guo, Z.; Jacobsen, C. Enrichment of mayonnaise with a high fat fish oil-in-water emulsion stabilized with modified datem c14 enhances oxidative stability. *Food Chem.* **2021**, *341*, 128141. [CrossRef] [PubMed]

14. Agyei-Amponsah, J.; Macakova, L.; DeKock, H.L.; Emmambux, M.N. Effect of substituting sunflower oil with starch-based fat replacers on sensory profile, tribology, and rheology of reduced-fat mayonnaise-type emulsions. *Starch—Stärke* **2021**, *73*, 2000092. [CrossRef]

15. Akhtar, M.; Murray, B.S.; Dickinson, E. Perception of creaminess of model oil-in-water dairy emulsions: Influence of the shear-thinning nature of a viscosity-controlling hydrocolloid. *Food Hydrocoll.* **2006**, *20*, 839–847. [CrossRef]

16. Liu, K.; Tian, Y.; Stieger, M.; van der Linden, E.; van de Velde, F. Evidence for ball-bearing mechanism of microparticulated whey protein as fat replacer in liquid and semi-solid multi-component model foods. *Food Hydrocoll.* **2016**, *52*, 403–414. [CrossRef]

17. Upadhyay, R.; Chen, J. Smoothness as a tactile percept: Correlating 'oral' tribology with sensory measurements. *Food Hydrocoll.* **2019**, *87*, 38–47. [CrossRef]

18. Le Calvé, B.; Saint-Léger, C.; Babas, R.; Gelin, J.-L.; Parker, A.; Erni, P.; Cayeux, I. Fat perception: How sensitive are we? *J. Texture Stud.* **2015**, *46*, 200–211. [CrossRef]

19. Sarkar, A.; Kanti, F.; Gulotta, A.; Murray, B.S.; Zhang, S. Aqueous lubrication, structure and rheological properties of whey protein microgel particles. *Langmuir* **2017**, *33*, 14699–14708. [CrossRef]

20. Liu, K.; Stieger, M.; van der Linden, E.; van de Velde, F. Effect of microparticulated whey protein on sensory properties of liquid and semi-solid model foods. *Food Hydrocoll.* **2016**, *60*, 186–198. [CrossRef]

21. Torres, O.; Andablo-Reyes, E.; Murray, B.S.; Sarkar, A. Emulsion microgel particles as high-performance bio-lubricants. *ACS Appl. Mater. Interfaces* **2018**, *10*, 26893–26905. [CrossRef] [PubMed]

22. Hu, J.; Andablo-Reyes, E.; Soltanahmadi, S.; Sarkar, A. Synergistic microgel-reinforced hydrogels as high-performance lubricants. *ACS Macro Lett.* **2020**, *9*, 1726–1731. [CrossRef]

23. Liu, H.; Xu, X.M.; Guo, S.D. Rheological, texture and sensory properties of low-fat mayonnaise with different fat mimetics. *LWT-Food Sci. Technol.* **2007**, *40*, 946–954. [CrossRef]

24. Stokes, J.R.; Boehm, M.W.; Baier, S.K. Oral processing, texture and mouthfeel: From rheology to tribology and beyond. *Curr. Opin. Colloid Interface Sci.* **2013**, *18*, 349–359. [CrossRef]

25. de Vicente, J.; Stokes, J.R.; Spikes, H.A. Soft lubrication of model hydrocolloids. *Food Hydrocoll.* **2006**, *20*, 483–491. [CrossRef]

26. Selway, N.; Chan, V.; Stokes, J.R. Influence of fluid viscosity and wetting on multiscale viscoelastic lubrication in soft tribological contacts. *Soft Matter* **2017**, *13*, 1702–1715. [CrossRef] [PubMed]

27. Stokes, J.R.; Macakova, L.; Chojnicka-Paszun, A.; de Kruif, C.G.; de Jongh, H.H.J. Lubrication, adsorption, and rheology of aqueous polysaccharide solutions. *Langmuir* **2011**, *27*, 3474–3484. [CrossRef]

28. Appelqvist, I.A.M.; Cochet-Broch, M.; Poelman, A.A.M.; Day, L. Morphologies, volume fraction and viscosity of cell wall particle dispersions particle related to sensory perception. *Food Hydrocoll.* **2015**, *44*, 198–207. [CrossRef]

29. Malone, M.E.; Appelqvist, I.A.M.; Norton, I.T. Oral behaviour of food hydrocolloids and emulsions. Part 1. Lubrication and deposition considerations. *Food Hydrocoll.* **2003**, *17*, 763–773. [CrossRef]

30. Blok, A.E.; Bolhuis, D.P.; Kibbelaar, H.V.M.; Bonn, D.; Velikov, K.P.; Stieger, M. Comparing rheological, tribological and sensory properties of microfibrillated cellulose dispersions and xanthan gum solutions. *Food Hydrocoll.* **2021**, *121*, 107052. [CrossRef]

31. Li, Y.; Joyner, H.S.; Carter, B.G.; Drake, M.A. Effects of fat content, pasteurization method, homogenization pressure, and storage time on the mechanical and sensory properties of bovine milk. *J. Dairy Sci.* **2018**, *101*, 2941–2955. [CrossRef] [PubMed]

32. He, Q.; Hort, J.; Wolf, B. Predicting sensory perceptions of thickened solutions based on rheological analysis. *Food Hydrocoll.* **2016**, *61*, 221–232. [CrossRef]

33. Su, H.-P.; Lien, C.-P.; Lee, T.-A.; Ho, J.-H. Development of low-fat mayonnaise containing polysaccharide gums as functional ingredients. *J. Sci. Food Agric.* **2010**, *90*, 806–812. [CrossRef]
34. Krzeminski, A.; Prell, K.A.; Busch-Stockfisch, M.; Weiss, J.; Hinrichs, J. Whey protein–pectin complexes as new texturising elements in fat-reduced yoghurt systems. *Int. Dairy J.* **2014**, *36*, 118–127. [CrossRef]
35. Ribes, S.; Peña, N.; Fuentes, A.; Talens, P.; Barat, J.M. Chia (*Salvia hispanica* L.) seed mucilage as a fat replacer in yogurts: Effect on their nutritional, technological, and sensory properties. *J. Dairy Sci.* **2021**, *104*, 2822–2833. [CrossRef]
36. Yildiz-Akgül, F. Enhancement of torba yoghurt with whey protein isolates. *Int. J. Dairy Technol.* **2018**, *71*, 898–905. [CrossRef]
37. Nastaj, M.; Sołowiej, B.G.; Gustaw, W.; Peréz-Huertas, S.; Mleko, S.; Wesołowska-Trojanowska, M. Physicochemical properties of high-protein-set yoghurts obtained with the addition of whey protein preparations. *Int. J. Dairy Technol.* **2019**, *72*, 395–402. [CrossRef]
38. Clare, D.A.; Lillard, S.J.; Ramsey, S.R.; Amato, P.M.; Daubert, C.R. Calcium effects on the functionality of a modified whey protein ingredient. *J. Agric. Food Chem.* **2007**, *55*, 10932–10940. [CrossRef]
39. Phan-Xuan, T.; Durand, D.; Nicolai, T.; Donato, L.; Schmitt, C.; Bovetto, L. Tuning the structure of protein particles and gels with calcium or sodium ions. *Biomacromolecules* **2013**, *14*, 1980–1989. [CrossRef]
40. Mehalebi, S.; Nicolai, T.; Durand, D. The influence of electrostatic interaction on the structure and the shear modulus of heat-set globular protein gels. *Soft Matter* **2008**, *4*, 893–900. [CrossRef]
41. Hongsprabhas, P.; Barbut, S. Protein and salt effects on ca2+-induced cold gelation of whey protein isolate. *J. Food Sci.* **1997**, *62*, 382–385. [CrossRef]
42. Ju, Z.; Kilara, A. Aggregation induced by calcium chloride and subsequent thermal gelation of whey protein isolate. *J. Dairy Sci.* **1998**, *81*, 925–931. [CrossRef]
43. Shen, X.; Guo, Q. Fabrication of robust protein-based foams with multifunctionality by manipulating intermolecular interactions. *Green Chem.* **2021**, *23*, 8187–8199. [CrossRef]
44. Valim, M.D.; Cavallieri, A.L.; Cunha, R.L. Whey protein/arabic gum gels formed by chemical or physical gelation process. *Food Biophys.* **2009**, *4*, 23–31. [CrossRef]
45. Oztop, M.H.; Rosenberg, M.; Rosenberg, Y.; McCarthy, K.L.; McCarthy, M.J. Magnetic resonance imaging (mri) and relaxation spectrum analysis as methods to investigate swelling in whey protein gels. *J. Food Sci.* **2010**, *75*, E508–E515. [CrossRef] [PubMed]
46. Fan, L.; Ge, A.; Chen, X.D.; Mercadé-Prieto, R. The role of non-covalent interactions in the alkaline dissolution of heat-set whey protein hydrogels made at gelation ph 2–11. *Food Hydrocoll.* **2019**, *89*, 100–110. [CrossRef]
47. Chang, C.; Li, J.; Li, X.; Wang, C.; Zhou, B.; Su, Y.; Yang, Y. Effect of protein microparticle and pectin on properties of light mayonnaise. *LWT-Food Sci. Technol.* **2017**, *82*, 8–14. [CrossRef]
48. Liu, J.; Shim, Y.Y.; Shen, J.; Wang, Y.; Ghosh, S.; Reaney, M.J. Variation of composition and functional properties of gum from six canadian flaxseed (*Linum usitatissimum* L.) cultivars. *Int. J. Food Sci. Tech.* **2016**, *51*, 2313–2326. [CrossRef]
49. Janssen, A.M.; Terpstra, M.E.J.; De Wijk, R.A.; Prinz, J.F. Relations between rheological properties, saliva-induced structure breakdown and sensory texture attributes of custards. *J. Texture Stud.* **2007**, *38*, 42–69. [CrossRef]
50. Bicerano, J.; Douglas, J.F.; Brune, D.A. Model for the viscosity of particle dispersions. *J. Macromol. Sci. Polym. Rev.* **1999**, *39*, 561–642. [CrossRef]
51. Lopez, C.G.; Voleske, L.; Richtering, W. Scaling laws of entangled polysaccharides. *Carbohydr. Polym.* **2020**, *234*, 115886. [CrossRef] [PubMed]
52. Ozer, B.H.; Bell, A.E.; Grandison, A.S.; Robinson, R.K. Rheological properties of concentrated yoghurt (labneh). *J. Texture Stud.* **1998**, *29*, 67–79. [CrossRef]
53. Nasirpour-Tabrizi, P.; Azadmard-Damirchi, S.; Hesari, J.; Khakbaz Heshmati, M.; Savage, G.P. Production of a spreadable emulsion gel using flaxseed oil in a matrix of hydrocolloids. *J. Food Process. Preserv.* **2020**, *44*, e14588. [CrossRef]
54. Dickinson, E. Hydrocolloids at interfaces and the influence on the properties of dispersed systems. *Food Hydrocoll.* **2003**, *17*, 25–39. [CrossRef]
55. Fuhrmann, P.L.; Kalisvaart, L.C.M.; Sala, G.; Scholten, E.; Stieger, M. Clustering of oil droplets in o/w emulsions enhances perception of oil-related sensory attributes. *Food Hydrocoll.* **2019**, *97*, 105215. [CrossRef]
56. Imai, E.; Hatae, K.; Shimada, A. Oral perception of grittiness: Effect of particle size and concentration of the dispersed particles and the dispersion medium. *J. Texture Stud.* **1995**, *26*, 561–576. [CrossRef]
57. Breen, S.P.; Etter, N.M.; Ziegler, G.R.; Hayes, J.E. Oral somatosensatory acuity is related to particle size perception in chocolate. *Sci. Rep.* **2019**, *9*, 7437. [CrossRef]
58. Hossain, M.K.; Keidel, J.; Hensel, O.; Diakité, M. The impact of extruded microparticulated whey proteins in reduced-fat, plain-type stirred yogurt: Characterization of physicochemical and sensory properties. *LWT* **2020**, *134*, 109976. [CrossRef]
59. Dickinson, E. On the road to understanding and control of creaminess perception in food colloids. *Food Hydrocoll.* **2018**, *77*, 372–385. [CrossRef]
60. Upadhyay, R.; Aktar, T.; Chen, J. Perception of creaminess in foods. *J. Texture Stud.* **2020**, *51*, 375–388. [CrossRef]
61. Wagoner, T.B.; Çakır-Fuller, E.; Shingleton, R.; Drake, M.; Foegeding, E.A. Viscosity drives texture perception of protein beverages more than hydrocolloid type. *J. Texture Stud.* **2020**, *51*, 78–91. [CrossRef] [PubMed]
62. Nishinari, K.; Fang, Y.; Rosenthal, A. Human oral processing and texture profile analysis parameters: Bridging the gap between the sensory evaluation and the instrumental measurements. *J. Texture Stud.* **2019**, *50*, 369–380. [CrossRef]

63. Funami, T.; Nakauma, M. Instrumental food texture evaluation in relation to human perception. *Food Hydrocoll.* **2022**, *124*, 107253. [CrossRef]
64. Conti-Silva, A.C.; Ichiba, A.K.T.; Silveira, A.L.d.; Albano, K.M.; Nicoletti, V.R. Viscosity of liquid and semisolid materials: Establishing correlations between instrumental analyses and sensory characteristics. *J. Texture Stud.* **2018**, *49*, 569–577. [CrossRef] [PubMed]

 *foods*

Article

Lubrication and Sensory Properties of Emulsion Systems and Effects of Droplet Size Distribution

Qi Wang, Yang Zhu *, Zhichao Ji and Jianshe Chen

Laboratory of Food Oral Processing, School of Food Science and Biotechnology, Zhejiang Gongshang University, Hangzhou 310018, China; qiwong1995@163.com (Q.W.); jjzzcc1998@163.com (Z.J.); jschen@zjgsu.edu.cn (J.C.)
* Correspondence: zhuyang@zjgsu.edu.cn

Abstract: The functional and sensory properties of food emulsion are thought to be complicated and influenced by many factors, such as the emulsifier, oil/fat mass fraction, and size of oil/fat droplets. In addition, the perceived texture of food emulsion during oral processing is mainly dominated by its rheological and tribological responses. This study investigated the effect of droplet size distribution as well as the content of oil droplets on the lubrication and sensory properties of o/w emulsion systems. Friction curves for reconstituted milk samples (composition: skimmed milk and milk cream) and Casein sodium salt (hereinafter referred to as CSS) stabilized model emulsions (olive oil as oil phase) were obtained using a soft texture analyzer tribometer with a three ball-on-disc setup combined with a soft surfaces (PDMS) tribology system. Sensory discrimination was conducted by 22 participants using an intensity scoring method. Stribeck curve analyses showed that, for reconstituted milk samples with similar rheological properties, increasing the volume fraction of oil/fat droplets in the size range of 1–10 μm will significantly enhance lubrication, while for CSS-stabilized emulsions, the size effect of oil/fat droplets reduced to around 1 μm. Surprisingly, once the size of oil/fat droplets of both systems reached nano size ($d_{90} = 0.3$ μm), increasing the oil/fat content gave no further enhancement, and the friction coefficient showed no significant difference ($p > 0.05$). Results from sensory analysis show that consumers are capable of discriminating emulsions, which vary in oil/fat droplet size and in oil/fat content ($p < 0.01$). However, it appeared that the discrimination capability of the panelist was significantly reduced for emulsions containing nano-sized droplets.

Keywords: oral lubrication; oral tribology; food emulsions; droplet size distribution; food sensory

Citation: Wang, Q.; Zhu, Y.; Ji, Z.; Chen, J. Lubrication and Sensory Properties of Emulsion Systems and Effects of Droplet Size Distribution. *Foods* **2021**, *10*, 3024. https://doi.org/10.3390/foods10123024

Academic Editors: Harjinder Singh and Alejandra Acevedo-Fani

Received: 31 October 2021
Accepted: 29 November 2021
Published: 6 December 2021

1. Introduction

Awareness of the critical importance of diet on human health and wellness is growing among consumers, especially due to well-reported evidence showing excessive calorie intake could lead to occurrence of obesity, high blood pressure, and cardiovascular diseases [1,2]. The food industry has, therefore, been actively seeking new techniques and food formulations of calorie reduction, such as low-fat or fat-free versions of traditional food products that maintain their sensory properties [3–5]. However, removing or reducing the amount of oil/fat in food systems is not as easy as it seems to be, simply due to the multiple roles of oil/fat in food formulations, particularly when the oil/fat is in a dispersed status and functions as a structural component of the matrix. Any small change in the mass fraction and droplet size distribution of oil droplets, the type and concentration of emulsifier, will alter the physico-chemical properties of the emulsion [3]. For example, fat droplets can affect the appearance (optical properties) [6], flavor characteristics (molecular distribution) [7,8], texture properties (rheology, tribology) [9], and shelf life (stability) [10] of food emulsions. In relation to texture and mouthfeel, oil/fat provides at least three basic functional roles: (1) as a filler ingredient to alter microstructure and rheology of such systems; (2) adsorption on the tongue surface for enhanced oil/fat sensation; and (3) as a particulate component to improve lubricating properties [11,12].

In past decades, rheology has been taken granted as a 'gold standard' instrumental technique for mapping or predicting the perceived texture and mouthfeel of liquid products [13–15]. However, as liquid and semisolid food are squeezed and rubbed between the tongue and hard plate during oral processing, the traditional rheology approach was found limited in explaining the perception of such as smoothness, creaminess, greasiness, etc. Instead, growing evidence shows that the lubrication behavior in the oral cavity plays a dominating role in influencing oral sensation and sensory perception [16–18]. In a typical tribological measurement, consisting of two surfaces lubricated by a Newtonian fluid, a Stribeck curve can be obtained to give the relationship between the friction coefficient and the thickness of the lubricating film [19–21]. If one of the surfaces is soft and deformable (i.e., tongue against hard palate), it is a soft-contact lubrication. Three different lubrication regimes have been established for a typical Stribeck curve: the boundary regime at low thickness of the lubricating film, the mixed regime at medium thickness, and the hydrodynamic regime at high thickness. The underlying mechanisms of lubrication regimes have been well described by a number of researchers [17,22–24].

Since the sensory mouthfeel of a food emulsion is closely linked to its rheological and tribological properties, it is now commonly accepted that oil/fat volume fraction, size of oil/fat droplets, and the viscosity of the emulsion will have direct impacts on the lubrication behavior and the perception of fat-related sensory attributes of such systems [25]. Malone et al. confirmed many years ago that correlations existed between sensory-perceived slipperiness and friction coefficients for biopolymer solutions [26]. A recent study by authors' group investigated the lubrication properties of emulsions stabilized by different emulsifiers [27], observing a good correlation between the sensation of finger and oral smoothness and the measured friction coefficient, which suggests that smoothness perception is mainly driven by tactile sensation. Chojnicka-Paszun et al. demonstrated a significant correlation between creamy attributes and the friction coefficient measured on soft rubber surfaces [28]. They suggested that creaminess is best predicted by the friction obtained at low speeds (comparable with the speed in the mouth) that corresponded to the boundary lubrication regime. However, the literature research findings are not always straightforward and consistent. For example, some researchers noted that the oil/fat volume fractions seem to have little effect on the lubrication behavior in the boundary regime, while increasing the oil/fat volume fractions of o/w emulsion system, resulting in an earlier shift from boundary to mixed regime [10,27]. A study by Laguna et al. showed that in vitro lubrication tests failed to distinguish milk samples containing different mass fractions of fat while untrained panelists were able to [25]. In addition to the oil/fat mass fraction, it has been found that the coalescence of fat droplets during oral processing may led to a decrease in friction coefficient and an increased perception of fat-related sensory attributes, such as creaminess [29]. An increased, effective oil/fat volume fraction and the increased viscosity could be the reasons behind this observation [7]. The presence of saliva may increase the friction of the system through the flocculation phenomenon for positively charged droplets or promote the adherence of saliva proteins to the substrate for negatively charged emulsions, both resulting in a stronger correlation of perceived graininess and fattiness to tribological properties [30–32].

Most of the previous tribological studies examined the effect of the oil/fat volume fraction on the lubrication behavior of liquid dairy products or model emulsion systems. However, the size effect of dispersed fat droplets on the lubrication behavior and sensory perception appears to be somewhat contradictory in the literature. It has been speculated that decreasing the oil droplet size at constant oil volume fraction could increase the perceived creaminess [33], while others discovered contradictory phenomena, showing that the clustered emulsion has significantly higher creaminess intensity than single droplets [32]. Therefore, in this study, we choose two o/w emulsion systems as major research objects (commercial dairy products composed of skimmed milk and milk cream; model emulsion composed of CSS and olive oil), with the objective to reveal the effect of the size of the fat droplet on the in vitro lubrication behavior of emulsions and then to investigate possible applications of the tribological approach in assessing the texture perception of food emulsions.

2. Materials and Methods

2.1. Sample Preparation

Pasteurized skimmed milk with fat content of 0% (*w/w*) (Avonmore, China Resources Wufeng Distribution (Shenzhen) Co., Ltd., Shenzhen, China) and milk cream with fat content of 35% (President, Angliss (Shanghai) Food Co., Ltd., Shanghai, China) were mixed in order to obtain reconstituted milk samples with specified fat contents of 0.3, 0.7, 2.0, 3.5, and 7.5 wt.%, respectively. The reconstituted milk samples (o/w emulsion) were first sheared using an Ultra-Turrax (Polytron, Kinematica AG, Lucerne, Switzerland) at 3000 rpm for 120 s (first stage) and homogenized using a high-pressure homogenizer (HPH—second stage). Samples with different fat droplet size distribution were obtained by adjusting the operating pressures (bar) and time (s) of the high-pressure homogenizer (AH-BASIC, ATS, Canada). The thickened samples were obtained by adding 1.4 wt.% commercial thickener (Ourdiet Swallow®, Guangzhou Ourdiet Biotechnology Co., Ltd., Guangzhou, China) into the prepared reconstituted milk samples at room temperature under magnetic stirring, then treated by Shear homogenizer at 3000 rpm for 120 s, accelerating the dissolution of thickener. In accordance with the specification of the manufacturer, the main ingredients of the thickener are xanthan gum (60%) and maltodextrin (28%).

The abbreviation used in the following text: reconstituted milk samples with different fat content homogenized under an Ultra-Turrax were labeled as 'Fat content-Shear', but samples which were homogenized under high pressure were labeled as 'Fat content-HPH'. Samples with the addition of thicker were labeled as 'Fat content-thicken-Shear/HPH'.

Model oil-in-water emulsions, consisting of 7.5 wt.% olive oil and 92.5 wt.% aqueous phase containing 1 wt.% Casein sodium salt (Sigma Aldrich, Auckland, New Zealand), were prepared by pre-homogenizing the ingredients using an Ultra-Turrax. Emulsions of different fat droplet size were obtained by adjusting the operating pressures (bar) and time (s) of the high-pressure homogenizer.

2.2. Droplet Size Studies

The droplet size distribution of all the samples was measured using a laser diffraction particle size analyzer (Mastersizer 3000, Malvern Instruments, Ltd., Worcestershire, UK). Freshly prepared emulsion samples were gently stirred and diluted by adding small droplets (about 30 µL) into a measurement chamber containing water, until the instrument gave an optimum obscuration rate around 10%. The optical model used to resolve the droplet size distribution from scattering data used a refractive index of 1.33 for the aqueous phase and 1.43 for the fat/oil phase (emulsions). Analysis was set for a regular round droplet and measured in quintuplicate for determinations for all samples. An averaged size distribution was obtained for each sample, and both the Sauter mean $d_{3,2}$ and the 90th percentile (d_{90}) were used to compare differences in the surface-weighted droplet size and distribution of coarser droplet, respectively.

2.3. Flow Behavior and Apparent Viscosity

The rheological properties of reconstituted milk and model oil-in-water emulsions were measured using a rotational rheometer (Discovery Hybrid Rheometer-2, TA Instrument, New Castle, DE, USA). All the measurments were carried out by using a cone-and-plate geometry (diameter: 40 mm, angle: 2.017°, operating gap: 55 µm) under the flow ramp, with shear rate ranging between 0.01 and 1000 s^{-1}. All the samples were equilibrated at 25 °C before measurement. After being loaded, each sample was equilibrated again for 60 s at 25 °C before test was started. All measurements were performed in triplicate. The results indicated that samples of both types are typical non-Newtonian and, therefore, the apparent viscosity at 50 s^{-1} was used for the analysis of the tribological data.

2.4. Tribology Measurement

Tribological measurements were carried out using a Soft TA-Tribometer (STAT). This device was developed based on a commercial texture analyzer TA, XTPlus (Stable Micro Systems, Surrey, UK), with a specially designed fixture attached to it. Details of an experimental setup have been given previously [34], following the operating principles described before [21]. Figure 1 shows the schematic diagram of the used tribometer; briefly, a soft probe containing three PDMS hemispheres was used to represent the hard palate, and PDMS substrate was used to represent the human tongue surface. The temperature of 25 °C was chosen for tribological measurements, to be in line with the temperature at which the samples were prepared, and Stribeck curves were obtained using a set of sliding speed between 0.1 and 20 mm/s. A normal force of 0.0394 N was applied for tribology measurement. For each measurement, about 6 mL of test sample was transferred onto the surface of a silicone elastomer using a disposable pipet, giving a thickness of fluid film around 2.5 mm. Care must be taken to ensure no air bubble was created during sample transition.

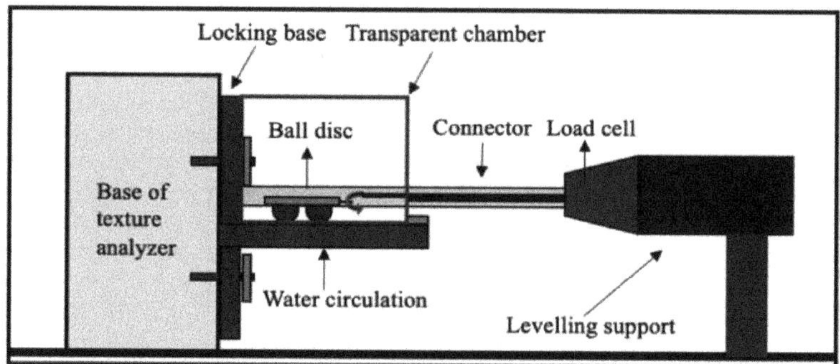

Figure 1. An illustration of the experiment setup 'Soft Texture Analyzer Tribometer'.

Test data (force, distance, and speed, as well as time) were recorded automatically by the Exponent software (Exponent, Stable Microsystems, version 6.1.9.0, London, UK). Dividing the friction force (N) by the surface load (N) gives the apparent friction coefficient. However, for convenience, all apparent friction coefficients were expressed as friction coefficient in this work, a general practice, which is widely accepted in the literature. Results were the average ± the standard deviation of at least three replicate runs conducted for each experiment.

2.5. Sensory Analysis

Reconstituted milk and thickened samples were evaluated by 22 pre-trained panelists (11 males, 11 females, mean age: 23 years) at the Laboratory of Food Oral Processing, School of Food Science and Biotechnology, Zhejiang Gongshang University. All experiments were performed in compliance with the relevant laws, and an ethical approval was given for this study by the Research Ethics Committee of the Zhejiang Gongshang University (Approval Code: 2020041011). As a selection criterion, all participants were frequent milk drinkers, both of skimmed and full-fat milk. Assessors were asked to refrain from eating and drinking at least 2 h prior to sensory analysis, except for water. They received instructions regarding the evaluation procedure in both written and verbal formats, prior to sample evaluation. All assessors were comprehended about the triangle test and the rating system. A signed consent form was obtained from each subject before the experiment officially started. All experiments were carried out in a specially designated laboratory, where temperature and light intensity can be adjusted. The sensory test consists of two different parts:

2.6. Triangle Test and Intensity Scoring with Sensory Descriptors

In the first session, assessors were presented with two sets of milk samples simultaneously, each consisting of three samples of reconstituted milk or thickened fluids. Same mean droplet size, $d_{90} = 0.3$ µm, was maintained for each set of samples, but with varied fat content (either 2% or 7.5%).

In the second session, assessors were also presented with two sets of milk samples simultaneously, consisting of reconstituted milk and thickened fluids. Each set (with same fat content 7.5%, differed with droplet size,) consisted of three samples with the same fat content of 7.5 wt.%. Mean droplet size was set at either $d_{90} = 0.3$ µm or $d_{90} = 3.5$ µm, respectively. Two of the samples have the same fat droplet size, but the other one is different.

During sensory analysis, all assessors wore nose clips to prevent odor interference. Assessors were asked to sit still in a natural and comfortable manner, while they were guided to deposited 2 mL sample onto the top surface of the tongue. The assessors were asked to move the tongue tip against the palate in a parallel motion (to mimic a sliding movement), while ensuring no swallowing of the sample occurred. They were then asked to spit the sample into the wash basin after 3 s. Between each sample, bottled water was used for mouth cleansing. Assessors were allowed to repeatedly test the same sample, but trials should be no more than three times. After assessing all three samples, the assessor was then requested to determine which one was the odd sample. Assessors were also asked to discriminate samples using the attributes provided on the paper ballot and/or provide other sensory feelings they may experience. A six-point analytical rating scale with descriptions of degrees of each attribute was also provided, and subjects gave a mark for each sensory attribute on the scale. The ratings were then converted to a number between 0 (left) and 5 (right) (0 = not at all and 5 = very). Sensory attributes for assessment include smoothness, creaminess, and thickness. Smoothness was assumed to be proportional to the perceived friction force from the hard palate moving against the tongue surface. Creaminess was defined as the perception of 'oiliness' in the mouth and the degree of mouth coating. Thickness was defined as the sensed resistance to flow in the mouth.

2.7. Statistical Analyze

Statistical analysis was performed by SPSS software Version 25 (IBM, Chicago, IL, USA). The Shapiro–Wilk normality test was used to determine the normality; otherwise, a nonparametric test was used to analyze the differences between groups. Analysis of variance (ANOVA) and a nonparametric test (Mann–Whitney U Test) were used to determine the significance of difference between the friction coefficient and averaged sensory attributes.

3. Results and Discussion

3.1. Emulsion Characteristics

As the droplet size distribution might influence the texture and sensory perception, it is one of most important parameters that describe emulsion systems. The droplet size distribution data, for further analysis, was commonly presented by the form of its intensity and number as well as volume [35,36]. In this study, we presented the data with volume intensity for a clear observation. The droplet size distributions of reconstituted and thickened milk samples obtained after different homogenization procedures are shown in Figure 2, where volume density is plotted against the droplet size for selected milk samples (specific compositions of tested samples are shown in Table 1). As expected, the volume fraction of reconstituted milk samples (0.3, 0.7, 2, 3.5, 7.5 wt.% fat) homogenized using a shear homogenizer showed bimodal distribution, with its first peak occurring between 0.01–1 µm and its second peak between 1–10 µm. Thickened samples with fat content of 3.5 and 7.5 wt.% showed multimodal distribution with a third peak at 10–100 µm (Figure 2A,B). Figure 2C shows the fat droplet distribution of reconstituted and thickened milk samples after high-pressure homogenization. Typical monomodal distribution was observed for these samples, though a small tail was observable for samples containing

3.5% and 7.5% fat. The measured mean droplet size shows no statistical difference ($p > 0.05$) among these samples (see Figure 2C).

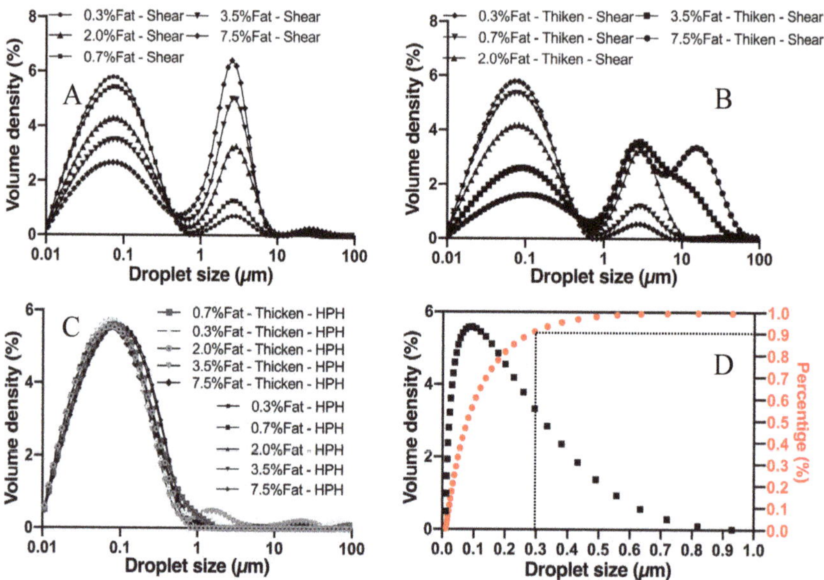

Figure 2. Droplet size distributions from the Malvern Mastersizer for reconstituted milk and thickened samples: (**A**) reconstituted milk homogenized under a shear homogenizer, (**B**) thickened reconstituted milk homogenized under a shear homogenizer, (**C**) reconstituted milk and thickened samples homogenized under HPH, and (**D**) droplet size accumulation curve.

Table 1. Composition of reconstituted milk and thickened samples.

Sample Name	Protein Concentration (wt.%)	Fat Concentration (wt.%)	Xanthan Concentration (wt.%)
0.3%Fat-Shear/HPH	3.74	0.30	
0.7%Fat-Shear/HPH	3.57	0.70	
2.0%Fat-Shear/HPH	3.51	2.00	None
3.5%Fat-Shear/HPH	3.44	3.50	
7.5%Fat-Shear/HPH	3.25	7.50	
0.3%Fat-T-Shear/HPH	3.74	0.30	
0.7%Fat-T-Shear/HPH	3.57	0.70	
2.0%Fat-T-Shear/HPH	3.51	2.00	0.86
3.5%Fat-T-Shear/HPH	3.44	3.50	
7.5%Fat-T-Shear/HPH	3.25	7.50	

It is not yet fully certain what causes bimodality in Figure 2A,B. However, according to Laguna et al., the first peak in both the milk and thickened milk samples could correspond to free casein micelles, and the second one represents the fat globules [25]. Commercial dairy colloids, such as whole fat milk and milk cream, stabilized by milk globulin, are generally very stable, with almost no coalescence of the droplets in the presence of sufficient emulsifiers [37,38]. In the case of reconstituted milk, visual observations showed that no destabilization occurred over the timescales of the experimental tests. However, the addition of xanthan gum could lead to structural alteration to the colloidal system. It has been reported that, at a certain concentration, xanthan gum may promote flocculation of fat droplets and inhibit emulsification of fat by a depletion mechanism [39]. Thus, we intend to believe that the third peak in Figure 2B could be caused by the clusters of the fat droplets.

Figure 3 shows the flow curves of reconstituted milk and thickened samples, where the apparent viscosity of the milk sample is shown as a function of the applied shear rates. The rheology measurement was conducted over shear rates between 1 and 100 s^{-1}, of similar order to the shear rate present in the tribological contact. Both types of systems exhibited a typical shear-thinning behavior, a decrease in apparent viscosity with increased shear rate, which is in good agreement with a previous study [40]. A milk sample, with a different fat mass homogenized using Ultra-Turrax (Shear), showed a weak shear-thinning effect and had a low viscosity (<0.01 Pa·s); an increase in fat mass fraction from 0.3 to 7.5% only had a marginal effect on viscosity increase (Figure 3A), which is in agreement with previous reports [25,28]. However, milk samples homogenized by HPH had significantly higher apparent viscosity compared with shear-homogenized samples. The former also had an obvious shear-thinning behavior, indicating that there are attractive interactions between the fat globules, probably caused by depletion forces induced by the casein micelles [41]. The flow curves of thickened milk are shown in Figure 3B. It is interesting to see that flow curves are very much overlapping; the addition of commercial thickener (xanthan) increased the viscosity of emulsion from 0.005~0.5 Pa·s to 0.1~6 Pa·s and controlled flow characteristics of the system. Compared with the unthickened samples in Figure 3A, the homogenization methods and fat content seem to have very limited effect on the rheology behavior of thickened samples. Table 2 shows droplet size and apparent viscosity at 50 s^{-1} for emulsion systems with and without thickening. One can see that the apparent viscosity only showed a slight increase (but with no statistical significance, $p > 0.05$) for a nearly two-magnitude reduction of fat droplet size.

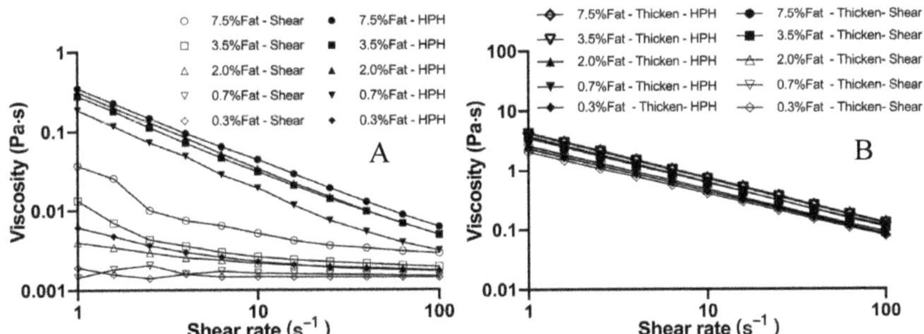

Figure 3. Flow curves for reconstituted milk (**A**) and thickened samples (**B**) with fat mass fractions of 0.3, 0.7, 2.0, 3.5, and 7.5 wt.%.

Table 2. Apparent viscosity (Pa·s) measured at 50 s^{-1} of milk samples with constant fat content and various droplet size distribution.

	7.5 wt.% Milk		7.5 wt.% Thickened Milk	
	Fat Droplet Size (d$_{90}$, μm)	**Viscosity (50 s^{-1})** (Pa·s)	**Fat Droplet Size** (d$_{90}$, μm)	**Viscosity (50 s^{-1})** (Pa·s)
	4.617	0.00915	20.733	0.153
	2.370	0.00855	8.686	0.163
	0.990	0.00939	4.070	0.151
	0.299	0.0111	1.460	0.169
			0.301	0.213
Significance	None		None	

3.2. Tribology: Lubrication Behavior of Milk and Thickened Milk Samples

Figure 4 shows the friction curves for all samples, in which the friction coefficient is plotted as a function of the sliding speed. We can see that two lubrication regimes, the boundary regime and the mixed regime, are clearly identifiable in most cases. The boundary regime at low sliding speeds has a friction coefficient nearly independent of the speed. At this regime, the load is supported by the asperity contact due to the absence of hydrodynamic pressure. At higher sliding speeds, the two surfaces start to separate from each other, and the lubrication effect starts to kick in, where a reduced friction coefficient becomes obvious for all sample systems.

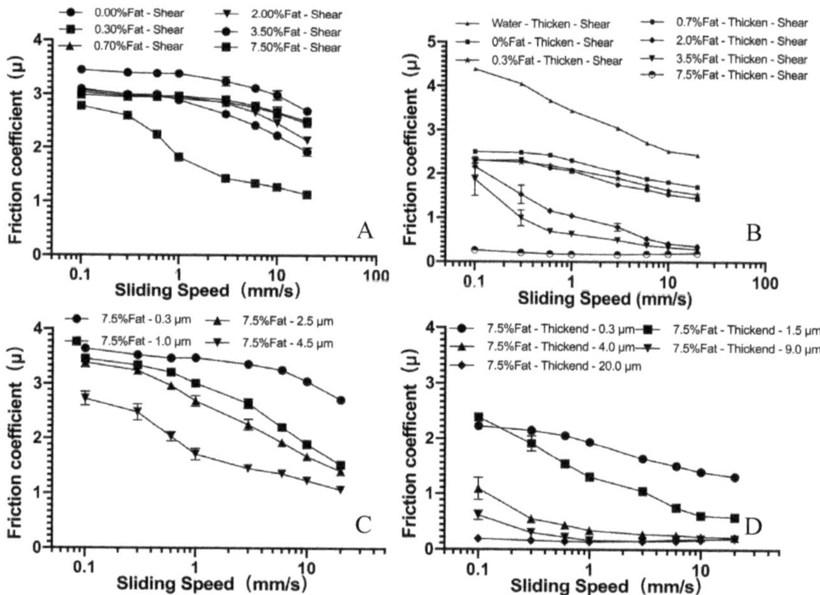

Figure 4. Stribeck curves for reconstituted milk and thickened samples, (**A**,**B**) with fat mass fractions between 0.3 and 7.5 wt.%, (**C**,**D**) with constant fat mass fraction of 7.5% and varied fat droplet size distribution. Lubrication tests were conducted at 25 °C and under a constant surface load of 0.0394 N.

3.2.1. Effect of Oil/Fat Mass Fraction on Lubrication Behavior

Figure 4A shows the Stribeck analysis for the milk samples. It is interesting to note that skimmed milk has the highest friction coefficient. It is also interesting to note that the presence of a small amount of fat droplets (i.e., 0.3%) can lead to a significant reduction of the friction coefficient at the boundary regime. This seems to suggest that a thin layer of emulsion film is sufficient to create a significant lubricating effect at low sliding speed [28]. However, it is also clear that the thin layer is not sufficient at this regime to create a floating pressure against the load for systems containing 3.5% fat and lower, probably due to the low viscosity of such systems [21]. For the highest fat content, it was found that the friction coefficient was significantly lower than that of all other systems, over the whole investigated range of sliding speed.

At higher sliding speeds, the friction coefficient started to decrease due to the capability of emulsion lubricant to separate two surfaces with a higher hydrodynamic pressure. Nonetheless, as discussed in Section 3.1, the changes in viscosity were only fractional. Therefore, we intend to believe that lubricant viscosity is not the main cause for the observed reduction of friction coefficient. The transition from boundary to mix regime at lower sliding speeds for high fat content sample is caused by the entrainment of fat droplets between the two surfaces, and the lubrication effect became evident.

Figure 4B shows the friction curves for thickened milk samples containing various amount of fat droplets, where thickened water was also shown as a reference. A pronounced lubricating effect was observed for such systems, with (1) a much smaller friction coefficient; and (2) a much earlier transition (at a lower sliding speed) from the boundary regime to the mixed regime. Most shocking, friction reduction was observed for the one containing 7.5 wt.% fat. Its friction curve remains very low and flat over the whole range of sliding speed (from 0.1 mm/s to 20 mm/s), convincingly indicating the formation of an effective lubricating film between the two moving surfaces. It is, therefore, reasonable to conclude that, compared with the reconstituted milk system, a thickened system with an increased viscosity leads to a higher hydrodynamic pressure, which may promote the entrainments of fat droplets into the lubricating interface, creating a significant reduction of the friction coefficient. The above results seem to be in line with those obtained by Chojnicka-Paszun and de Jongh, who suggested that the addition of xanthan gum might mask surface roughness and improve lubrication [42]. Moreover, the study conducted by Bongaerts confirms the importance of lubricant viscosity on friction behavior; high viscosity is no doubt the main reason enhancing the lubricating effect [19], and this conclusion is consistent with the friction curve in Figure 4A,B (the sample had various viscosity under the same fat content). However, the friction curves were various among different samples even when the viscosity of the tested system is similar (Figure 3B), suggesting that the tribological property of emulsion is not only determined by viscosity.

3.2.2. Effect of Oil/Fat Droplet Size on Lubrication Behavior

In addition to fat content, the fat droplet size has also been investigated for its effect on lubrication behavior, and the results are shown in Figure 4C,D. Bear in mind that these systems have rather similar rheological behavior and comparable shear viscosity (see Figure 3). However, it is striking to see hugely different Stribeck curves for these systems. The friction coefficient varies almost one order of magnitude at all experimental speeds. Larger droplet size seems to be beneficial and more efficient in friction reduction. For systems with similar viscosity (see Table 2), samples with larger fat droplets had a lower friction coefficient than those with smaller droplets as well as a prolonged boundary regime. This seems to further suggest that larger fat droplets are easier to be squeezed and entrained into the lubricant gap, and, as such, coalescence and accompanying film formation might occur more easily [32]. Investigations by Laguna et al. indicated that fat droplets could coalesce within the tribological contact surfaces, reducing the traction coefficient until a sufficiently high shear is established to disrupt any boundary layers [25]. Their result seems to suggest that, under inferior sliding speeds (compared to 1–1000 mm/s), boundary lubrication mediated mechanisms may not exist.

The results shown above demonstrate that increasing the mass content of fat droplets, with both a droplet distribution between 0.01 μm and 10 μm and droplet clustering (10–100 μm), can improve the lubricating effect. In order to identify the optimal distribution range of fat droplets for an enhanced lubricating effect, a series of milk and thickened samples with matching droplet size distribution have been fabricated using HPH with droplet size $d_{90} = 0.3$ μm (droplet size distributions are given in Figure 2D). Friction curves for the above-mentioned samples are shown in Figure 5, with two distinguishable patterns for the two different types of systems. The sample systems with a thickener addition show a much lower friction coefficient than their counterpart samples. However, for the same type of samples, the results seem to suggest that further reduction of droplet size may have little effect on the lubrication behavior. This observation seems contradictory to the previous report that dairy products with higher fat content should exhibit a superior lubricating effect [10,28]. Data in Figure 5 show that both systems tested under a controlled surface load of 0.0392 N do not appear to demonstrate a significantly lower friction coefficient with an increasing mass fraction of fat in both the boundary and mixed regimes, with no significant statistical difference ($p > 0.5$). For the effect of the speed of surface movement, the friction coefficient only shows a slight decrease with the increase of sliding speed.

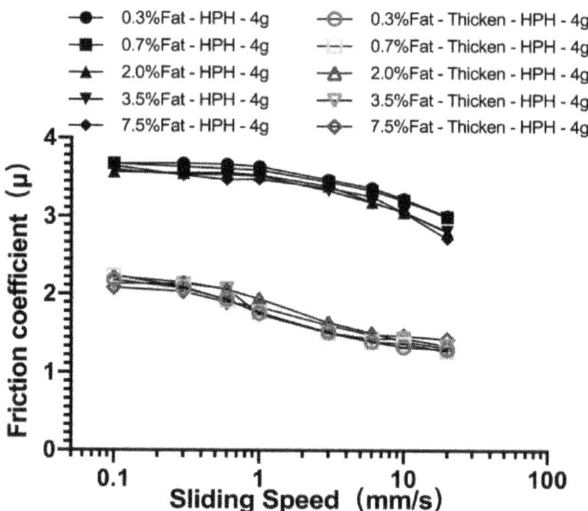

Figure 5. Stribeck curves for reconstituted milk and thickened samples with a constant fat droplet distribution and fat mass fractions between 0.3 and 7.5 wt.%. Lubrication tests were conducted at 25 °C and under surface load of 0.0394 N.

Based on the above results, one may conclude that the size of fat droplets is closely associated with this frictional phenomenon. The effect of the presence of fat droplets appears to be rather surprising: the presence of a small amount of fat content, i.e., 0.3%, leads to a big friction reduction, yet a further increase in fat content of up to 7.5% gives no further benefit. Given this, we hypothesize that the presence of fat droplets even in a limited number leads to the entrainment between moving surfaces and may help to promote lubricating effect. We further speculate that once enough droplets for entrainment exist between the two surfaces, a further increase in the number of fat droplets gives no further benefit in surface lubrication. This could be the reason that there was no further friction coefficient reduction for a system with the fat content increased from 0.3% to 7.5%. A similar observation was obtained by previous studies [25,43]. However, these authors believed that the identical tribology curve was the result of using PDMS versus PDMS as contact material. As the current study also used PDMS as medium, and the friction curves were clearly distinguished for skimmed milk and milk containing 0.3% fat, we intend to believe that substrate material is not the dominating cause, while droplet size is.

3.2.3. Relating Lubrication Behavior of Model Emulsion with Oil Droplet Size

The lubricating mechanism of milk fat droplet size was also conducted on vegetable oil (olive). Prior to lubrication study, the flow behavior of CSS-stabilized emulsions has to be investigated; the results are shown in Figure 6, so we can see that the apparent shear viscosity against the shear rate exhibits a near-Newtonian flow behavior for all studied systems. A mild influence of oil content and droplet size on shear viscosity is observable. However, all emulsion systems have a relatively low viscosity, particular at the higher end of shear rates. These results seem to be in contradiction with previous studies, where almost all emulsion systems appear to have a significant increase in apparent viscosity with the increase in fat content or fat droplet size [27,44]. The reason could be due to the different shear rate range for observation. At a low shear rate when there is no disruption of internal structure, the shear viscosity will generally show a very high value. Once internal structure is disrupted at a high shear rate, viscosity increase will usually be diminished.

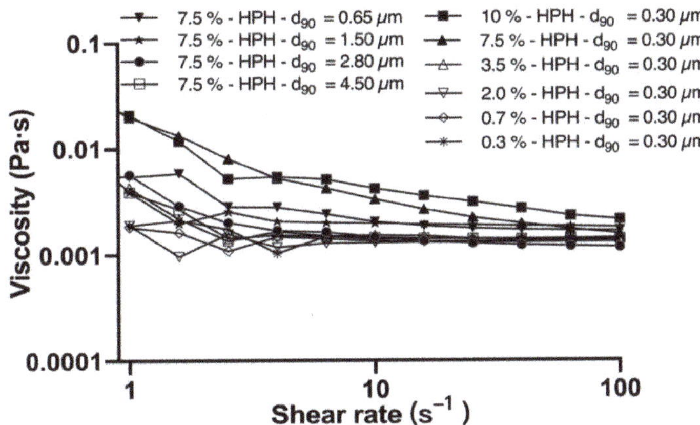

Figure 6. Flow *curves for CSS stabilized emulsions with oil* mass fractions of 0.3, 0.7, 2.0, 3.5, 7.5, and 10 wt.%.

The lubrication behavior of the above emulsions has been carefully studied, and the results are given in Figure 7, where the friction coefficient is plotted against the sliding speed. As show in Figure 7A, friction curves show a high similarity for all emulsion systems of the same droplet size distribution, yet with a different oil content (varying from 0.3% to 10%) ($p > 0.05$). All systems show a monotonous decrease with the increase of speed; however, no clear transition from boundary to mixed regime or from mixed to hydrodynamic regime could be observed. Compared to the work of Upadhyay, the measured friction was correlated negatively with the oil content (1–30%) and the friction coefficient was significantly different ($p < 0.05$), despite four types of emulsifiers being used [27]. It should be noted that the span of oil content in the study by Upadhyay was much higher compared with our work. A high concentration of droplets may alter the way of droplet entrainment during lubrication and, therefore, causes a considerable change to lubrication behavior.

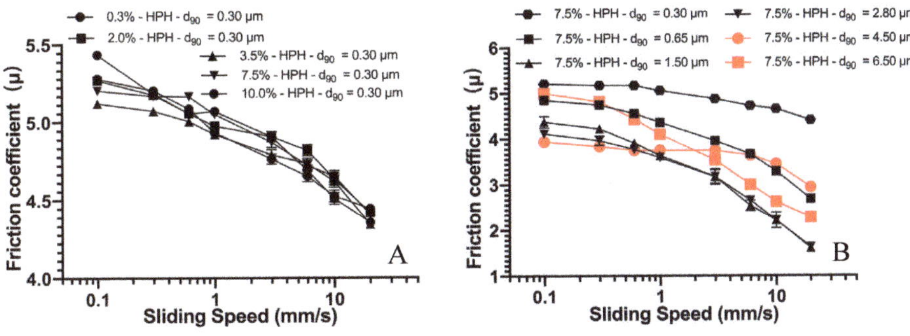

Figure 7. Stribeck curves for *CSS*-stabilized o/w emulsions with oil mass fractions between 0.3 and 10 wt.% with the same fat droplet distribution (**A**), with constant oil mass fractions 7.5% and varied oil droplet size distribution (**B**). Lubrication tests were conducted at 25 °C and under surface load of 0.0392 N.

The friction curves of emulsion with the same oil content but different droplet size distribution is shown in Figure 7B. It is interesting to see that, despite all emulsion systems showing a gradual decrease in the friction coefficient with the increase in sliding speed, the droplet size appears to have a significant influence on the friction coefficient. The friction coefficient shows significant decrease, when increasing the droplet size over the whole range of the sliding speed, even when the volume fraction was the same. From the data in Figure 7B, one may find that emulsions with a droplet size distribution d_{90} value at 1.5 and 2.8 μm have the optimal lubricating effect. Once the droplet size is larger than 4.5 microns, inferior lubrication seems to become the case. The exact reason is not yet clear. However, one possible explanation could be that large droplets may have difficulty for entrainment, in particular when droplets are close in size to the thickness of the lubricating film. Another possibility could be the limited number of available droplets for entrainment (since the oil content remained the same for all emulsion systems). Nevertheless, our results again suggest that appropriate droplet entrainment is the key for emulsion lubrication.

To take the results a step further, we combined the droplet size distribution and friction results together for discussion (as shown in Figure 8). As described in Figure 7A, the lubricating effect of nano-sized emulsion seems independent of fat/oil content; in other words, adjusting the volume fraction of fat/oil droplets in the white area in Figure 8A should not interfere with the friction curve. When the fat content is kept constant, increasing the percentage of large oil droplets (show as the patterned area in Figure 8A,B) facilitated the lubricating effect. However, once the droplet size reached d_{90} = 2.8, further increasing the droplet size in the system (blue area in Figure 8C) seems to have little contribution to the lubricating effect (Figure 8C).

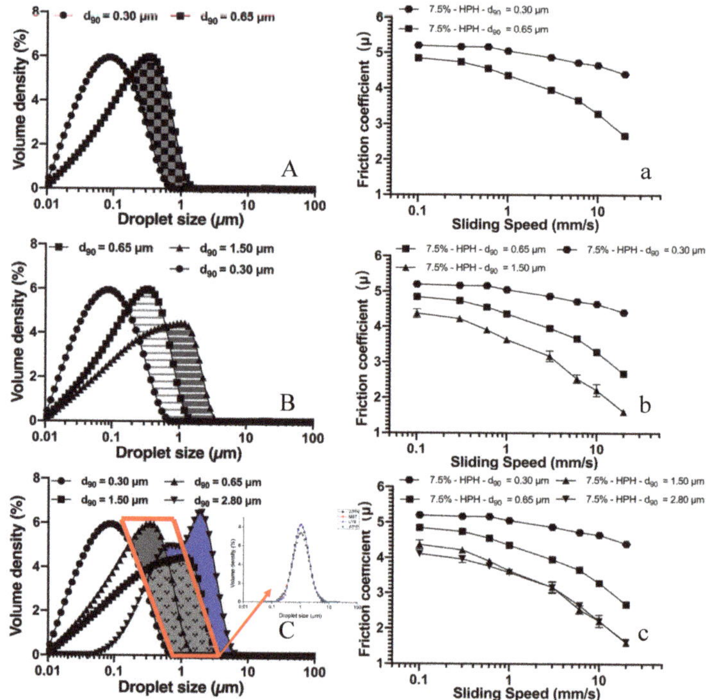

Figure 8. A schematic diagram showing the range of 'lubricating fat droplets' and its friction curves. Figures with capitalized letter shows the droplets distribution of emulsion sample with constant oil mass fraction but varied droplets size (**A–C**), figures with minuscules letter shows its friction curve (**a–c**).

Based on the results shown above, a schematic representation of the 'lubricating fat droplets' range was built, referred to as the red box in Figure 8C. The lubricating effect of samples with fat droplet size, mainly distributed on the left and right sides of the red box, appears to be poorer compared to samples whose fat/oil droplets are distributed within the red box. The distribution of the speculated range of oil droplets, which can facilitate the lubricating effect, is in accordance with the work by Upadhyay and Chen, with normalized oil droplet distribution (show in the insert picture in Figure 8C) and an increase in the oil mass fraction from 1% to 30% demonstrating a significantly lowered friction coefficient, although a different emulsifier was used [27].

3.3. Sensory Analysis

Discrimination and rating test analysis were used to collect sensory data; the three most-used sensory attributes and paraphrases were selected and provided to consumers. For a triangle test with total 22 responses, the minimum number of correct responses required for significance at $p < 0.001$ and $p < 0.01$ is 15 and 12, respectively (ISO 4120:2021 Sensory analysis—Methodology—Triangle test). Table 3 shows that the number of panelists who were able to discriminate between milk or thickened milk samples with constant fat content or fat droplet size was statistically significant.

Table 3. Number of correct answers for sensory analysis using a discrimination test for milk and thickened milk samples.

Sample	Number of Correct Answers	Total Responses
2%/7.5% fat milk with same fat droplet size ($d_{90} = 0.3$ μm)	12 *	22
7.5% fat milk with different fat droplet size ($d_{90} = 0.3/3.5$ μm)	13 *	22
2%/7.5% fat thickened milk with same fat droplet size ($d_{90} = 0.3$ μm)	20 ***	22
7.5% fat thickened milk with different fat droplet size ($d_{90} = 0.3/4$ μm)	17 ***	22

* Significance at 5%, *** Significance at 0.1%.

As it can be observed in Table 3, for the discrimination ability by the panelists of reconstituted milk with different fat content but with the same fat droplet size ($d_{90} = 0.3$ μm), the number of correct responses just meets the significant level. In addition, for the other three sets, panelists showed a higher discrimination capability. The selected samples were successfully differentiated in terms of thickness, creaminess, and smoothness ($p < 0.05$). Figure 9 shows the average sensory attribute intensity scores with a standard error for all samples.

A sensory discrimination of samples with the same droplet size distribution but different oil content is shown in Figure 9A,C. The mean intensity rating of 'smoothness' did not show significant difference. This seems to agree with the lubrication results, where the friction-overlapped curves were observed for these samples (Figure 5). However, the emulsion of high oil content gives a significantly higher perception of 'thickness' and 'creaminess', with statistical significance. It suggests that perceived 'thickness' and 'creaminess' are the dominating sensory attributes used by panelists for discrimination of these emulsion samples.

Figure 9B shows sensory results for milk samples with the same amount of oil volume fraction (7.5%) but rather different droplet size distribution ($d_{90} = 0.3$ μm and 3.5 μm, respectively). Significant differences in perceived intensity of 'smoothness' ($p < 0.05$) were reported by the panelists, despite no significant differences in perceiving 'thickness' and 'smoothness'. As expected, milk samples presenting high ranking in the 'smoothness' attributes depicted the lowest friction coefficient, as it was assumed to be negatively related to the frictional force caused by contact between the two surfaces [45]. However, this trend was not seen for further-thickened samples. As seen in Figure 9D, panelists appear to be unable to differentiate between the smoothness of the two samples, despite the significant difference in their friction behavior.

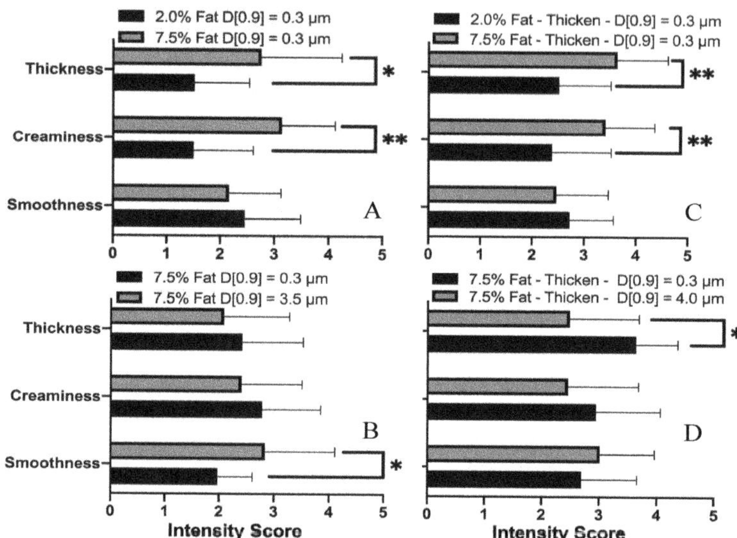

Figure 9. Ranking of selected sensory attributes obtained for samples with constant droplets size but with different fat content (**A,C**), or with same fat content but various fat droplet size (**B,D**). * Significance at 5%, ** Significance at 1%.

A phenomenon worth noting is that compared to a sample with fat content of 2%, panelists rated 'creaminess' significantly higher for samples with fat content of 7.5% (Figure 9A,C); reduced fat droplet size also showed an enhanced 'creaminess' sensation (Figure 9B,D) for samples with fat content of 7.5%, although it was not significant. It should also be noted that 'creaminess' is a very complex, multimodal sensory attribute involving olfactory, gustatory, and tactile cues [12]. Kokini first developed a texture-dominating model which describes 'creaminess' as a function of 'smoothness' and 'thickness' [45]. However, Chen and Eaton later indicated that the perception of 'creaminess' is strongly influenced by olfactory sensation [46]. Therefore, rather than a single stimulation, the perceived 'creaminess' should be a combined result or an integrated result of multiple sensations. Nevertheless, except for the negative correlation with droplet size-related data, a sample with a higher 'creaminess' score also showed a higher apparent viscosity (although not significant), which indicates a positive correlation between the 'creaminess' score and the rheology data, as observed by Sonne [47].

The results from the sensory test suggest that the panelists were able to differentiate samples with the same droplet size distribution but with different levels of oil/fat content, even though the perceived 'smoothness' may not play a dominant role in the discrimination. For thickened samples, almost all panelists were able to identify the difference within two sets of samples, mostly likely based on the perceived 'thickness' and 'creaminess'. Moreover, it was noticed that the xanthan addition could cause an undesirable 'slimy' feeling, which may interfere with the sensation of 'smoothness' attribute as reported by a few participants. This may explain why milk samples with a similar 'smoothness' rating have a very different in vitro lubrication behavior.

We should also bear in mind that saliva is a significant factor in the oral processing of food and plays an important role in texture perception. During oral processing, oil droplets may experience flocculation due to the participation of salivary proteins. Even though the oral destabilization of food emulsions is beyond the scope of the current study, one should be aware that the different environment between in vitro tribological tests and real oral sensory analysis could also be an important reason for some discrepancy observed between the two sets of results.

4. Conclusions

Droplet size and oil/fat content are the two most important controlling factors for the design and production of food emulsion, playing an equally important role as surfactants, particular in dairy and other o/w emulsion products. How such factors affect the lubrication behavior and then oral sensation is a core question for oral lubrication research of food emulsions. For this reason, this study has carefully reconstituted sets of milk samples with carefully controlled oil/fat content and droplet size distribution as well as shear viscosity; their lubrication behavior was carefully analyzed. Sensory analysis has also been conducted, in trying to reveal the sensory meaning of the lubrication data. Obtained Stribeck curves gave a clear discrimination for emulsions that vary in oil/fat droplet size (but with constant fat content) and for emulsions that vary in oil content (but with the same droplet size). Besides, results from this study demonstrated that droplet size played an important role in the flow and friction behavior of o/w emulsion systems. An increase in the volume fraction of oil/fat droplets within a certain range exhibits a significant effect on the lubricating effect. Results from sensory analysis showed a majority of panelists were able to discriminate milk samples with different lubrication behavior.

Author Contributions: Conceptualization, J.C. and Y.Z.; methodology, Q.W.; validation, J.C., Y.Z. and Q.W.; formal analysis, Y.Z.; investigation, Q.W. and Z.J.; resources, Q.W. and Z.J.; data curation, Q.W. and Z.J.; writing—original draft preparation, Q.W.; writing—review and editing, Y.Z. and J.C.; visualization, Q.W.; supervision, Y.Z. and J.C.; project administration, J.C.; funding acquisition, J.C. and Y.Z. All authors have read and agreed to the published version of the manuscript.

Funding: This research was funded by National Natural Science Foundation of China (Grant No. 31871885, 32001833).

Institutional Review Board Statement: The study was conducted according to the guidelines of the Declaration of Helsinki, and approved by the Institutional Review Board (or Ethics Committee) of the Zhejiang Gongshang University (protocol code 2020041011 and 10 April 2020).

Informed Consent Statement: Informed consent was obtained from all subjects involved in the study.

Conflicts of Interest: The authors declare no conflict of interest.

References

1. Bluher, M. Obesity: Global epidemiology and pathogenesis. *Nat. Rev. Endocrinol.* **2019**, *15*, 288–298. [CrossRef]
2. McClements, D.J.; Barrangou, R.; Hill, C.; Kokini, J.L.; Lila, M.A.; Meyer, A.S.; Yu, L.L. Building a Resilient, Sustainable, and Healthier Food Supply Through Innovation and Technology. *Annu. Rev. Food Sci. Technol.* **2021**, *12*, 1–28. [CrossRef] [PubMed]
3. Chung, C.; Degner, B.; McClements, D.J. Designing reduced-fat food emulsions: Locust bean gum-fat droplet interactions. *Food Hydrocoll.* **2013**, *32*, 263–270. [CrossRef]
4. Di Cicco, F.; Oosterlinck, F.; Tromp, H.; Sein, A. Comparative study of whey protein isolate gel and polydimethylsiloxane as tribological surfaces to differentiate friction properties of commercial yogurts. *Food Hydrocoll.* **2019**, *97*, 105204. [CrossRef]
5. Laiho, S.; Williams, R.P.W.; Poelman, A.; Appelqvist, I.; Logan, A. Effect of whey protein phase volume on the tribology, rheology and sensory properties of fat-free stirred yoghurts. *Food Hydrocoll.* **2017**, *67*, 166–177. [CrossRef]
6. McClements, D.J. Colloidal basis of emulsion color. *Curr. Opin. Colloid Interface Sci.* **2002**, *7*, 451–455. [CrossRef]
7. Benjamins, J.; Vingerhoeds, M.H.; Zoet, F.D.; de Hoog, E.H.A.; van Aken, G.A. Partial coalescence as a tool to control sensory perception of emulsions. *Food Hydrocoll.* **2009**, *23*, 102–115. [CrossRef]
8. Mosca, A.C.; Rocha, J.A.; Sala, G.; van de Velde, F.; Stieger, M. Inhomogeneous distribution of fat enhances the perception of fat-related sensory attributes in gelled foods. *Food Hydrocoll.* **2012**, *27*, 448–455. [CrossRef]
9. Liu, K.; Stieger, M.; van der Linden, E.; van de Velde, F. Fat droplet characteristics affect rheological, tribological and sensory properties of food gels. *Food Hydrocoll.* **2015**, *44*, 244–259. [CrossRef]
10. Li, Y.; Joyner, H.S.; Carter, B.G.; Drake, M.A. Effects of fat content, pasteurization method, homogenization pressure, and storage time on the mechanical and sensory properties of bovine milk. *J. Dairy Sci.* **2018**, *101*, 2941–2955. [CrossRef]
11. Jones, S.A.; Roller, S. *Handbook of Fat Replacers*, 1st ed.; CRC Press: Boca Raton, FL, USA, 1996. [CrossRef]
12. Upadhyay, R.; Aktar, T.; Chen, J. Perception of creaminess in foods. *J. Texture Stud.* **2020**, *51*, 375–388. [CrossRef] [PubMed]
13. Cutler, A.N.; Morris, E.R.; Taylor, L.J. Oral perception of viscosity in fluid foods and model systems. *J. Texture Stud.* **1983**, *14*, 377–395. [CrossRef]
14. Dickie, A.M.; Kokini, J.L. An improved model for food thickness from non-Newtonian fluid mechanics in the mouth. *J. Food Sci.* **1983**, *48*, 57–61. [CrossRef]

15. Stanley, N.L.; Taylor, L.J. Rheological basis of oral characteristics of fluid and semi-solid foods: A review. *Acta Psychol.* **1993**, *84*, 79–92. [CrossRef]
16. Chen, J.; Stokes, J.R. Rheology and tribology: Two distinctive regimes of food texture sensation. *Trends Food Sci. Technol.* **2012**, *25*, 4–12. [CrossRef]
17. Sarkar, A.; Andablo-Reyes, E.; Bryant, M.; Dowson, D.; Neville, A. Lubrication of soft oral surfaces. *Curr. Opin. Colloid Interface Sci.* **2019**, *39*, 61–75. [CrossRef]
18. Shewan, H.M.; Pradal, C.; Stokes, J.R. Tribology and its growing use toward the study of food oral processing and sensory perception. *J. Texture Stud.* **2020**, *51*, 7–22. [CrossRef] [PubMed]
19. Bongaerts, J.H.H.; Fourtouni, K.; Stokes, J.R. Soft-tribology: Lubrication in a compliant PDMS–PDMS contact. *Tribol. Int.* **2007**, *40*, 1531–1542. [CrossRef]
20. Taylor, B.L.; Mills, T.B. Using a three-ball-on-plate configuration for soft tribology applications. *J. Food Eng.* **2020**, *274*, 109838. [CrossRef]
21. Wang, Q.; Wang, X.; Chen, J. A new design of soft texture analyzer tribometer (STAT) for in vitro oral lubrication study. *Food Hydrocoll.* **2020**, *110*, 106146. [CrossRef]
22. de Vicente, J.; Stokes, J.R.; Spikes, H.A. Soft lubrication of model hydrocolloids. *Food Hydrocoll.* **2006**, *20*, 483–491. [CrossRef]
23. de Vicente, J.; Stokes, J.; Spikes, H.A. The Frictional Properties of Newtonian Fluids in Rolling–Sliding soft-EHL Contact. *Tribol. Lett.* **2005**, *20*, 273–286. [CrossRef]
24. Selway, N.; Stokes, J.R. Insights into the dynamics of oral lubrication and mouthfeel using soft tribology: Differentiating semi-fluid foods with similar rheology. *Food Res. Int.* **2013**, *54*, 423–431. [CrossRef]
25. Laguna, L.; Farrell, G.; Bryant, M.; Morina, A.; Sarkar, A. Relating rheology and tribology of commercial dairy colloids to sensory perception. *Food Funct.* **2017**, *8*, 563–573. [CrossRef] [PubMed]
26. Malone, M.E.; Appelqvist, I.A.M.; Norton, I.T. Oral behaviour of food hydrocolloids and emulsions. Part 1. Lubrication and deposition considerations. *Food Hydrocoll.* **2003**, *17*, 763–773. [CrossRef]
27. Upadhyay, R.; Chen, J. Smoothness as a tactile percept: Correlating 'oral' tribology with sensory measurements. *Food Hydrocoll.* **2019**, *87*, 38–47. [CrossRef]
28. Chojnicka-Paszun, A.; de Jongh, H.H.J.; de Kruif, C.G. Sensory perception and lubrication properties of milk: Influence of fat content. *Int. Dairy J.* **2012**, *26*, 15–22. [CrossRef]
29. Dresselhuis, D.M.; de Hoog, E.H.A.; Stuart, M.A.C.; Vingerhoeds, M.H.; van Aken, G.A. The occurrence of in-mouth coalescence of emulsion droplets in relation to perception of fat. *Food Hydrocoll.* **2008**, *22*, 1170–1183. [CrossRef]
30. Douaire, M.; Stephenson, T.; Norton, I.T. Soft tribology of oil-continuous emulsions. *J. Food Eng.* **2014**, *139*, 24–30. [CrossRef]
31. Dresselhuis, D.M.; Klok, H.J.; Stuart, M.A.C.; de Vries, R.J.; van Aken, G.A.; de Hoog, E.H.A. Tribology of o/w emulsions under mouth-like conditions: Determinants of friction. *Food Biophys.* **2007**, *2*, 158–171. [CrossRef]
32. Fuhrmann, P.L.; Kalisvaart, L.C.M.; Sala, G.; Scholten, E.; Stieger, M. Clustering of oil droplets in o/w emulsions enhances perception of oil-related sensory attributes. *Food Hydrocoll.* **2019**, *97*, 105215. [CrossRef]
33. Lett, A.M.; Norton, J.E.; Yeomans, M.R. Emulsion oil droplet size significantly affects satiety: A pre-ingestive approach. *Appetite* **2016**, *96*, 18–24. [CrossRef]
34. Chen, J.; Liu, Z.; Prakash, S. Lubrication studies of fluid food using a simple experimental set up. *Food Hydrocoll.* **2014**, *42*, 100–105. [CrossRef]
35. Smułek, W.; Siejak, P.; Fathordoobady, F.; Masewicz, Ł.; Guo, Y.; Jarzębska, M.; Kitts, D.D.; Kowalczewski, P.Ł.; Baranowska, H.M.; Stangierski, J.; et al. Whey proteins as a potential co-surfactant with Aesculus hippocastanum L. as a stabilizer in nanoemulsions derived from hempseed oil. *Molecules* **2021**, *26*, 5856. [CrossRef] [PubMed]
36. Jarzębski, M.; Siejak, P.; Smułek, W.; Fathordoobady, F.; Guo, Y.; Pawlicz, J.; Trzeciak, T.; Kowalczewski, P.Ł.; Kitts, D.D.; Singh, A.; et al. Plant extracts containing saponins affects the stability and biological activity of hempseed oil emulsion system. *Molecules* **2020**, *25*, 2696. [CrossRef] [PubMed]
37. Hinderink, E.B.A.; Kaade, W.; Sagis, L.; Schroën, K.; Berton-Carabin, C.C. Microfluidic investigation of the coalescence susceptibility of pea protein-stabilised emulsions: Effect of protein oxidation level. *Food Hydrocoll.* **2020**, *102*, 105610. [CrossRef]
38. Van Aken, G.A.; Zoet, F.D. Coalescence in highly concentrated coarse emulsions. *Langmuir* **2000**, *16*, 7131–7138. [CrossRef]
39. McClements, D.J. Food emulsions: Principles, practices, and techniques: Second edition. In *Food Emulsions: Principles, Practices, and Techniques*, 2nd ed.; CRC Press: Boca Raton, FL, USA, 2004.
40. Nguyen, P.T.M.; Bhandari, B.; Prakash, S. Tribological method to measure lubricating properties of dairy products. *J. Food Eng.* **2016**, *168*, 27–34. [CrossRef]
41. Grotenhuis, E.; Tuinier, R.; De Kruif, C.G. Phase stability of concentrated dairy products. *J. Dairy Sci.* **2003**, *86*, 764–769. [CrossRef]
42. Chojnicka-Paszun, A.; de Jongh, H.H.J. Friction properties of oral surface analogs and their interaction with polysaccharide/MCC particle dispersions. *Food Res. Int.* **2014**, *62*, 1020–1028. [CrossRef]
43. Godoi, F.C.; Bhandari, B.R.; Prakash, S. Tribo-rheology and sensory analysis of a dairy semi-solid. *Food Hydrocoll.* **2017**, *70*, 240–250. [CrossRef]
44. Joyner, H.; Pernell, C.W.; Daubert, C. Impact of Oil-in-Water emulsion composition and preparation method on emulsion physical properties and friction behaviors. *Tribol. Lett.* **2014**, *56*, 143–160. [CrossRef]
45. Kokini, J.L.; Kadane, J.B.; Cussler, E.L. Liquid texture perceived in the mouth. *J. Texture Stud.* **1977**, *8*, 195–218. [CrossRef]

46. Chen, J.; Eaton, L. Multimodal mechanisms of food creaminess sensation. *Food Funct.* **2012**, *3*, 1265–1270. [CrossRef] [PubMed]
47. Sonne, A.; Busch-Stockfisch, M.; Weiss, J.; Hinrichs, J. Improved mapping of in-mouth creaminess of semi-solid dairy products by combining rheology, particle size, and tribology data. *LWT Food Sci. Technol.* **2014**, *59*, 342–347. [CrossRef]

 *foods*

Article

Structural and Physicochemical Characteristics of Oil Bodies from Hemp Seeds (*Cannabis sativa* L.)

Francesca Louise Garcia, Sihan Ma, Anant Dave and Alejandra Acevedo-Fani *

Riddet Institute, Massey University, Private Bag 11222, Palmerston North 4442, New Zealand; francescalouise.garcia@gmail.com (F.L.G.); S.Ma@massey.ac.nz (S.M.); anant.dave@fonterra.com (A.D.)
* Correspondence: a.acevedo-fani@massey.ac.nz

Citation: Garcia, F.L.; Ma, S.; Dave, A.; Acevedo-Fani, A. Structural and Physicochemical Characteristics of Oil Bodies from Hemp Seeds (*Cannabis sativa* L.). *Foods* **2021**, *10*, 2930. https://doi.org/10.3390/2930 foods10122930

Academic Editor: Dennis Fiorini

Received: 29 October 2021
Accepted: 19 November 2021
Published: 26 November 2021

Publisher's Note: MDPI stays neutral with regard to jurisdictional claims in published maps and institutional affiliations.

Abstract: The structural and physicochemical characteristics of oil bodies from hemp seeds were explored in this study. Oil bodies from several plant-based sources have been previously studied; however, this is the first time a characterisation of oil bodies from the seeds of industrial hemp is provided. The morphology of oil bodies in hemp seeds and after extraction was investigated using cryo-scanning electron microscopy (cryo-SEM), and the interfacial characteristics of isolated oil bodies were studied by confocal laser scanning microscopy (CLSM). Proteins associated with oil bodies were characterised using sodium dodecyl sulphate polyacrylamide gel electrophoresis (SDS-PAGE). The effect of pH and ionic strength on colloidal properties of the oil bodies was investigated. Oil bodies in hemp seeds appeared spherical and sporadically distributed in the cell, with diameters of 3 to 5 μm. CLSM images of isolated oil bodies revealed the uniform distribution of phospholipids and proteins at their interface. Polyunsaturated fatty acids were predominant in the lipid fraction and linoleic acid accounted for ≈61% of the total fatty acids. The SDS-PAGE analysis of washed and purified oil bodies revealed major bands at 15 kDa and 50–25 kDa, which could be linked to membrane-specific proteins of oil bodies or extraneous proteins. The colloidal stability of oil bodies in different pH environments indicated that the isoelectric point was between pH 4 and 4.5, where oil bodies experienced maximum aggregation. Changes in the ionic strength decreased the interfacial charge density of oil bodies (ζ-potential), but it did not affect their mean particle size. This suggested that the steric hindrance provided by membrane-specific proteins at the interface of the oil bodies could have prevented them from flocculation at low interfacial charge density. The results of this study provide new tertiary knowledge on the structure, composition, and colloidal properties of oil bodies extracted from hemp seeds, which could be used as natural emulsions or lipid-based delivery systems for food products.

Keywords: hempseeds; physicochemical properties; oil bodies; oleosomes; structure; delivery systems

1. Introduction

Hemp (*Cannabis sativa* L.) cultivars with less than 0.3% or 0.2% of delta-9-tetrahydrocan nabinol (THC) are actively grown for several industrial purposes including food applications. Production of hemp seeds from these varieties is rapidly increasing because of their great nutritional and functional value. The seeds from hemp are an excellent source of dietary lipid, protein, and fibre. The oil extracted from the hemp seeds contains more than 90% of polyunsaturated fatty acids (PUFAs) including two essential fatty acids (EFAs), which are linoleic acid and α-linolenic acid [1,2]. The high PUFA and EFA contents of hemp seed oil can improve health and development when included in the diet as EFAs cannot be synthesised by the human body. Moreover, hemp seed oil has an ω-6/ω-3 fatty acid ratio of around 2.5:1 that falls within the optimal balance of 2:1 and 3:1; this ratio is considered essential for the metabolism of PUFAs [1]. In addition, hemp seed oil contains bioactive compounds, such as tocopherols, flavonoids, chlorophyll pigments, and phenolic compounds [3]. Because of its high nutritive value, hemp seed oil shows great potential as a bioactive food ingredient. The oil is usually extracted by cold-pressing to preserve its bioactive compounds. Solvent extraction (usually carried out with hexane) is also used to increase the yield and reduce the cost of extraction [2]. However, this makes the oil inedible as it may result in oil degradation, having contaminants from residual solvents.

Within the seed, the oil is contained in small spherical intracellular organelles called oil bodies [4,5]. An oil body consists of a hydrophobic triacylglyceride (TAG) core surrounded by a complex membrane consisting of a monolayer of phospholipids embedded with proteins, called oleosins, that play a role in their synthesis and stability [5,6]. Among different species, the size of seed oil bodies varies but falls within a narrow range of 0.5–2.5 μm [4,7]. This narrow size range provides a large surface area for the attachment of lipase during seed germination for rapid mobilisation of reserve TAGs [5]. The isoelectric point of oil bodies generally ranges from 5.7 to 6.6 (which is close to the isoelectric point of oleosin), indicating that they have a net negative charge at their surface at neutral pH [7,8]. This is due to the interaction of the negatively charged phospholipids with the positively charged residues of the surface proteins [7]. The variation in the isoelectric point of oil bodies from different species may be attributed to the differences in the extraction process used for isolation since the chemicals used for extraction can remove the non-specifically associated proteins on the oil body surface [9–12].

Oil bodies are present in abundant amounts in plant tissues and can be considered as natural oil-in-water emulsions because of their unique structure [13,14]. The presence of oleosin at the interface provides oil bodies with steric hindrance, protecting the phospholipids against the action of phospholipases and also protecting oil bodies from coalescence or aggregation [6]. Moreover, the interaction between phospholipids and oleosins at the interface gives oil bodies a negative surface charge that leads to electrostatic repulsions [15]. This unique interface allows oil bodies to maintain their individuality and have remarkable stability inside the cell where they are stored for long periods until they are utilised for germination [4,5].

Hemp seed oil bodies can be a highly valuable novel ingredient for the food industry because of their potential as natural emulsions or nature-assembled delivery systems of PUFAs, EFAs, and liposoluble bioactive compounds. To date, no studies have been reported on the extraction of oil bodies from hemp seeds. Since the properties of oil bodies after extraction vary from source to source, the characterisation of new materials is necessary before their applications in food products. This study provides for the first time an in-depth characterisation of oil bodies from hemp seeds. The microstructure, composition, and the colloidal stability of oil bodies under different pH and ionic strength conditions were investigated to gain better understanding of their functional performance in food-like environments.

2. Materials and Methods

2.1. Materials

Unless stated otherwise, all chemicals were purchased from Sigma-Aldrich Ltd. (St. Louis, MO, USA), and the reagents were made up in Milli-Q water (Milli-Q apparatus; Millipore Corp., Bedford, MA, USA). Low-THC hulled hemp seed kernels (Supplier: Sunshine (Tianjin) Produce Limited, Country of Origin: China, AAA Grade, heat-treate) were purchased from a local supermarket (Davis Trading Company Ltd., Palmerston North, New Zealand). The content of THC of hemp seeds was less than 10 ppm, as declared by the supplier. The THC content in hemp seeds for food applications must be less than 0.35%, regulated by the Ministry of Health and New Zealand Legislation. All experiments were repeated at least twice and the means of the results from two or more experiments are reported.

2.2. Methods

2.2.1. Preparation of Hemp Milk and Isolation of Oil Bodies

Dehulled hemp seeds were soaked overnight in Milli-Q water at a seed-to-water ratio of 1:4 (w/w). Then, the mixture was ground using a colloid mill at a frequency of 30 Hz (Bematek Systems, Inc., Model CZ-60-PB, Salem, MA, USA). To remove the solids, we passed the resulting mixture through a clarifier separator (GEA Westfalia, Model CTC-03-107, New Jersey, NJ, USA) and the "hemp milk" was collected for cream extraction.

Hemp seed oil bodies were isolated from hemp milk by centrifugation at $10,000 \times g$ for 20 min at 20 °C. The resulting cream layer referred to as "oil bodies" or "oil body fraction" was then carefully removed from the aqueous layer for experimental studies and added with sodium azide (0.02% w/w) to inhibit the growth of microorganisms.

2.2.2. Microstructure

The morphology of oil bodies in hemp seed and after isolation were examined by cryo-scanning electron microscopy (cryo-SEM), whereas the interfacial properties of the oil body fraction was determined by confocal laser scanning microscopy (CLSM).

Cryo-SEM

For the analysis of oil bodies in the cross-section of the seed, dehulled hemp seeds were placed in the sample holder and flash-frozen by dipping in liquid nitrogen. After 10 min of contact with liquid nitrogen, the frozen sample was transferred to the cryo-unit sectioning chamber and placed under vacuum. The temperature of the sample was lowered to −120 °C followed by fracturing using a cold knife.

For the sample imaging, the temperature of the sample was gradually raised to −100 °C for 20 min for sublimation. The fractured surface of the seeds was then coated with a thin platinum coating (10 mA for 240 s), and the samples were transferred to the imaging chamber. SEM images of fractured sections were recorded at 6 to 20 kV on a Joel JSM 6500F Field Emission Scanning Microscope. A total of five different sections were examined, and images were captured at different resolutions.

Similarly, the morphology of the hemp seed oil body fraction was examined by cryo-SEM as per the described method but following the sample preparation protocol of Efthymiou et al. [16]. Briefly, a drop of oil body suspension was placed between two small copper grids, and the excess sample was gently soaked dry by a filter paper. The sample sandwiched between the grids was then placed in a jet freezer assembly (Baltec JFD 030) and frozen rapidly to −186 °C using a continuous propane stream for 30 min. Upon freezing, the copper grid from one of the sides was removed, and the sample was placed on the sample holder maintained at −180 °C of the cryo-SEM apparatus and was transferred to the cryo-unit sectioning chamber and placed under vacuum. Sample imaging was performed as previously described with modifications. Instead of a platinum coating, the surface of the sample was coated with a thin palladium coating and SEM images of the fractured sections were recorded at 5 kV at −120 °C on a Joel JSM 6500F Field Emission Scanning Microscope equipped with an energy-dispersive detector and a Gatan Alto2500 cryo-unit.

Confocal Laser Scanning Microscopy (CLSM)

The microstructure of hemp seed oil bodies was investigated using a confocal laser scanning microscope (Model Leica SP5, DM6000B, supplied by Leica Microsystems, Heidelberg, Germany) with a 63 mm oil immersion objective lens. Nile Red (1 mg/mL in acetone) and Fast Green FCF (1 mg/mL in Milli-Q water) were used to selectively stain neutral lipids and proteins, respectively, on the basis of the staining protocols described by Gallier et al. [17]. Briefly, 200 µL of the diluted oil body fraction (0.1% v/v fat) was mixed with 12 µL of Nile Red and 6 µL of Fast Green FCF. The stained sample was placed on a concave microscope slide, covered with a cover slip (0.17 mm thick), and immediately examined using the confocal laser scanning microscope. For phospholipids staining, Lissamine™ rhodamine B dye (Rd-DHPE, 1 mg/mL in chloroform) was used. Briefly, 100 μL of the diluted oil body fraction (0.1% *v/v* fat) was added with 5 μL of Lissamine™ rhodamine B (Rd-DHPE). The stained sample was placed on a concave microscope slide, covered with a cover slip (0.17 mm thick), and examined using the confocal laser scanning microscope. The images were processed using the software ImageJ.

2.2.3. Compositional Analysis of Hemp Seed Oil Bodies

Proximate Composition

The compositional analysis of the oil body fraction, such as moisture content, fat content, crude protein content, and ash content, was performed at the Nutrition Laboratory of Massey University, Palmerston North, New Zealand. The following methods were used for analysis: vacuum oven drying (AOAC 990.19, 990.20) for moisture, Mojonnier method (flour, baked, extruded products, AOAC 922.06) for fat content, Dumas method (AOAC 968.06) for crude protein content, and furnace drying at 550 °C (AOAC 942.05 for feed, meat) for ash content.

Fatty Acid Composition

The total lipids from the hemp seed oil body fraction were extracted using the B&D method [18]. The neutral lipids and polar lipids were separated by solid phase extraction (SPE) as described by Avalli and Contarini [19]. In brief, 400 mg lipid sample was dissolved in 1 mL of chloroform–methanol (2:1, *v/v*), and 0.75 mL of the mixture was loaded into hexane conditioned HybridSPE®—Phospholipid cartridge (bed weight 30 mg, 1 mL, Sigma Aldrich Ltd., St. Louis, MO, USA). The neutral lipids were eluted with 3 mL of hexane–diethyl ether (8:2, *v/v*) and 3 mL of hexane–diethyl ether (1:1, *v/v*). The recovered fraction for neutral lipids was dried for further fatty acids analysis. The polar lipids were eluted with 4 mL of methanol and followed with 2 mL of methanol and 2 mL of chloroform–methanol–water (3:5:2, *v/v/v*). The recovered fraction for polar lipids was dried for further fatty acids analysis.

Total fatty acids analysis in both lipid fractions were quantified by gas chromatography (GC), following the method described by Zhu et al. [20]. A sample of 20 mg of lipid fraction (neutral or polar) and 1 mL methyl tricosanoate/hexane (1 mg/mL) (internal standard) were mixed. The composition of the total fatty acids was determined by capillary GC on a Supelcowax™ 10 (30 m × 0.35 mm × 0.50 lm) capillary column (Supelco Park, Bellefonte, PA, USA) installed in an Agilent 7890 gas chromatography system equipped with a flame ionisation detector and a split/splitless injector (Agilent Technologies, Santa Clara, CA, USA). The initial oven temperature was 140 °C, which was maintained for 1 min before being increased to 180 °C at a rate of 10 °C/min, and then maintained at 180 °C for 1 min. The oven temperature was then increased to 210 °C at a rate of 2 °C/min, and then maintained at 210 °C for 10 min. Helium was used as the carrier gas at 72 cm/s running at a constant flow. The injector was set at 250 °C, and the detector was set at 260 °C. The split ratio was 20:1.

2.2.4. Protein Composition of Hemp Seed Oil Bodies by Sodium Dodecyl Sulphate Polyacrylamide Gel Electrophoresis (SDS-PAGE)

The oil body fraction was collected by separating hemp milk at 10,000× *g* centrifugation for 20 min. The oil bodies were washed by two methods, and the final samples were referred to as "phosphate buffer-washed oil bodies" (WC) and "purified oil bodies" (POB), respectively.

The oil body fraction was suspended in phosphate buffer (10 mM, pH 7.5) followed by centrifugation (10,000× *g* for 20 min). After washing it twice, the cream phase was collected and suspended with phosphate buffer, and then mixed with hexane (1:1) and re-centrifuged at 10,000× *g* for 20 min to remove free oil. The WC was suspended in phosphate buffer for analysis.

The oil body fraction was purified by the method described by Tzen et al. [21]. Briefly, the oil body fraction was suspended in phosphate buffer (5 mM, pH 7.5) containing 0.1% (*w/w*) Tween 20 and 0.2 M sucrose. The suspension was diluted 1:1 using phosphate buffer (10 mM, pH 7.5) and centrifuged at 10,000× *g* for 20 min. The cream layer obtained was suspended in 9 M urea solution, and then the suspension was diluted 2:1 using phosphate buffer (10 mM, pH 7.5) followed by centrifugation at 10,000× *g* for 20 min. The cream phase was re-suspended in phosphate buffer (10 mM, pH 7.5) containing 0.6 M sucrose and washed by hexane as described above. The final POB was re-suspended in phosphate buffer for analysis.

Sodium dodecyl sulphate polyacrylamide gel electrophoresis (SDS-PAGE) was performed under both non-reducing and reducing conditions as per the protocol described by Manderson et al. [22]. Hemp milk (M), phosphate buffer-oil bodies (WC), and purified oil bodies (POB) samples were mixed with PAGE sample reducing buffer and non-reducing buffer to a final protein concentration of 2 mg/mL. For the reducing condition, dithiothreitol was used in reducing sample buffer (200 mM), and the reducing samples were heated at 95 °C for 10 min. Then, 10–15 µL of samples were loaded onto Mini-Protean gels (Bio-Rad Laboratories, Richmond, CA, USA) and run at 120 V for about 120 min. The molecular imager Gel Doc XR system (Bio-Rad Laboratories, Richmond, CA, USA) was used for gel scanning, and ImageLab software was used for image analysis.

2.2.5. Physicochemical Characterisation and Stability of Oil Bodies
Particle Size

The particle size of hemp seed oil bodies was measured by static light scattering using a Mastersizer 2000 (Hydro MU, Malvern, Worcestershire, UK). Samples were dispersed in Milli-Q water or a sodium dodecyl sulphate (SDS) solution (1.2% *w/v*). The data were reported in Sauter-average diameter ($d_{3,2}$) and volume-mean diameter ($d_{4,3}$). The refractive index of hemp seed oil (1.475) and water (1.33) was used in the protocol. The mean particle size was calculated as the average of triplicate measurements.

ζ-Potential

The ζ-potential of hemp seed oil bodies was determined using a Zetasizer Nano ZS (Malvern Instruments Ltd., Malvern, United Kingdom) equipped with a 4 mW He/Ne laser at a wavelength output of 633 nm. Samples were diluted to a final particle concentration of 0.01% (*v/v*) in Milli-Q water and put in an electrophoresis cell (Model DTS1070, Malvern Instruments Ltd.) at 25 °C. The ζ-potential was read at least 10 times for each sample, and the ζ-potential values were calculated by Smoluchowski approximation. Mean values were calculated from triplicate measurements.

Effect of pH and Ionic Strength on Colloidal Stability

The stability of oil bodies in the food matrix should be understood once they are incorporated in the food matrix of commercial products. Their stability may be affected by changes in their environment such as variations in pH, ionic strength, and/or temperature. To determine the influence of pH on the colloidal stability, we suspended hemp seed oil bodies in Milli-Q water at an oil body fraction-to-water ratio of 1:3 (*w/w*). The suspensions were adjusted to pH 3–10 using 0.1 M HCl or 0.1 M NaOH. For the effect of ionic strength, the hemp seed oil body fraction was suspended at different concentrations of NaCl (0, 62.5, 125, 250, 500, and 1000 mM). The particle size and ζ-potential of the samples were determined as described in 'particle Size' and 'ζ-potential' sections, respectively.

3. Results and Discussion

3.1. Morphology of Hemp Seed

Figure 1 shows the microstructure of hemp seed by Cryo-SEM. The hemp seeds were filled with oil bodies (shown in white arrows) that were sporadically distributed inside the cell (cell wall indicated in green arrows) (Figure 1). Oil bodies appeared spherical with diameters ranging from 3 to 5 µm with minor irregularities in the shape. Aside from the cell walls and the oil bodies, no other structural elements could be identified from the micrographs. Moreover, some cells seemed to be devoid of any oil bodies in the plane of imaging, as seen in Figure 1A (black arrows). In Figure 1B, oil bodies appeared to have an outer covering that may or may not be part of the oil body membrane or the seed membrane. In addition, some cells also showed semi-hemispherical cavities (black arrows, Figure 1B), which may have contained an oil body previously but were dislodged during fracturing of the sample.

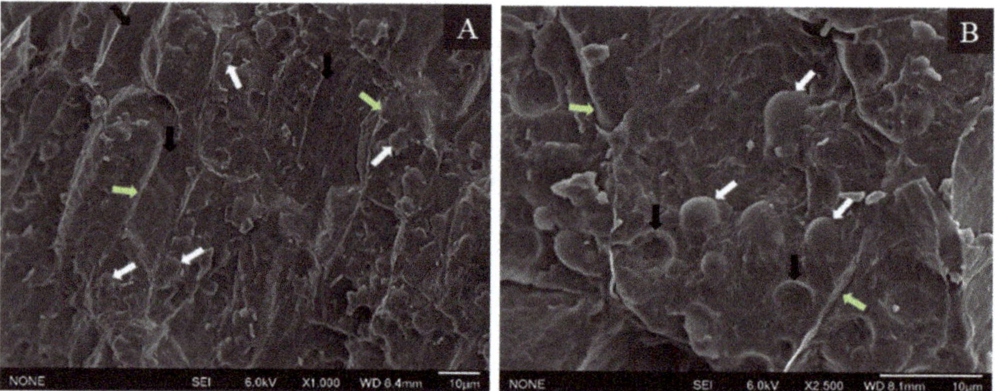

Figure 1. Cryo-SEM images of the cross-section of hemp seeds at 1000× (**A**) and 2500× (**B**) magnification. See the text for the description of the arrows.

Although there are no reported studies on the microstructure of oil bodies in hemp seeds, other studies have examined the structural characteristics of other oleaginous seeds and tissues. For instance, oil bodies with an apparent diameter of 1 to 3 µm were found to be densely packed in *E. plantagineum* seeds [13]. The TEM images showed that oil bodies were sporadically distributed within the seed, with most cells being densely packed with oil bodies. More recently, Paterlini et al. [23] also reported the morphology of *Jathropha peiranoi* seeds, where the cells of endosperm and embryo tissues were rich in oil bodies, containing about 13–15 grouped spherical structures of 2.5–4.5 µm, occupying the cell volume.

The dense cell environment causes the compression of oil bodies, leading to close packing [24]. The ability to withstand compression forces and maintain integrity indicates high flexibility of the hemp seed oil bodies as has been noted previously [25,26]. The close packing of oil bodies in the cell matrix also makes them appear asymmetrical. The same was observed for oil bodies in maize germ [25]. On the contrary, oil bodies in sunflower seeds have more spherical shape, partially attributed to the greater space available due to higher moisture content in the seeds [25].

The cryo-SEM images of hemp seeds also showed some cells with fewer oil bodies or completely devoid of any oil bodies (Figure 1). Such an asymmetric distribution of oil bodies has also been reported for coconut endosperm [26]. These regions in the cell without any oil bodies could be occupied by other intracellular organelles. It is possible that the images with asymmetric distribution of oil bodies represent a cross-section of cellular regions occupied by other intracellular organelles.

3.2. Microstructure of Hemp Seed Oil Bodies

The morphology of isolated oil bodies in aqueous suspension was further examined through cryo-SEM imaging. Oil bodies appeared to be spherical in shape (white arrows, Figure 2A), with the diameter ranging from 2 to 5 µm (1000× magnification). The size of the oil bodies in suspension agreed with the size of oil bodies in the seed (Figure 1). At 10,000× magnification (Figure 2B), the oil bodies appeared to have an irregular spherical shape and a rough surface. Furthermore, thread-like structures or connections between oil droplets were observed (black arrows, Figure 2A), likely caused by artefacts arising during the sublimation of ice during sample preparation.

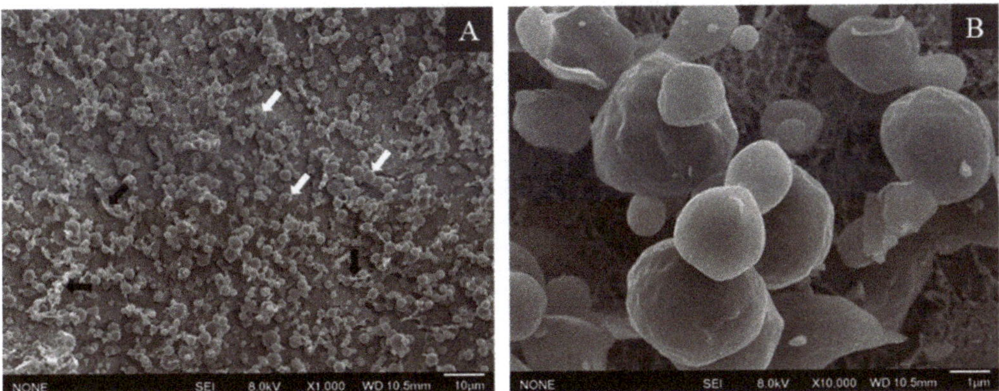

Figure 2. Cryo-SEM images of hemp seed oil bodies in an aqueous solution at 1000× (**A**) and 10,000× (**B**) magnification. For description of white and black arrows, please see text.

The microstructure of oil bodies was examined by confocal laser scanning microscopy using the dyes Nile Red, FG-FCF, and Lissamine™ rhodamine B (Rd-DHPE) to stain neutral lipids, proteins, and phospholipids, respectively. The CLSM images showed that hemp seed oil bodies existed as spherical droplets in aqueous solutions, and their diameter ranged from 3 to 7 µm without any noticeable flocculation or aggregation. They contained a large intense red fluorescent core region that represented the neutral lipids or TAGs (Figure 3A) surrounded by a uniform covering of proteins at the interface (stained green by FG-FCF), regardless of the oil body size (Figure 3B). No protein aggregates were observed in the continuous phase of the suspension. Lastly, staining with Rd-DHPE showed uniform intense red fluorescence on the surface of the oil droplets, confirming the presence of phospholipids (Figure 3B).

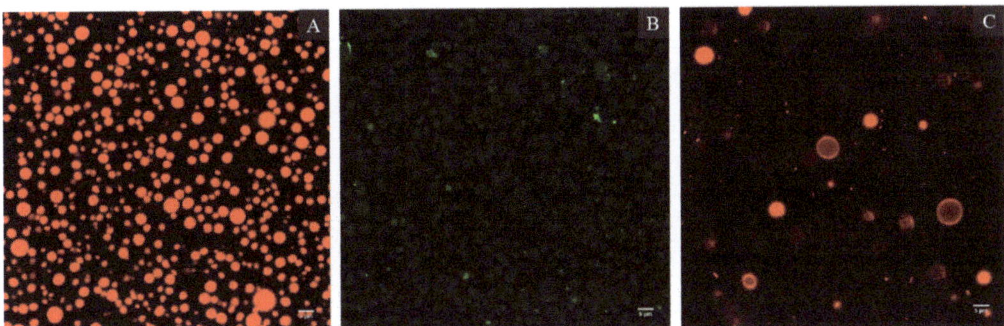

Figure 3. Confocal Laser Scanning Microcopy images of hemp seed oil bodies showing neutral lipids stained by Nile Red (**A**), proteins stained by Fast Green-FCF (**B**) dyes, and the interfacial distribution of phospholipids stained by Lissamine™ rhodamine B dye (**C**).

Both Cryo-SEM and CLSM imaging confirmed that oil bodies were not aggregated in aqueous suspension. No protein aggregates (stained by FG-FCF) were observed in the aqueous phase, indicating that the protocol was able to isolate intact oil bodies with minimal amounts of extraneous proteins. The hemp seed oil body interface appeared to have a uniform distribution of phospholipids as indicated by the intense red fluorescence on the oil body surface upon staining with Rd-DHPE. Overall, hemp seed oil bodies conform to the classic interfacial structure of seed bodies proposed by [5] and previously confirmed in oil bodies from nuts [17,27] and fruits [28] as well.

3.3. Proximate Composition of Hemp Seed Oil Bodies

The proximate composition of the hemp seed oil body fraction is shown in Table 1; this fraction had $79.3 \pm 2.8\%$ fat and $1.5 \pm 0.2\%$ protein, respectively. The protein content was within the range of 0.6 to 3.4% (*w/w*) reported for oil bodies from seeds of various species [7]. This could represent the proteins located at the oil body interface and possibly some extraneous proteins that were present in the oil body fraction after isolation. The phospholipid content in the oil body fraction was not determined in the study.

Table 1. Proximate composition of hemp seed oil bodies.

Parameter	Composition (%, Wet Basis)
Moisture	22.80 ± 2.09
Fat	79.30 ± 2.78
Crude protein	1.50 ± 0.22
Ash	0.20 ± 0.01

3.4. Fatty Acid Composition of Lipids in Hemp Seed Oil Bodies

Table 2 shows the fatty acid composition of the hemp seed oil body fraction. The composition of lipids in oil bodies contained $\geq 90\%$ of unsaturated fatty acids composed mainly of long-chain fatty acids. Linoleic acid ($C_{18:2}$) accounted for 61% of the total fatty acids. Other major fatty acids of lipids from hemp seed oil bodies were oleic acid ($C_{18:1}$, 15.1%) and α-linolenic acid ($C_{18:3}$, 11.7%). Saturated fatty acids such as palmitic acid ($C_{16:0}$, 6.1%) and stearic acid ($C_{18:0}$, 2.9%) were also found in minor concentrations. Teh and Birch [3], Oomah et al. [29], and Alonso-Esteban et al. [30] reported that linoleic acid was found as the major fatty acid in hemp seed oil (53–57%). The fatty acid composition found in oil bodies was similar to that previously reported for hemp oil. These similarities had been previously reported for other seeds and oil bodies fractions [13]. The ω-6/ω-3 ratio found in the oil body fraction was 2.6:1, which is in good aggrement with previous studies on oil extracted from hemp seed (2.5:1) [1]. The lipid composition of hemp seed oil can be comparable to walnut oil due to their high linoleic and α-linolenic acid content [31].

Table 2. Fatty acid composition (%) of lipids in hemp seed oil bodies, and their neutral and polar fractions.

Fatty Acid	Total Fat Percentage (%)		
	Total Lipids	Neutral Lipids	Polar Lipids
$C_{16:0}$ Palmitic acid	6.10 ± 0.02	5.95 ± 0.03	8.46 ± 0.05
$C_{18:0}$ Stearic acid	2.91 ± 0.03	2.86 ± 0.06	2.98 ± 0.07
$C_{18:1n9}$ Oleic acid	15.10 ± 0.06	15.00 ± 0.15	13.44 ± 0.20
$C_{18:2n6}$ Linoleic acid	60.74 ± 0.32	61.07 ± 0.20	47.36 ± 0.47
$C_{18:3n6}$ γ-Linolenic acid	0.039 ± 0.005	0	1.23 ± 0.04
$C_{18:3n3}$ α-Linolenic acid	11.66 ± 0.33	11.70 ± 0.23	23.62 ± 0.96
$C_{20:0}$ Arachidic acid	1.98 ± 0.01	1.99 ± 0.06	1.59 ± 0.01
$C_{20:1}$ Eicosenoic acid	1.00 ± 0.03	0.96 ± 0.02	0.68 ± 0.10
$C_{20:2}$ Eicosadienoic acid	0.486 ± 0.035	0.467 ± 0.015	0.644 ± 0.082

The composition of fatty acids in the neutral lipids fraction was very similar to the total lipids fraction of hemp seed oil bodies. This was expected because hemp seeds lipids mostly consist of TAG, which are neutral lipids, and about 1.5–2% are non-TAG lipids or unsaponifiable fraction [32]. Regarding the polar lipid fraction, the major fatty acids found were $C_{16:0}$, $C_{18:1}$, $C_{18:2}$, and $C_{18:3}$, following a similar trend to those in neutral lipids fraction but with higher levels of $C_{18:3}$.

Polar lipids are a diverse class of compounds that are divided into two main subclasses, phospholipids and sulfolipids [33]. Antonelli et al. [34] reported the polar lipid composition of hemp seed oil, and they found that 51% of the phospholipids were phosphatidylcholines (PC), and $C_{16:0}$, $C_{16:1}$, $C_{18:2}$, and $C_{18:3}$ were the most abundant fatty acid combinations. They also reported that about 72% of the sulfolipids were sulfoquinovosylmonoacylglycerols (SQDG), and the fatty acid combinations $C_{16:0}$, $C_{18:1}$, and $C_{18:3}$ were the most abundant. Therefore, the major fatty acids found in the polar lipid fraction of hemp seed oil bodies are likely linked to the presence of PC and SQDG at high concentrations.

3.5. Protein Composition

The protein composition of the oil body fraction was determined using SDS-PAGE under reducing and non-reducing conditions (Figure 4). The gel patterns showed the polypeptide bands present in hemp milk (M), phosphate-washed oil bodies (WC), and purified oil bodies washed with urea (POB).

The hemp milk (lane M) in both gels showed more than 30 protein bands ranging in MW from 250 to 5 kDa. The most intense bands identified were around MW 52 kDa (in non-reducing conditions) and MW 18, 20, and 34 kDa (under reducing conditions). These most likely corresponded to edestin, the major globulin of hemp seeds protein. Edestin is a hexamer with MW of about 300 kDa that is composed of six identical sub-units linked by non-covalent interactions, each unit having five cysteine residues [35–37]. Two of five cysteine residues were disulphide-linked through basic sub-units (BS) with MW about 18–20 kDa, and three acidic sub-units (AS) of MW around 34 kDa [38], which were visible in the gels made under reducing conditions (disulphide-linked Stibbards units are dissociated under these condition) (Figure 4). The polypeptide band at 52 kDa under non-reducing conditions could correspond to the AS-BS units linked by disulphide bonds.

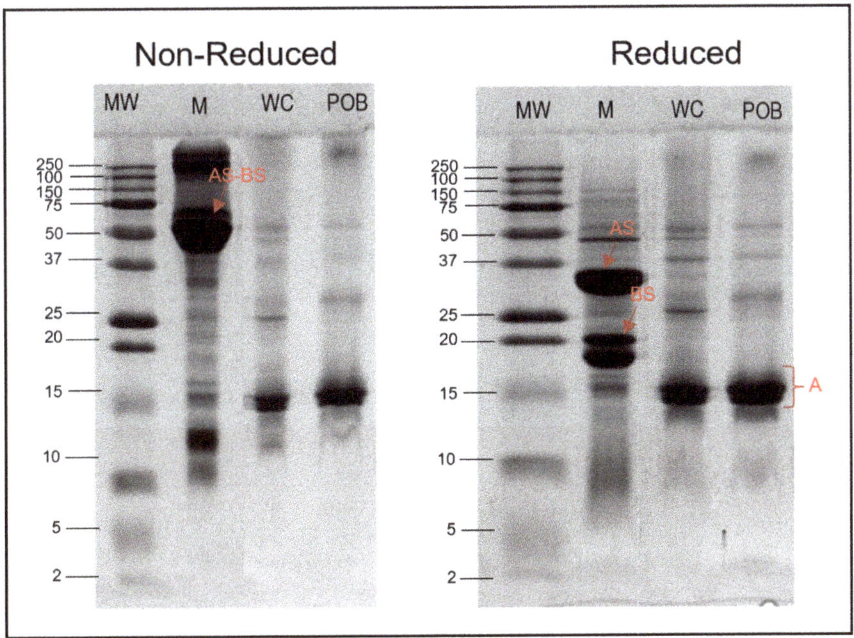

Figure 4. Composition of proteins in hemp seed oil bodies in reducing and non-reducing conditions. MW: molecular weight marker; M: hemp milk; WC: phosphate buffer-washed oil bodies; POB: purified oil bodies washed with urea.

The number of polypeptide bands found in WC lane were significantly less than in M lane. Only around five bands were observed with molecular weight ranging from 55 to 25 kDa. These differences indicated that there are remarkably less proteins in the oil body fraction after isolation from hemp milk. The bands identified in WC lane may be linked to either membrane-specific proteins of oil bodies (oleosins, caleosins, steroleosins) [39], or to extraneous proteins carried over after isolation from hemp milk. Nikiforidis et al. [40] reported that oil bodies after isolation contain their own structural proteins (also called membrane-specific proteins) and exogenous proteins belonging to the matrix where oil bodies were isolated from, with these proteins varying from source to source.

Similarly, the number of polypeptide bands found in POB lane were even less than those found in the WC lane (in reducing and non-reducing conditions), suggesting that proteins remaining in POB should be only those tightly linked to the membrane of oil bodies. The major band in POB at around 15 kDa could be related to oleosins. Maize germ oleosins have been previously identified by SDS-PAGE analysis, being located at 15 and 16 kDa bands [41]. Similarly, oleosins have been identified from oil bodies isolated from rice bran [42] and sesame seeds [43].

3.6. Physicochemical Properties of Hemp Seed Oil Bodies

3.6.1. Particle Size and ζ-Potential

The size distribution, average size, and charge of hemp seed oil bodies were characterised after dispersion in water or 1.2% SDS (Figure 5). The average diameter of oil bodies dispersed in water was $d_{4,3}$: 4.9 $\pm$ 0.7 μm and $d_{3,2}$: 3.1 $\pm$ 0.1 μm (Figure 5). These values agree with the size of oil bodies measured by cryo-SEM images (Figure 2) and CLSM images (Figure 3). This indicates that the integrity and structure of oil bodies in the seed are maintained upon extraction and that they exist as spherical droplets in solution. The size of oil bodies is higher than the range reported for other plant species which ranged from 0.5 to 2.5 μm; the size of oil bodies is generally influenced by biological factors [4,7]. Oil bodies that are smaller in size usually exhibit a low TAG to interfacial protein ratio, while those with larger sizes have lower interfacial protein content [4,40].

The presence of SDS in the dispersion did not cause a change in the $d_{4,3}$ and $d_{3,2}$ of oil bodies, which were 5.1 $\pm$ 0.8 μm and 2.9 $\pm$ 0.2 μm, respectively. SDS disrupts the hydrophobic interactions between membrane-specific proteins on the oil body surface and the extraneous proteins; thus, the mean particle diameter represented the real size of individual oil bodies [11]. Irrespective of the presence of SDS in the samples, the size distributions of the oil bodies showed a monomodal peak, indicating that most oil bodies had similar sizes. The similarities of size distributions of oil bodies with and without SDS also suggests the absence of flocculation or aggregation, which agrees with the results from CLSM imaging (Figure 3). The CLSM images showed the absence of protein aggregates (stained by FG-FCF) in the aqueous phase, suggesting that the protocol was able to isolate intact oil bodies with minimal amounts of extraneous proteins.

The ζ-potential of hemp seed oil bodies at pH 7 was found to be -32.8 ± 5.1 mV. The ζ-potential was measured at pH 7 in all samples to avoid fluctuations due to variations in the pH of hemp oil body extracts (results not shown). Along with the particle size results, the surface charge of oil bodies demonstrates their stability in aqueous suspension. At neutral pH, the oil bodies also exhibited a high magnitude of ζ-potential (absolute value above 30 mV), which indicates the good electrostatic stability of the oil bodies dispersed in water. A ζ-potential magnitude of 30 mV is an indication of good electrostatic stability of emulsions [44].

Parameter	Value	Value (SDS)
$d_{4,3}$ (µm)	4.9 ± 0.7	5.1 ± 0.8
$d_{3,2}$ (µm)	3.1 ± 0.1	2.9 ± 0.2
ζ-potential (mV) at pH 7	-32.8 ± 5.1	-

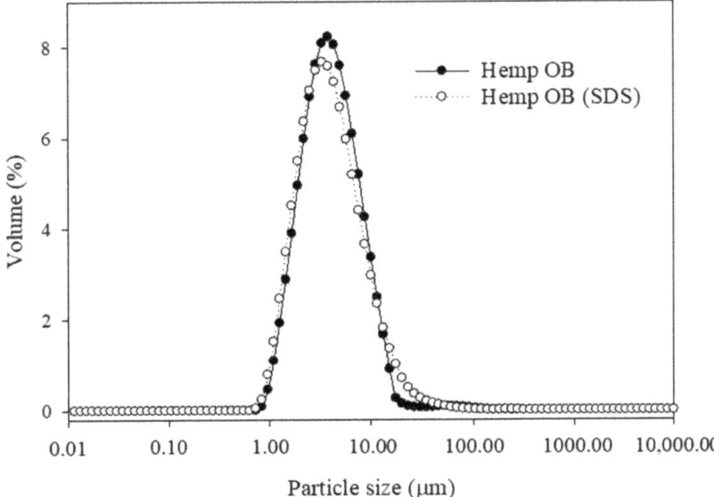

Figure 5. Particle size ($d_{4,3}$ and $d_{3,2}$) and ζ-potential of hemp seed oil bodies suspended in water and SDS.

3.6.2. Effect of pH and Ionic Strength on Colloidal Stability

The influence of pH and ionic strength on the colloidal stability of oil body fractions extracted from hemp seeds was determined. The mean particle diameter ($d_{4,3}$) and ζ-potential were measured in oil body dispersions at different pH (3–10) and ionic strength (0–1000 mM) conditions, and the results are shown in Figure 6. Extreme pH values (highly acidic and highly alkaline) did not affect the mean particle diameter ($d_{4,3}$) of hemp seed oil bodies (Figure 6A). The particle size increased between pH 4 to 5, reaching a maximum value of 14.4 ± 2.9 µm at pH 5. However, oil bodies maintained their size that ranged from 3.3 ± 1.2 µm to 5.1 ± 0.7 µm at alkaline pH values. Particle size data also indicate oil body aggregation at pH values close to the point of zero charge. Similarly, other authors observed that the average particle size of in oil body fractions was relatively small at pH values far from the isoelectric point [14]. Aggregation of oil bodies at different pH values is dependent on electrostatic interactions: a strong electrostatic repulsion prevents the oil bodies from aggregating at pH values away from their isoelectric point [11]. On the other hand, at pH values close to the isoelectric point, there is weak electrostatic repulsion that cannot overcome the attractive forces, thereby causing oil bodies to flocculate [8,45,46]. As such, the flocculation of oil bodies at pH values close to the isoelectric point could be attributed to the weak electrostatic repulsion due to the zero net charge of the oleosins at the interface [8,25].

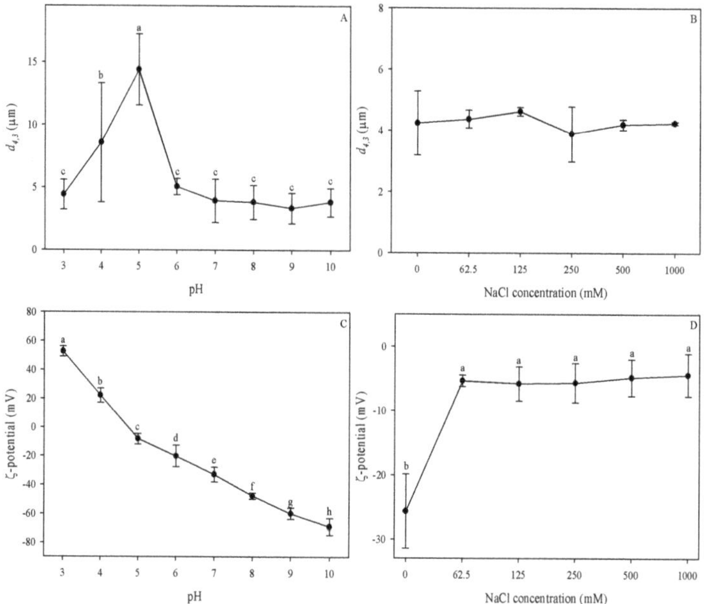

Figure 6. Effect of pH (**A,C**) and ionic strength (**B,D**) on the particle size (**A,B**) and ζ-potential (**C,D**) of hemp seed oil bodies. Means that do not share a letter in the same graph are statistically significant at $\alpha = 0.05$. Error bars indicate standard deviations.

The effect of pH on the ζ-potential of hemp seed oil bodies is illustrated in Figure 6C. As the pH increased, the surface charge changed from positive to negative, and hemp seed oil bodies were found to be negatively charged at neutral pH. The point of zero charge was observed between pH 4 and 4.5; at pH values below this, oil bodies were positively charged, and at pH values above this, they were negatively charged. The highest ζ-potential value of -68.7 ± 5.9 mV was observed at pH 10. The effect of pH on the surface charge of the hemp seed oil body fraction is consistent with other types of oil bodies, wherein the ζ-potential changes from positive to negative with the increase in pH and had a zero value at the isoelectric point [14,46]. This behaviour of oil bodies under different pH conditions is similar to that observed in protein-stabilised emulsions, where droplets are positively charged at low pH, have zero charge at the isoelectric point, and are negatively charged at high pH [44]. Hence, it can be inferred that the membrane-specific proteins (e.g., oleosins) of oil bodies are mostly responsible for these changes in surface charge of oil bodies at various pH values [14,44]. The isoelectric point determined for hemp seed oil bodies was comparable to those reported for maize and soybean oil bodies, which were between pH 4 and 5 [11,14,46].

The changes in the particle size ($d_{4,3}$) and ζ-potential of the oil bodies at different ionic strengths are shown in Figure 6B,D. Variations of the NaCl concentration in hemp seed oil bodies dispersions did not have a significant effect on the $d_{4,3}$. In contrast, the increase in NaCl concentration resulted in a significant decrease in the ζ-potential of the oil bodies. After the addition of NaCl, the magnitude of the ζ-potential decreased from -25.8 ± 5.8 to around -5 mV at 62.5 to 1000 mM NaCl concentration. No significant difference was observed for the ζ-potential at all the NaCl concentrations studied. Generally, increasing the ionic strength of the solution causes a reduction of the charge density of colloidal suspensions due to electrostatic screening effects [44]. The charges are screened until a point where flocculation cannot be prevented, thereby causing the formation of aggregates. In the case of hemp seed oil bodies, a reduction in the magnitude of the ζ-potential was observed with the addition of NaCl at the lowest concentration; it remained similar as

Foods **2021**, 10, 2930

the salt concentration increased. The negative charge of oil bodies, mainly attributed to the phospholipids and oleosins at the interface, seems to be largely affected by small changes in the ionic strength causing a decrease of the negative charge density of the oil bodies interface. Similarly, other types of oil bodies such as maize, soybean, and safflower exhibited a reduction in ζ-potential with the addition of salt [11,14,47]. However, contrary to the decrease in electric potential, no significant change in the mean particle diameter was observed despite the increase in the ionic strength of the solution. The resistance of oil bodies to flocculation can be attributed to the oleosins at the interface, which not only provide electrostatic repulsion but also steric hindrance that enhances their stability [8]. This is supported by the presence of phospholipids that promote the firm anchorage of the oleosins, strengthening the oil body interface [48]. Emulsions that are sterically stabilised are more resistant to variations in pH and ionic strength compared to electrostatically stabilised emulsions [44]. Thus, the reduction in the charge density was not enough to cause the flocculation of oil bodies due to steric effects provided by the oleosins at the oil body surface.

4. Conclusions

Overall, this work determined the composition, structural, and physicochemical properties of oil bodies from hemp seed. The dense packing oil bodies within the cell along with other cell components showed a high degree of flexibility of hemp seed oil bodies. In aqueous solutions, hemp seed oil bodies showed a high net negative charge at neutral pH, which prevented flocculation. The CLSM images showed that the oil bodies consisted of a lipid core surrounded by an interface consisting of proteins and phospholipids. The lipids consisted of long-chain fatty acids, with linoleic acid comprising 61% of the total fatty acids and an ω-6/ω-3 ratio of 2.6:1. The fatty acid composition of polar lipids was dominated by $C_{16:0}$, $C_{18:1}$, $C_{18:2}$, and $C_{18:3}$, most likely due to the presence of phospholipids and sulfolipids in great proportions. The investigation of interfacial proteins suggests that the oleosins most likely had a molecular weight of approximately 15 kDa and did not show disulfide bonding. After isolation of oil bodies, they exhibited minimal flocculation in an aqueous suspension, as evidenced by their high ζ-potential and monomodal size distribution when dispersed in water and SDS. Hemp seed oil bodies were negatively charged at neutral pH and aggregated at pH values close to their isoelectric point, exhibiting a similar behaviour to protein-stabilised emulsions at different pH values. With the addition of salt, a reduction in ζ-potential was observed, but no further increase in particle diameter was seen with increasing salt concentration, indicating that steric effects of the oleosins at the interface were able to resist the reduction in electrical potential. Crude hemp seed oil bodies can provide a naturally stable emulsion and have the potential to be used in a variety of food applications.

Author Contributions: F.L.G.: investigation, formal analysis, visualisation, writing—original draft. S.M.: investigation, writing—original draft. A.D.: conceptualisation, validation, methodology, writing—review and editing, supervision. A.A.-F.: conceptualisation, validation, methodology, writing—review and editing, supervision. All authors have read and agreed to the published version of the manuscript.

Funding: There is no funding grant number for this particular project.

Data Availability Statement: The datasets used and/or analysed during the current study are available from the corresponding author on request.

Acknowledgments: Authors thank Harjinder Singh for his valuable scientific contributions and support with editing, David Flynn (McDiarmid Institute, School of Chemistry and Physics, Victoria University of Wellington) for assistance with Cryo-SEM imaging, Matthew Savoian (Manawatu Microscopy and Imaging Centre of Massey University) for support in CLSM imaging, and Dan Wu (Massey University) for her technical assistance. Author Francesca Garcia also thanks the New Zealand Ministry of Foreign Affairs and Trade (MFAT) for the scholarship and Massey University School of Food and Advanced Technology for the postgraduate research grant.

Conflicts of Interest: The authors declare no conflict of interest.

References

1. Callaway, J.C. Hempseed as a nutritional resource: An overview. *Euphytica* **2004**, *140*, 65–72. [CrossRef]
2. Small, E. *Cannabis: A Complete Guide*; CRC Press, Taylor and Francis Group: Boca Raton, FL, USA, 2017.
3. Teh, S.-S.; Birch, J. Physicochemical and quality characteristics of cold-pressed hemp, flax and canola seed oils. *J. Food Compos. Anal.* **2013**, *30*, 26–31. [CrossRef]
4. Huang, A.H.C. Oil bodies and oleosins in seeds. *Annu. Rev. Plant Physiol. Plant Mol. Biol.* **1992**, *43*, 177–200. [CrossRef]
5. Tzen, J.T.C.; Huang, A.H.C. Surface structure and properties of plant seed oil bodies. *J. Cell Biol.* **1992**, *117*, 327–335. [CrossRef]
6. Huang, A.H.C. Structure of plant seed oil bodies. *Curr. Opin. Struct. Biol.* **1994**, *4*, 493–498. [CrossRef]
7. Tzen, J.T.C.; Cao, Y.Z.; Laurent, P.; Ratnayake, C.; Huang, A.H.C. Lipids, proteins, and structure of seed oil bodies from diverse species. *Plant Physiol.* **1993**, *101*, 267–276. [CrossRef] [PubMed]
8. Maurer, S.; Waschatko, G.; Schach, D.; Zielbauer, B.I.; Dahl, J.; Weidner, T.; Bonn, M.; Vilgis, T.A. The role of intact oleosin for stabilization and function of oleosomes. *J. Phys. Chem. B* **2013**, *117*, 13872–13883. [CrossRef]
9. Chen, Y.; Ono, T. Simple extraction method of non-allergenic intact soybean oil bodies that are thermally stable in an aqueous medium. *J. Agric. Food Chem.* **2010**, *58*, 7402–7407. [CrossRef]
10. De Chirico, S.; di Bari, V.; Foster, T.; Gray, D. Enhancing the recovery of oilseed rape seed oil bodies (oleosomes) using bicarbonate-based soaking and grinding media. *Food Chem.* **2018**, *241*, 419–426. [CrossRef]
11. Sukhotu, R.; Shi, X.; Hu, Q.; Nishinari, K.; Fang, Y.; Guo, S. Aggregation behaviour and stability of maize germ oil body suspension. *Food Chem.* **2014**, *164*, 1–6. [CrossRef]
12. Zhao, L.; Chen, Y.; Chen, Y.; Kong, X.; Hua, Y. Effects of pH on protein components of extracted oil bodies from diverse plant seeds and endogenous protease-induced oleosin hydrolysis. *Food Chem.* **2016**, *200*, 125–133. [CrossRef]
13. Gray, D.A.; Payne, G.; McClements, D.J.; Decker, E.A.; Lad, M. Oxidative stability of *Echium plantagineum* seed oil bodies. *Eur. J. Lipid Sci. Technol.* **2010**, *112*, 741–749. [CrossRef]
14. Iwanaga, D.; Gray, D.A.; Fisk, I.D.; Decker, E.A.; Weiss, J.; McClements, D.J. Extraction and characterization of oil bodies from soy beans: A natural source of pre-emulsified soybean oil. *J. Agric. Food Chem.* **2007**, *55*, 8711–8716. [CrossRef]
15. Huang, A.H.C. Oleosins and oil bodies in seeds and other organs. *Plant Physiol.* **1996**, *110*, 1055–1061. [CrossRef] [PubMed]
16. Efthymiou, C.; Williams, M.A.K.; McGrath, K.M. Revealing the structure of high-water content biopolymer networks: Diminishing freezing artefacts in cryo-SEM images. *Food Hydrocoll.* **2017**, *73*, 203–212. [CrossRef]
17. Gallier, S.; Gordon, K.C.; Singh, H. Chemical and structural characterisation of almond oil bodies and bovine milk fat globules. *Food Chem.* **2012**, *132*, 1996–2006. [CrossRef]
18. Bligh, E.G.; Dyer, W.J. A rapid method of total lipid extraction and purification. *Can. J. Biochem. Physiol.* **1959**, *37*, 911–917. [CrossRef]
19. Avalli, A.; Contarini, G. Determination of phospholipids in dairy products by SPE/HPLC/ELSD. *J. Chromatogr. A* **2005**, *1071*, 185–190. [CrossRef] [PubMed]
20. Zhu, X.; Ye, A.; Verrier, T.; Singh, H. Free fatty acid profiles of emulsified lipids during in vitro digestion with pancreatic lipase. *Food Chem.* **2013**, *139*, 398–404. [CrossRef]
21. Tzen, J.T.; Peng, C.-C.; Cheng, D.-J.; Chen, E.C.; Chiu, J.M. A new method for seed oil body purification and examination of oil body integrity following germination. *J. Biochem.* **1997**, *121*, 762–768. [CrossRef]
22. Manderson, G.; Hardman, M.; Creamer, L. Effect of heat treatment on the conformation and aggregation of β-lactoglobulin A, B, and C. *J. Agric. Food Chem.* **1998**, *46*, 5052–5061. [CrossRef]
23. Paterlini, P.; Jaime, G.S.; Aden, C.; Olivaro, C.; Gómez, M.I.; Cruz, K.; Tonello, U.; Romero, C.M. Seeds characterization of wild species Jatropha peiranoi endemic of arid areas of Monte Desert Biome, Argentina. *Ind. Crop. Prod.* **2019**, *141*, 111796. [CrossRef]
24. Capuano, E.; Pellegrini, N.; Ntone, E.; Nikiforidis, C.V. In vitro lipid digestion in raw and roasted hazelnut particles and oil bodies. *Food Funct.* **2018**, *9*, 2508–2516. [CrossRef] [PubMed]
25. Nikiforidis, C.V.; Kiosseoglou, V.; Scholten, E. Oil bodies: An insight on their microstructure—Maize germ vs sunflower seed. *Food Res. Int.* **2013**, *52*, 136–141. [CrossRef]
26. Nikiforidis, C.V. Structure and functions of oleosomes (oil bodies). *Adv. Colloid Interface Sci.* **2019**, *274*, 102039. [CrossRef] [PubMed]
27. Gallier, S.; Tate, H.; Singh, H. In Vitro Gastric and Intestinal Digestion of a Walnut Oil Body Dispersion. *J. Agric. Food Chem.* **2013**, *61*, 410–417. [CrossRef]
28. Dave, A.C.; Ye, A.; Singh, H. Structural and interfacial characteristics of oil bodies in coconuts (*Cocos nucifera* L.). *Food Chem.* **2019**, *276*, 129–139. [CrossRef]
29. Oomah, B.D.; Busson, M.; Godfrey, D.V.; Drover, J.C.G. Characteristics of hemp (*Cannabis sativa* L.) seed oil. *Food Chem.* **2002**, *76*, 33–43. [CrossRef]
30. Alonso-Esteban, J.I.; González-Fernández, M.J.; Fabrikov, D.; Torija-Isasa, E.; Sánchez-Mata, M.d.C.; Guil-Guerrero, J.L. Hemp (*Cannabis sativa* L.) Varieties: Fatty Acid Profiles and Upgrading of γ-Linolenic Acid–Containing Hemp Seed Oils. *Eur. J. Lipid Sci. Technol.* **2020**, *122*, 1900445. [CrossRef]

31. Miraliakbari, H.; Shahidi, F. Lipid class compositions, tocopherols and sterols of tree nut oils extracted with different solvents. *J. Food Lipids* **2008**, *15*, 81–96. [CrossRef]

32. Liang, J.; Appukuttan Aachary, A.; Thiyam-Holländer, U. Hemp seed oil: Minor components and oil quality. *Lipid Technol.* **2015**, *27*, 231–233. [CrossRef]

33. Frentzen, M. Phosphatidylglycerol and sulfoquinovosyldiacylglycerol: Anionic membrane lipids and phosphate regulation. *Curr. Opin. Plant Biol.* **2004**, *7*, 270–276. [CrossRef]

34. Antonelli, M.; Benedetti, B.; Cannazza, G.; Cerrato, A.; Citti, C.; Montone, C.M.; Piovesana, S.; Laganà, A. New insights in hemp chemical composition: A comprehensive polar lipidome characterization by combining solid phase enrichment, high-resolution mass spectrometry, and cheminformatics. *Anal. Bioanal. Chem.* **2020**, *412*, 413–423. [CrossRef] [PubMed]

35. Tang, C.H.; Ten, Z.; Wang, X.S.; Yang, X.Q. Physicochemical and functional properties of hemp (*Cannabis sativa* L.) protein isolate. *J. Agric. Food Chem.* **2006**, *54*, 8945–8950. [CrossRef] [PubMed]

36. Hadnađev, M.; Dapčević-Hadnađev, T.; Lazaridou, A.; Moschakis, T.; Michaelidou, A.M.; Popović, S.; Biliaderis, C.G. Hempseed Meal Protein Isolates Prepared by Different Isolation Techniques. Part I. physicochemical properties. *Food Hydrocoll.* **2018**, *79*, 526–533. [CrossRef]

37. Potin, F.; Lubbers, S.; Husson, F.; Saurel, R. Hemp (*Cannabis sativa* L.) Protein Extraction Conditions Affect Extraction Yield and Protein Quality. *J. Food Sci.* **2019**, *84*, 3682–3690. [CrossRef]

38. Malomo, S.A.; Aluko, R.E. A comparative study of the structural and functional properties of isolated hemp seed (*Cannabis sativa* L.) albumin and globulin fractions. *Food Hydrocoll.* **2015**, *43*, 743–752. [CrossRef]

39. Napier, J.A.; Beaudoin, F.; Tatham, A.S.; Alexander, L.G.; Shewry, P.R. The Seed Oleosins: Structure, Properties and Biological Role. In *Advances in Botanical Research*; Academic Press: Cambridge, MA, USA, 2001; Volume 35, pp. 111–138.

40. Nikiforidis, C.V.; Matsakidou, A.; Kiosseoglou, V. Composition, properties and potential food applications of natural emulsions and cream materials based on oil bodies. *RSC Adv.* **2014**, *4*, 25067–25078. [CrossRef]

41. Nikiforidis, C.V.; Karkani, O.A.; Kiosseoglou, V. Exploitation of maize germ for the preparation of a stable oil-body nanoemulsion using a combined aqueous extraction-ultrafiltration method. *Food Hydrocoll.* **2011**, *25*, 1122–1127. [CrossRef]

42. Xu, D.; Gao, Q.; Ma, N.; Hao, J.; Yuan, Y.; Zhang, M.; Cao, Y.; Ho, C.-T. Structures and physicochemical characterization of enzyme extracted oil bodies from rice bran. *LWT* **2021**, *135*, 109982. [CrossRef]

43. Lin, L.-J.; Liao, P.-C.; Yang, H.-H.; Tzen, J.T.C. Determination and analyses of the N-termini of oil-body proteins, steroleosin, caleosin and oleosin. *Plant Physiol. Biochem.* **2005**, *43*, 770–776. [CrossRef]

44. McClements, D.J. *Food Emulsions: Principles, Practices, and Techniques*, 3rd ed.; CRC Press: Boca Raton, FL, USA, 2016.

45. Demetriades, K.; Coupland, J.N.; McClements, D.J. Physical properties of whey protein stabilized emulsions as related to pH and NaCl. *J. Food Sci.* **1997**, *62*, 342–347. [CrossRef]

46. Nikiforidis, C.V.; Kiosseoglou, V. Aqueous extraction of oil bodies from maize germ (*Zea mays*) and characterization of the resulting natural oil-in-water emulsion. *J. Agric. Food Chem.* **2009**, *57*, 5591–5596. [CrossRef]

47. Lan, X.; Qiang, W.; Yang, Y.; Gao, T.; Guo, J.; Du, L.; Noman, M.; Li, Y.; Li, J.; Li, H.; et al. Physicochemical stability of safflower oil body emulsions during food processing. *LWT* **2020**, *132*, 109838. [CrossRef]

48. Payne, G.; Lad, M.; Foster, T.; Khosla, A.; Gray, D. Composition and properties of the surface of oil bodies recovered from *Echium plantagineum*. *Colloids Surf. B Biointerfaces* **2014**, *116*, 88–92. [CrossRef] [PubMed]

Article

Microencapsulation of *Cyclocarya paliurus* (Batal.) Iljinskaja Extracts: A Promising Technique to Protect Phenolic Compounds and Antioxidant Capacities

Xiao Chen [1], Senghak Chhun [1], Jiqian Xiang [2], Pipat Tangjaidee [1], Yaoyao Peng [1] and Siew Young Quek [1,3,*]

[1] Food Science, School of Chemical Sciences, The University of Auckland, Auckland 1010, New Zealand; xche622@aucklanduni.ac.nz (X.C.); schh883@aucklanduni.ac.nz (S.C.); ptan226@aucklanduni.ac.nz (P.T.); yaoyao.peng@auckland.ac.nz (Y.P.)

[2] Enshi Tujia & Miao Autonomous Prefecture Academy of Agricultural Sciences, Enshi 445002, China; hmxjq@163.com

[3] Riddet Institute, Centre of Research Excellence in Food Research, Palmerston North 4474, New Zealand

* Correspondence: sy.quek@auckland.ac.nz

Abstract: This study aimed to protect phenolic compounds of *Cyclocarya paliurus* (Batal.) Iljinskaja (*C. paliurus*) using a microencapsulation technique. Ethanol and aqueous extracts were prepared from *C. paliurus* leaves and microencapsulated via microfluidic-jet spray drying using three types of wall material: (1) maltodextrin (MD; 10–13, DE) alone; (2) MD:gum acacia (GA) of 1:1 ratio; (3) MD:GA of 1:3 ratio. The powders' physicochemical properties, microstructure, and phenolic profiles were investigated, emphasizing the retentions of the total and individual phenolic compounds and their antioxidant capacities (*AOC*) after spray drying. Results showed that all powders had good physical properties, including high solubilities (88.81 to 99.12%), low moisture contents (4.09 to 6.64%) and low water activities (0.11 to 0.19). The extract type used for encapsulation was significantly ($p < 0.05$) influenced the powder color, and more importantly the retention of total phenolic compounds (TPC) and *AOC*. Overall, the ethanol extract powders showed higher TPC and *AOC* values (50.93–63.94 mg gallic acid equivalents/g and 444.63–513.49 μM TE/g, respectively), while powders derived from the aqueous extract exhibited superior solubility, attractive color, and good retention of individual phenolic compounds after spray drying. The high-quality powders obtained in the current study will bring opportunities for use in functional food products with potential health benefits.

Keywords: spray drying; gum acacia; maltodextrin; plant extract; functional ingredients; powder

Citation: Chen, X.; Chhun, S.; Xiang, J.; Tangjaidee, P.; Peng, Y.; Quek, S.Y. Microencapsulation of *Cyclocarya paliurus* (Batal.) Iljinskaja Extracts: A Promising Technique to Protect Phenolic Compounds and Antioxidant Capacities. *Foods* **2021**, *10*, 2910. https://doi.org/10.3390/foods10122910

Academic Editors: Harjinder Singh and Alejandra Acevedo-Fani

Received: 29 October 2021
Accepted: 22 November 2021
Published: 24 November 2021

Publisher's Note: MDPI stays neutral with regard to jurisdictional claims in published maps and institutional affiliations.

1. Introduction

Cyclocarya paliurus (Batal.) Iljinskaja (*C. paliurus*), also named "sweet tea tree", is an edible and medicinal plant native to southern China, predominantly Jiangsu, Hunan, Hubei, Henan, Guangxi, Fujian, and Anhui provinces [1]. *C. paliurus* leaves were approved by the Food and Drug Administration as the first Chinese health tea in 1999 [2]. This is because the extract of *C. paliurus* exhibits numerous health benefits, ranging from antihyperlipidemic and antihyperglycemic [3] to antioxidant [3,4] and antimicrobial activities [5,6]. Many other therapeutic effects were also observed for *C. paliurus*, for example, immunomodulation, antihypertensive activity, and improvement of mental efficiency [1,7,8]. These bioactivities are associated with the presence of abundant phenolic compounds in leaves, such as myricetin-3-*O*-β-D-glucuronide, kaempferol, quercetin, caffeic acid, quercetin-3-*O*-β-D-glucuronide, and quercetin-3-*O*-α-D-glucuronide [5,9]. Other phytochemical compounds, for example polysaccharides and triterpenoids, also show antidiabetic, anticancer, anti-inflammatory, and hepatoprotective properties [10–13].

Bioactive ingredients can be extracted from plants and further incorporated into different foods for developing functional products. However, the application of plant-derived phenolic compounds, especially polyphenols, is limited by their low stability as induced by pH, temperature, oxygen, light and enzymes [14]. Microencapsulation of plant extracts by spray-drying is among the viable solutions, in which bioactive components can be entrapped by wall materials and thereby well protected from any harsh environmental conditions. Moreover, microencapsulation of *C. paliurus* extracts into powders brings numerous other advantages such as ease of storage, transportation, and application in foods. Regarding the drying technique, conventional spray-drying is inferior to the monodisperse droplet technique in terms of controlling the physicochemical and functional properties of the resulting powders [15]. This is because traditional spray-drying exhibits the drawback of producing atomized droplets of diverse sizes, which means the droplets will experience different drying histories even under the same drying conditions [15]. By contrast, the micro-fluidic-jet spray dryer resembles a special nozzle that aids in the production of microcapsules with uniform morphology, shape and particle size [15,16]. Up to now, this novel technique has been successfully used in the production of spray-dried powders including fish oil emulsion [17], vitamin E and CoQ_{10} [18,19], cranberry juice [20], noni juice [16], and mandarin juices [21], but has yet been explored for plant extracts.

Maltodextrin (MD) and gum acacia (GA) are popular hydrocolloid-based materials for encapsulating sensitive components or fruit/vegetable juices, such as green tea polyphenols [22], spent coffee's phenolic extract [23], *Phaseolus lunatus* bioactive peptides [24], fermented noni juice [25], extracted soy isoflavones [26], and mandarin juice [21]. The wide use of GA is due to its high solubility, ideal emulsifying properties and good viscosity [27]. On the other hand, MD has fast reconstitution ability, which creates colorless and low viscous solutions without altering the original flavor properties of the core material [28]. Some studies have shown that blends of GA and MD are more efficient than a single carrier during microencapsulation, which significantly increases the stability of bioactive compounds in food systems [16,20].

Nowadays, tea infusion is one of the main food-level usages of *C. paliurus*. To diversify the application of *C. paliurus* in food products and ensure the stability of its bioactive extracts, microencapsulation of *C. paliurus* into a solid form is favorable, but no study has been conducted so far, to the best of the authors' knowledge. This study aimed to encapsulate the ethanol and aqueous extracts of *C. paliurus* via a monodisperse spray-dryer using MD and its combination with GA as wall materials. Physicochemical properties of spray-dried powders were studied, emphasizing on the retentions of total and individual phenolic compounds and their *AOC* after spray drying. This study filled the knowledge gap in the spray-drying of *C. paliurus* extracts while offering an alternative way to deliver bioactive components of *C. paliurus* to consumers.

2. Materials and Methods

2.1. Materials and Chemicals

Commercially dried *C. paliurus* leaves were supplied by the Enshi Tujia & Miao Autonomous Prefecture Academy of Agricultural Sciences, Hubei province, China. The leaves were ground into a powder using a coffee blender (Breville, Australia) for 3 min, and subjected to a 200 mesh US standard sieve to achieve a small and uniform particle size that facilitates the extraction. The fine powder samples were kept at $-20\ ^\circ$C until further requirement. MD [10–13], dextrose equivalent (DE) was purchased from Ingredion (Singapore) and GA was gifted by Hawkins Watts Limited (Auckland, New Zealand).

Folin-Ciocalteu's phenol reagent, gallic acid, 2,4,6-tri(2-pyridyl)-s-triazine (TPTZ), 2,2-diphenyl-1-picrylhydrazyl (DPPH), and 6-hydroxy-2,5,7,8-tetramethyl-chroman-2-carboxylic acid (Trolox) were acquired from Sigma-Aldrich (St. Louis, MO, USA). Authentic standards including kaempferol-3-*O*-glucoside, kaempferol-3-*O*-rhamnoside, myricetin-3-*O*-galactoside, and quercetin-3-*O*-rhamnoside were purchased from Extrasynthese (Genay, France). Quercetin-3-*O*-galactoside and quercetin-3-*O*-glucuronide were sourced from Cayman Chemical (Ann Arbor, MI, USA). Phenolic acid standards including 1,3-dicaffeoylquinic acid, 5-*O*-caffeoylquinic acid, 4,5-dicaffeoylquinic acid, 3,4-dicaffeoylquinic acid, and 1,5-dicaffeoylquinic acid were obtained from Biosynth Carbosynth®(Carbosynth Ltd., Berkshire, UK). Other phenolic compounds including quercetin-3-*O*-glucoside, caffeic acid, chlorogenic acid, and phloretin were acquired from Sigma-Aldrich (St. Louis, MO, USA).

2.2. Preparation of C. paliurus Extracts

Experiments were conducted to optimize the extraction parameters, including time, temperature, material to solvent ratio (w/v), and ethanol concentration (%), using orthogonal design (data not shown). The optimum conditions were then applied to prepare the ethanol and aqueous extracts as below. For the preparation of ethanol extract, 1 g of ground sample was placed into an Eppendorf tube, followed by a 30% ethanol solution supplement to achieve a material to solvent ratio of 1:30 (w/w). The extraction was carried out at 60 °C for 60 min with constant stirring at 550 rpm. Afterwards, the solution was vacuum filtered with a MS2 filter paper (MicroScience, New South Wales, Australia) and stored at −20 °C until further analysis. For the preparation of the aqueous extract, the procedure used was similar to that described above. The optimum conditions applied were material to solvent ratio of 1:30 (w/w) (1 g of ground sample, 30 mL of Mill-Q water), extraction at 90 °C, and extraction time of 60 min under constant stirring of 550 rpm.

2.3. Feed Solution and Spray-Drying Process

Preliminary experiments were conducted to find the suitable spray drying conditions and feed solution composition. According to the results, three formulations (Table 1) were chosen from each extract for preparing the feed solutions. The wall materials or carrier matrices were blended with either the ethanol or aqueous extracts of *C. paliurus* to achieve a core to wall ratio of 1:3 (w/w). The sample was thoroughly mixed at room temperature by stirring at 700 rpm using a magnetic hot plate until all solids were fully dissolved.

Table 1. Wall material formulations for spray-drying.

No.	Core (Extract)	Sample Abbreviations	Wall Materials (%) [a]	
			MD	GA
1	WE	WE-MD-1	100	0
2	WE	WE-MD/GA-2	50	50
3	WE	WE-MD/GA-3	75	25
4	EE	EE-MD-1	100	0
5	EE	EE-MD/GA-2	50	50
6	EE	EE-MD/GA-3	75	25

WE, aqueous extract; EE, ethanol extract; MD, maltodextrin (10–13 DE); GA, gum acacia. [a] The composition of wall materials; core to wall material used in the current study was 1:3 (w:w).

A micro-fluidic-jet spray-dryer supplied by Dong Concept Pty Ltd. (Nantong, China) was used to spray dry feed solutions [21]. The drying parameters were set as follows based on results from the preliminary experiments: inlet temperature of 180 °C, pressure at 0.5 ± 0.02 kg/cm^2, cooling airflow rate at 250 ± 2 L/min, nozzle driver frequency of 10kHz, dispersed air flow rate at 10 L/min, reservoir temperature of 20 °C, and outlet temperature of 81 °C. A polytetrafluoroethylene nozzle with a diameter of 75μm was utilized to obtain monodisperse droplets during the drying process. Powders collected from the dryer were immediately transferred into sample tubes which were then flushed with nitrogen, sealed by parafilm and stored in a desiccator kept at 4 °C before further analysis.

2.4. Characterization of Spray-Dried Powders

2.4.1. Moisture Content and Water Activity

The water activity (a_w) was measured using a water activity analyzer (Novasina, AW SPRINT, Switzerland) at 25 °C. The moisture content of powder samples was determined by the gravimetry method, by drying the samples in an oven (Heraeus D-63450, Germany) at 105 °C until a consistent weight was maintained.

2.4.2. Color Analysis

The color of powder samples was evaluated using three indices (a*, b*, and L*) obtained from a colorimeter (Minolta CR-300, Tokyo, Japan). The a* value denotes the variation from red (60) to green (-60), the L* value indicates lightness from blank (0) to white (100), and the b* value represents the variation from yellow (60) to blue (-60).

2.4.3. Bulk Density

Bulk ($\rho bulk$) density was obtained as described in our previous study [21]. Briefly, spray-dried *C. paliurus* powders were accurately weighted and gently put into a 10 mL cylinder. The powder volume (V_0) was recorded and the $\rho bulk$ was calculated using Equation (1).

$$\rho_{bulk} \text{ (g/mL)} = \frac{powder\ mass}{V_0} \tag{1}$$

2.4.4. Hygroscopicity

Powder hygroscopicity was measured according to Rigon & Noreña [29] with minor modifications. Briefly, 300 mg powder was put into an aluminium cap and then transferred into a chamber containing saturated NaCl with roughly 75% relative humidity at room temperature. The sample mass was accurately taken after 48 h of storage. The hygroscopicity was calculated as the mass of moisture (g) absorbed by 100 g powder (g/100 g).

2.4.5. Solubility

The solubility of powders was measured as described in our previous study [21]. In brief, 0.75 g sample was resuspended in 75 mL of Milli-Q water under vortex for 5 min. The suspension was centrifuged at $7000\times g$ for 10 min. An aliquot of 20 mL supernatant was transferred into a pre-weighed beaker and oven-dried (Heraeus D-63450, Hanau, Germany) at 105 °C until a consistent weight was achieved. Solubility (%) was determined by the mass difference following Equation (2).

$$\text{Solubility (\%)} = \frac{(M_1 - M_0)}{0.2} \times 100 \tag{2}$$

where M_0 is the baker mass, and M_1 is the total mass of the sample and beaker after drying.

2.5. Particle Morphology

The powder morphology was visualized using a Hitachi Scanning Electron Microscope (Mode TM3030Plus, Tokyo, Japan) as described [16,25]. Briefly, the powder sample was placed on a carbon adhesive tape stuck on an aluminium stub. The sample was sputter-coated with a golden layer and then transferred into the specimen chamber for observation. The surface, overall appearance, and cross-section of the powder were observed under $500-800\times$ magnification.

2.6. Total Phenolic Content (TPC) Analysis

TPC of the powder samples was obtained using the Folin-Ciocalteu assay [16,20]. Gallic acid (GA) solutions were prepared at a series of concentrations to construct a standard curve ($R^2 > 0.99$). Results were expressed as mg GA equivalents/g dried powders [20].

The retention of TPC after spray drying was calculated using Equation (3).

$$\text{TPC retention (\%)} = \frac{\text{TPC in powder}}{\text{Total solids content in powder}} \div \frac{\text{TPC in feed solution}}{\text{Total solids content in feed solution}} \times 100 \tag{3}$$

2.7. Antioxidant Capacity (AOC) Analysis

2.7.1. DPPH Assay

The DPPH assay was performed according to the procedure described by Lei et al. with minor modifications [5]. Briefly, the sample (10 µL) was transferred into a 96-well plate and 200 μL of a methanolic DPPH solution (40 mg/mL) were added. The reaction took place in darkness for 60 min. Afterwards, the sample absorbance was measured at 517 nm using a plate reader (EnSpire Multimode, Perkin Elmer, Waltham, MA, USA). The standards and the reagent blank were ethanolic Trolox (at 10, 25, 50, 100, 200, 300, and 400 μM) and absolute ethanol, respectively. Results were expressed as Trolox equivalents per g dried powders (μM TE/g dw).

The retention of *AOC* as evaluated by the *DPPH* assay was calculated using Equation (4).

$$AOC \text{ retention } (\%) = \frac{AOC_{DPPH} \text{ in powder}}{\text{Total solids content in powder}} \div \frac{AOC_{DPPH} \text{ in feed solution}}{\text{Total solids content in feed solution}} \times 100 \qquad (4)$$

2.7.2. Ferric Reducing Antioxidant Powder (FRAP) Assay

An aliquot of 10μL sample was mixed with 200 μL of FRAP working solution [23]. The working solution was made up of the following constituents in a proportion of 10:1:1 (*v/v*), i.e., sodium acetate buffer (0.3 M, pH 3.6), 20mM $FeCl_3$ solution, and 10 mM TPTZ diluted in 40 mM HCl. After incubation in the dark for 60 min, the sample was submitted to the EnSpire Multimode plate reader (Perkin Elmer, USA) for absorbance analysis at 593nm. A calibration curve ($R^2 > 0.99$) was constructed using ethanolic Trolox solutions of 10, 25, 50, 100, 200, 300, and 400 μM. The *AOC* were expressed as Trolox equivalents (TE) per g dried powders (μM TE/g dw).

The retention of *AOC* as evaluated by the FRAP assay was calculated using Equation (5).

$$AOC \text{ retention } (\%) = \frac{AOC_{FRAP} \text{ in powder}}{\text{Total solids content in powder}} \div \frac{AOC_{FRAP} \text{ in feed solution}}{\text{Total solids content in feed solution}} \times 100 \qquad (5)$$

2.8. Individual Phenolic Compound Analysis

Phenolic compounds in the feed solutions and spray-dried powders were identified and quantitated by an Agilent high-performance liquid chromatography (HPLC) system equipped with a Phenomenex C_{18} column (5 μm, 4.6 × 250 mm, Torrance, CA, USA) and a diode array detector. Mobile phases of (A) acetonitrile with 0.1% (*v/v*) formic acid and (B) water with 0.1% (*v/v*) formic acid were used at the flow rate of 1mL/min under 25 °C for compound separation. The gradient elution was conditioned as follows: 0min, 5% B; 10 min, 15% B; 20 min, 25% B; 30 min, 35% B; 35 min, 45% B; 40 min, 35% B; 45 min, 15% B; and 50 min, 5% B. The detector wavelength was set as 280 and 320 nm.

The stock solution containing 14 phenolic standards was used to identify and quantitate analytes in the present study, including quercetin-3-*O*-rhamnoside, quercetin-3-*O*-glucuronide, quercetin-3-*O*-glucoside, 4,5-dicaffeoylquinic acid, quercetin-3-*O*-galactoside, kaempferol-3-*O*-glucoside, kaempferol-3-*O*-rhamnoside, myricetin-3-*O*-galactoside, 1,5-dicaffeoylquinic acid, caffeic acid, 5-*O*-caffeoylquinic acid, 3,4-dicaffeoylquinic acid, chlorogenic acid, and 1,3-dicaffeoylquinic acid. The stock solation was diluted into a series of concentrations to construct standard curves (see Supplementary Table S1). The retention of individual phenolic compounds after spray drying was calculated using Equation (6).

$$\text{Phenolic retention } (\%) = \frac{\text{Content in powder}}{\text{Total solids content in powder}} \div \frac{\text{Content in feed solution}}{\text{Total solids content in feed solution}} \times 100 \qquad (6)$$

2.9. Statistical Analysis

All experiments were conducted in triplicate, and data were presented as mean ± standard deviation. One-way ANOVA and Duncan's multiple range test was carried out to evaluate statistical significance using SPSS Statistics 25 from IBM (Armonk, NY, USA). The principal component analysis (PCA) was conducted using SIMCA 16.0 software from Sartorius (Umeå, Sweden).

3. Results and Discussion

3.1. Physical Properties of Spray-Dried C. paliurus Extracts

3.1.1. Water Activity and Moisture Content

Moisture content indicates the water available in a food system and is an important factor determining the powder products' stability and shelf life. The food industry prefers a value ranging from 1–6% for a stable storage purpose [20]. From Table 2, except for the EE-MD/GA-3 sample (6.64%), all powders fell in this range, i.e., from 4.09–5.95%, suggesting good stability of spray-dried *C. paliurus* extracts during storage. The use of MD as wall material without GA decreased moisture residue in the final product, which was evident for the *C. paliurus* powders derived from the ethanol extract ($p < 0.05$). A similar influence was found during the microencapsulation of *Sideritis stricta* tea [30], a ai berry [31], and cranberry juice [20]. Since GA has high number of hydrophilic pockets associated with water-binding effect, more considerable water residue could be retained in the GA-encapsulated powders [20]. In addition, residue protein (2%) in GA could also increase the percentage of bound water inside the molecules, leading to a lower water diffusion rate during processing [32].

Table 2. Physical properties of spray-dried *C. paliurus* extracts.

Properties	WE-MD-1	WE-MD/GA-2	WE-MD/GA-3	EE-MD-1	EE-MD/GA-2	EE-MD/GA-3
Moisture Content (%)	4.09 ± 0.23 [a]	4.45 ± 0.21 [a]	4.73 ± 0.78 [a]	4.63 ± 0.12 [a]	5.95 ± 0.77 [b]	6.64 ± 1.16 [b]
Water Activity	0.14 ± 0.01 [bc]	0.18 ± 0.00 [d]	0.19 ± 0.02 [d]	0.11 ± 0.00 [a]	0.13 ± 0.02 [b]	0.16 ± 0.01 [c]
Hygroscopicity (g moisture/ 100 g solids)	3.14 ± 0.40 [b]	3.00 ± 0.20 [b]	3.70 ± 0.92 [b]	2.07 ± 0.34 [a]	2.07 ± 0.25 [a]	2.16 ± 0.78 [a]
Solubility (%)	99.12 ± 0.72 [d]	94.65 ± 1.40 [c]	92.04 ± 1.68 [b]	88.81 ± 1.15 [a]	89.85 ± 3.11 [ab]	89.83 ± 1.42 [ab]
Bulk Density (g/mL)	0.37 ± 0.01 [c]	0.38 ± 0.01 [d]	0.38 ± 0.02 [cd]	0.32 ± 0.01 [ab]	0.33 ± 0.01 [b]	0.30 ± 0.00 [a]

Different letters ([a,b,c,d]) in the same row indicate significant differences at the level of $p < 0.05$. WE, aqueous extract; EE, ethanol extract; MD, malto-dextrin (10–13 DE); GA, gum acacia. The numbers indicate the wall materials ratio; 1 = only MD, 2 = MD:GA of 1:1 ratio, 3 = MD:GA of 1:3 ratio.

Water activity (a_w) represents the free water available for biochemical and microbial reactions. As an essential index for spray-dried powders, water activity largely influences product safety and shelf life. Powders with an a_w below 0.6 could be microbiologically stable due to the limited availability of free water molecules for microbial proliferation [33]. The current samples exhibited the a_w values ranging from 0.11 to 0.19, regardless of wall materials and extract types. These a_w values were much lower than the powders produced from other plant-originated materials, e.g., watermelon [33], cranberry juice [20], and aai berry [31]. Our results indicated that the spray-dried *C. paliurus* extracts were microbiologically stable. Furthermore, the use of MD as wall material alone significantly lowered the water activity of the final products ($p < 0.05$), in line with the phenomenon of moisture-content reduction as discussed above.

3.1.2. Hygroscopicity

Hygroscopicity indicates the vulnerability of a powder product to absorb moisture when exposed to high relative humidity conditions. In the present study, all samples showed low hygroscopicity ranging from 2.07–3.70% in an environment having 75% relative humidity (Table 2). The values were much lower than those reported for fruit juice powders produced from noni (19.4–22.7%) [16], aai berry (11.8–19.7%) [31], and mandarin (17.5–18.53%) [21]. For food powders, higher hygroscopicity is frequently associated with core material rich in sugars and organic acids with low glass transition temperatures, which increase the powder's tendency to water adsorption [34]. The hygroscopicity results obtained from this research generally agreed with that of the spray-dried grape polyphenols extract encapsulated by MD:GA of 6:4 under the core to coating ratio of 1:2 (hygroscopicity, 2.5%), as reported by Tolun et al. [35]. They also found that the type of wall materials applied has a profound effect on the powder hygroscopicity. For example, MD with a relatively higher DE (17–20) could raise the hygroscopicity value of the resulting powder, probably due to the presence of more hydroxyl moieties.

In the current study, the powders produced from both the ethanol and aqueous extracts of *C. paliurus* showed no significant variations in hygroscopicity regardless of the wall material used. However, the hygroscopicity (3.00–3.70%) of the aqueous extract powders was significantly ($p < 0.05$) higher than all the ethanol extract samples (2.07–2.16%). These results indicate that the nature of the extract used for feed solution preparation could have a more significant effect on the powders' hygroscopicity than the wall materials applied. Nevertheless, the effect of wall material on powder hygroscopicity cannot be dismissed. In a previous study, we applied similar wall materials, i.e., GA and MD (two types, 17–20 DE and 10–13 DE), for encapsulation of noni juice, and found the wall material type significantly affected the hygroscopicity of the powders produced [16]. The trend of powder hygroscopicity was observed as MD (17–20 DE) > GA > MD (10–13 DE). In the current study, we have applied a combination of MD and GA in different ratios instead of using a single wall material. The interaction of MD and GA would modify the hydrogen bonds of the hydroxyl groups in the crystalline and amorphous regions of macromolecules, affecting the moisture absorption ability of the wall materials [16]. This may be the reason for the different results obtained using single wall material compared to when used in combination.

3.1.3. Solubility

Solubility, reflecting the reconstitution capacity of powders, is an essential indicator for assessing the quality of spray-dried powders. High solubility is preferred by the food industry for the manufacturing of different food products. Overall, the solubility of the powder samples is high within the range of 88.8% to 99.12% (Table 2). The aqueous extract microencapsulated using MD (WE-MD-1) gave the maximum solubility (99.12%) among the powders. In comparison with previous studies, the solubility of spray-dried powders derived from *C. paliurus* extracts was significantly superior to that of the mandarin juice (73.82–75.30%) [21], aai berry incorporated in Tapioca starch (32.08%) [31], and green tea leaf extracts (63–72.66%) [36]. The obtained solubility values were also comparable to that of the spray-dried powders containing mao fruit juice (89.29–96.87%) [37], and the aqueous extracts of mountain tea (97.3–98.6%) [30] and blackberry (88.2–97.4%) [29]. Comparing the two different types of extracts, the solubility of powders produced from the aqueous extract were significantly ($p < 0.05$) higher than those of the ethanol extracts (i.e., 92.04 to 99.12% vs. 88.81 to 89.85%). Silva et al. reported a similar finding for green tea microcapsules. They explained that chemicals with hydrophobic groups might present in the original ethanol extract, leading to lower solubility [36]. Their argument was supported by a study on a medicinal plant called Devil's Club (*Oplopanax horridus*) where several hydrophobic constituents have been identified in the ethanol extract including *trans*-nerolidol, falcarindiol, oplopandiol acetate, and oplopandiol [38].

3.1.4. Bulk Density

Bulk density is a crucial parameter for powder products as it is associated with mixing, packaging, transportation, and storage [30]. In the present study, the bulk densities ranged from 0.30 to 0.38 g/mL for all samples, comparable to that of the particles derived from the mountain tea water extract (0.38 g/mL) [30]. In contrast, the bulk density of mandarin juice powders (0.56–0.61 g/mL) [21] and avocado powders (0.41–0.51 g/mL) [32] were much higher. The differences observed may be ascribed to the particle size and surface morphology of the powders. For example, a large particle size with a wrinkled surface would result in a greater interspace between particles, therefore occupying a high volume, giving lower bulk density [16]. The particle size distribution of spray-dried powders could also be responsible for this phenomenon. A homogenous spray-dried powder with uniform particle size decreases bulk density due to larger interparticle voids. In contrast, in a system of non-uniform particle size, bulk density will increase because the spaces between large interparticle voids are loaded with smaller microcapsules [39]. Furthermore, the type of wall material used for microencapsulation could influence the surface structure of particles and consequently affect the bulk density of powders produced. For example, Zhang et al. found that the noni juice powders encapsulated by GA-alone had a shriveled and rough surface with thicker indentations [16]. This surface morphology causes a large interspace between particles, giving a lower bulk density. However, the effect of wall materials was not apparent in the current study, probably due to the use of MD and GA in combination instead of using each individually.

3.1.5. Color Evaluation

Color gives the first sensory impression and quality about a food product. Table 3 shows the color attributes and appearance of different spray-dried microcapsules of *C. paliurus* ethanol and aqueous extracts. Generally, the L* value varied from 59.23 to 60.52. The b* and a* values showed a relatively wider range, from 14.82 to 19.07 and from 11.49 to 13.68, respectively. Both wall materials and extracts used for encapsulation showed no significant effect on the brightness (L*). In contrast, the a* and b* values showed significant statistical differences; the *C. paliurus* aqueous extract powders were less reddish and more yellow than those obtained from the ethanol extract. Moreover, the wall material type significantly affected the redness (a*) and the yellowness (b*) of the powders ($p < 0.05$), with partial substitution of MD with GA in the feed solution producing more reddish and browner powders. Şahin Nadeem et al. [30] reported similar observations for spray-dried mountain tea extract. They ascribed this to the natural color of the core material and the minor protein content (2%) and reducing sugars in GA, which were susceptible to the Maillard browning reactions during spray-drying.

Table 3. Color attributes and appearance of spray-dried *C. paliurus* extracts.

Samples	Colour Attributes			Powder Appearance
	L*	A*	B*	
WE-MD-1	60.23 ± 0.84 [a]	11.49 ± 0.15 [a]	19.07 ± 0.45 [d]	
WE-MD/GA-2	59.95 ± 1.53 [a]	11.56 ± 0.15 [a]	17.90 ± 0.52 [c]	
WE-MD/GA-3	59.75 ± 1.14 [a]	11.65 ± 0.07 [a]	18.90 ± 0.39 [d]	
EE-MD-1	59.23 ± 0.63 [a]	13.25 ± 0.19 [c]	17.12 ± 0.41 [b]	
EE-MD/GA-2	59.61 ± 0.94 [a]	12.94 ± 0.39 [b]	14.82 ± 0.94 [a]	

Table 3. *Cont.*

Samples	Colour Attributes			Powder Appearance
	L*	**A***	**B***	
EE-MD/GA-3	60.52 ± 0.62 [a]	13.68 ± 0.13 [d]	15.42 ± 0.26 [a]	

Different letters ([a,b,c,d]) in the same column indicate significant differences at the level of $p < 0.05$. WE, aqueous extract; EE, ethanol extract; MD, maltodextrin (10–13 DE); GA, gum acacia. The numbers indicate the wall materials ratio; 1 = only MD, 2 = MD:GA of 1:1 ratio, 3 = MD:GA of 1:3 ratio.

3.2. Morphology Observation

Figure 1 reveals the external and cross-section morphology of the spray-dried *C. paliurus* powders. Generally, all samples have a marginally spherical shape with size uniformity, surface wrinkles and dents, and no apparent cracks (rows A and B; columns a, b, and c). The wrinkle and dent formations on the particle surface may be caused by rapid water removal and cooling of the surface crust [35]. The intact particle surface is desirable as it minimizes gas permeability and reduces the surface area for oxidation, thereby enhancing the preservation and retention of the core material [16,35,36]. Moreover, the uniformity of particle shape and size observed for all samples was significantly different from the agglomerated and uneven morphology of powders prepared using the conventional spray-drying technique [30–32]. The consistent size and surface morphology was attributed to using a micro-fluidic-jet spray-dryer equipped with a special nozzle [15].

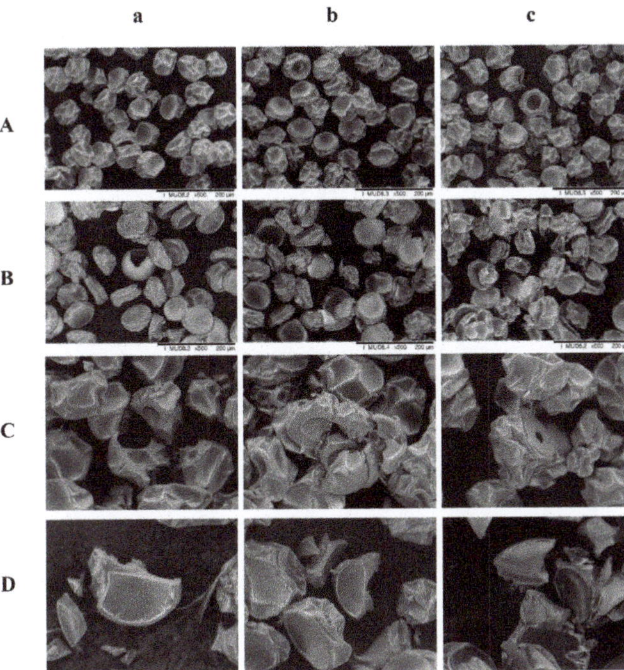

Figure 1. SEM micrograph of spray-dried *C. paliurus* extracts microencapsulated with (a) MD, (b) MD and GA at the ratio of 1:1, and (c) MD and GA at the ratio of 3:1. Rows (**A,B**) refers to the surface structure of powders containing aqueous and ethanol extract, respectively (500× magnification); rows (**C,D**) are the cross-section structure of the aqueous-extract and ethanol extract powders, respectively (800× magnification).

There were no clear correlations between surface morphology and wall materials used for the spray-dried powders from the aqueous extracts (rows A; columns a, b, and c). For the powders containing the ethanol extract (row B), the MD-GA particles (row B; columns b and c) displayed relatively more roughness and shaper indentations than the MD-alone powder (row B; columns a). Zhang et al. reported a similar phenomenon for the cranberry juice powders produced via a conventional mini spray dryer [20]. The rate of dry-shell formation determines the surface morphology due to the inflation and rupture of voids during the spray-drying process [40]. An early dry shell formation resulted in uneven moisture distribution within the particle, and consequently, a shriveled and rough surface [41]. Furthermore, the cross-section images (rows C and D; columns a, b, and c) also did not illustrate any obvious differences among the three samples with different wall materials compositions. However, the type of extract used for spray drying indicates an influence on the cross-section morphology of the final microcapsules. The surfaces of the ethanol extract particles (row D) were generally more homogenous with few smaller pores than the aqueous extract samples (row C). Whether this structure would be conducive to the retention of total phenolic compounds in the spray-dried powders will be further investigated in the following section.

3.3. Retentions of TPC and AOC after Spray Drying

TPC and AOC retentions were examined to evaluate the effect of spray-drying on the bioactive compounds in the *C. paliurus* extracts. The results are shown in Figure 2. Furthermore, the TPC and AOC values of the original aqueous and ethanol extracts are listed in Supplementary Table S2 for comparison.

The TPC of powder varied from 50.35 to 63.94mg GA equivalents/g, significantly lower than the value of the original aqueous and ethanol extracts (65.18 and 94.76 mg GA equivalents/dw). This indicated the degradation of some phenolic compounds during thermal treatment as induced by the spray drying process. Both wall material and extract types significantly ($p < 0.05$) influenced the TPC retention in the final products (Figure 2B). The highest TPC retention was in the WE-MD/GA-2 sample (90.31%), followed by the WE-MD-1 and WE-MD/GA-3 samples (84.69% and 81.02%, respectively). These results indicate that the MD-GA mixture in a 1:1 ratio worked well to protect the phenolic compounds of *C. paliurus* aqueous extract. The superiority of MD-GA blends was also observed for the microencapsulation of cranberry juice (TPC retention, 138–216%) [20] and mandarin juice (chlorogenic acid retention, 92.7–225%) [21], and was further confirmed by Idham et al. [42], who studied the microencapsulation efficiency of anthocyanins from *Hibiscus Sabdariffa* L. The chemical composition of MD and GA could have contributed to their superiority in preserving the TPC of the aqueous extract-derived powders. GA contains a trace amount of protein (2%) which can be covalently bound to the carbohydrate group in MD, forming a sound barrier for protecting core materials. In addition, both protein and GA also have film-forming properties. These factors may explain the good bioactive preserving properties of GA when used in combination with MD [20,25]. Furthermore, it is also possible for the hydroxyl moieties in the phenolic compounds to form hydrogen bonds with the trace protein in GA, which would further reinforce the protection of phenolic in the microcapsules, thereby enhancing the TPC retention [43]. Maltodextrin also provided promising results, as observed in the aqueous and ethanol extract powders (84.69% and 75.38%, respectively). These values were reasonable and comparable to the retention of anthocyanins (85%) in blackberry pulp powder [44] and *p*-hydroxybenzoic acid in the majority of mandarin powder samples (81.5–105.8%) [21], though lower than that of TPC retention in cranberry powder (138–216%) [20].

Comparing the extracts, TPC retention in the ethanol extract powders was significantly lower ($p < 0.05$) than the aqueous extract powders (Figure 2B). This could be ascribed to the different chemical compositions of the two extracts. The phenolic compounds in ethanol extract could be more sensitive to heat during spray drying [4], while the aqueous extract also contains polysaccharides [6,12] that might potentially protect phenolics to a certain extent. The lowest TPC retention value (56.23%) was observed for the powder of ethanol extract microencapsulated by MD and GA in 1:1 ratio (EE-MD/GA-2). The more wrinkled particle surface of the EE-MD/GA-2 sample (Figure 1, row B and Column b) also pointed to a relatively higher total surface area than the aqueous extract powders, which could be linked to the lower encapsulation efficiency of phenolic compounds. In addition, the loss of phenolic compounds could also be attributed to their thermal degradation during spray-drying treatment [45].

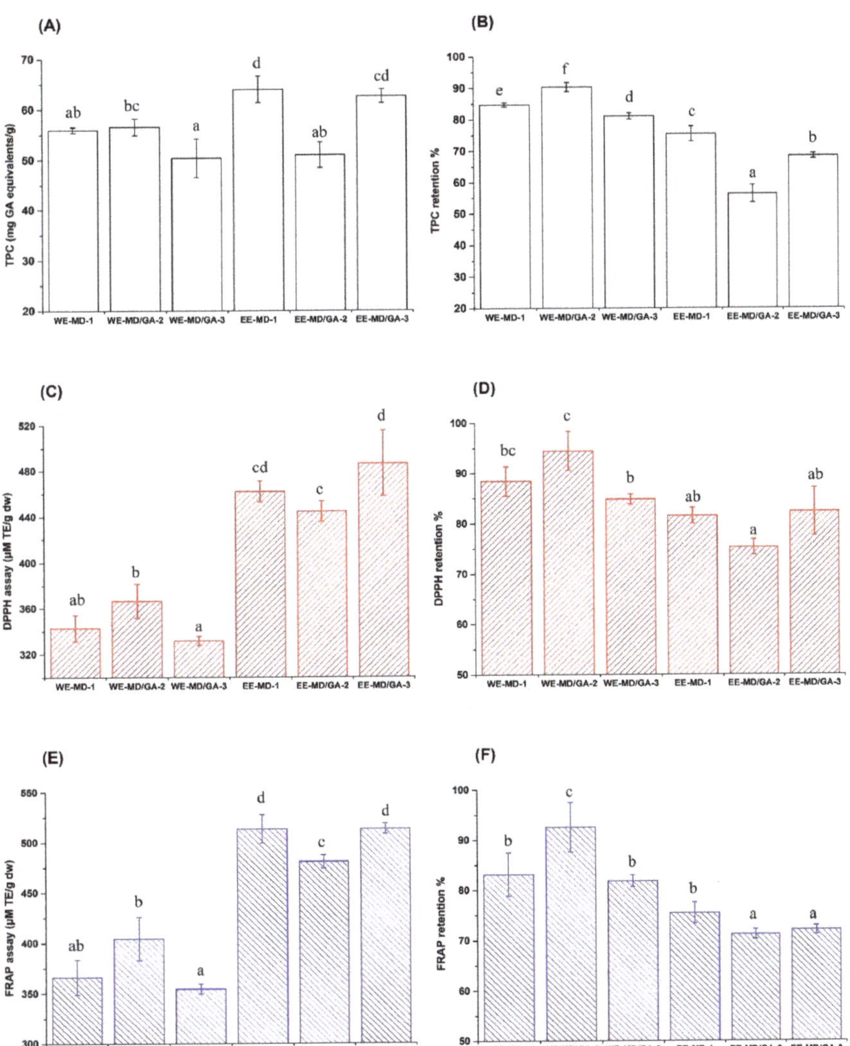

Figure 2. TPC and *AOC* of spray-dried *C. paliurus* extracts and their retentions after spray drying. (**A**) TPC of spray-dried powders; (**B**) The retention (%) of TPC after spray drying; (**C**) *AOC* of spray-dried powders evaluated by DPPH assay; (**D**) retention (%) of *AOC* after spray drying obtained by DPPH assay; (**E**) *AOC* of spray-dried powders evaluated by FRAP assay; (**F**) retention (%) of *AOC* after spray drying obtained by FRAP assay. (WE, aqueous extract; EE, ethanol extract; MD, malto-dextrin (10–13 DE); GA, gum acacia. The numbers indicate the wall materials ratio; 1 = only MD, 2 = MD:GA of 1:1 ratio, 3 = MD:GA of 1:3 ratio). Different letters (a, b, c, d, e, f) in the figure indicate significant differences at the level of $p < 0.05$.

Figure 2C–F show the antioxidant activities of spray-dried powders and their retentions after spray drying. Comparing the *AOC* of original extracts without encapsulation, the FRAP assay showed significantly higher values ($p < 0.05$) than those from the DPPH assay (i.e., 567.61 vs. 447.54 µM TE/g dw for aqueous extracts and 688.23 vs. 593.87 µM TE/g dw for ethanol extracts). As two assays have different reaction mechanisms, it is necessary to consider both when evaluating the antioxidant capacities, and this has been a practice in many previous studies [11,29,37]. Both DPPH and FRAP assays have similar *AOC* ranges for powder products (Figure 2C,E), i.e., from 331.69 to 486.72 and from 354.43 to 513.49 µM TE/g dw, respectively. These results were positively correlated with the TPC of most powders, indicating a significant contribution of phenolic compounds to the antioxidant properties. An exception was observed for the EE-MD/GA-2 and WE-MD/GA-3 samples; the EE-MD/GA-2 sample has significantly ($p < 0.05$) higher DPPH and FRAP antioxidant activities (444.63 and 480.85 µM TE/g dw, respectively) than the WE-MD/GA-3 sample (331.69 and 354.43 µM TE/g dw), but their TPC values had no statistical significance ($p > 0.05$). The increased *AOC* as displayed by the EE-MD/GA-2 sample might be attributed to its Maillard reaction products that showed antioxidant potential, such as 5-hydroxymethylfurfural [46]. After spray drying, the retention of *AOC* as analyzed by DPPH and FRAP assays varied from 75.27–94.42% and 71.18–92.52% for both extracts, respectively (Figure 2D,F). The aqueous extract encapsulated by the GA-MD mixture of 1:1 showed the highest *AOC* retention of 94.42%. The possible structural changes, degradation, re-synthesis and polymerization of phenolic compounds induced by the heat treatment may be responsible for the reduced antioxidant activities after spray drying [35].

3.4. Individual Phenolic Compounds in C. paliurus Powders

An HPLC-DAD system was used to analyze and quantitate individual phenolic compounds in the resulting powders to investigate the effect of wall materials on phenolic retention and to clarify the suitable sample type (aqueous or ethanol extract) for preparing the *C. paliurus* microcapsules. Table 4 showed the concentrations of 14 major phenolic compounds (seven each for phenolic acids and flavonols) in *C. paliurus* extracts before and after spray drying.

Chlorogenic acid remained predominant in both original extracts and spray-dried powders, accounting for 51–54% of the total quantified phenolic components (33.06–36.32 mg/g dw for powder samples; 40.8 and 45.5 mg/g dw for aqueous and ethanol extracts, respectively). The ethanol extract contained 23.52 mg/g dw chlorogenic acid, significantly higher than the aqueous one (21.96 mg/g dw). The *C. paliurus* powders contained 16.94–17.38 mg/g dw of chlorogenic acid, with the highest value found in the WE-MD/GA2 sample (18.55 mg/g dw). Quercetin-3-*O*-glucuronide was the second major phenolic component, occupied approximately 17% of the total in powder samples. The concentrations of quercetin-3-*O*-glucuronide in the original aqueous and ethanol *C. paliurus* extracts were 4.13 and 4.88 mg/g dw, respectively (Table 4), which was much higher than the findings of Zhou et al. (0.2–3.7 mg/g) [11]. This might be attributed to the plant diversity of *C. paliurus*, as the phenolic profiles of *C. paliurus* were found to be quantitatively and qualitatively different across geographical location [11]. The other major phenolic components in the powder samples included quercetin-3-*O*-rhamnoside, 5-*O*-caffeoylquinic acid, myricetin-3-*O*-galactoside, and 1,3-dicaffeoylquinic acid, with proportions of approximately 8.4%, 6.6%, 5.1%, and 3.1%, respectively. The total quantitative contribution of the remaining eight phenolic compounds was less than 10%, which included kaempferol-3-*O*-glucoside (0.73–0.95 mg/g), kaempferol-3-*O*-rhamnoside (0.71–1.06 mg/g), quercetin-3-*O*-glucoside (0.53–0.80 mg/g), 4,5-dicaffeoylquinic acid (0.36–0.51 mg/g), quercetin-3-*O*-galactoside (0.21–0.32 mg/g), 1,5-dicaffeoylquinic acid (0.15–0.17 mg/g), 3,4-dicaffeoylquinic acid (0.13–0.16 mg/g), and caffeic acid (0.11–0.16 mg/g).

Table 4. The content of major phenolic compounds in powders and their retentions after spray drying.

Phenolic Compounds	Content (mg/g dw)							
	WE	EE	WE-MD-1	WE-MD/GA-2	WE-MD/GA-3	EE-MD-1	EE-MD/GA-2	EE-MD/GA-3
quercetin-3-*O*-glucuronide	4.13 ± 0.01 [A]	4.88 ± 0.03 [B]	5.73 ± 0.26 [a]	6.26 ± 0.04 [b]	5.75 ± 0.15 [a]	5.98 ± 0.02 [ab]	6.02 ± 0.03 [ab]	6.05 ± 0.06 [ab]
R (%)			134.06 ± 6.15 [bc]	152.01 ± 0.84 [d]	138.68 ± 3.72 [c]	123.25 ± 0.38 [a]	122.54 ± 0.56 [a]	127.55 ± 1.22 [ab]
myricetin-3-*O*-galactoside	1.24 ± 0.01 [B]	0.96 ± 0.00 [A]	1.68 ± 0.06 [d]	1.84 ± 0.01 [e]	1.44 ± 0.03 [c]	1.47 ± 0.07 [cd]	1.31 ± 0.03 [ab]	1.24 ± 0.05 [a]
R (%)			128.80 ± 4.62 [ab]	149.87 ± 0.70 [cd]	115.63 ± 2.37 [a]	152.68 ± 6.82 [d]	136.65 ± 2.66 [bc]	136.92 ± 4.92 [bc]
5-*O*-caffeoylquinic acid	2.45 ± 0.01 [A]	2.48 ± 0.00 [A]	2.15 ± 0.02 [bc]	2.30 ± 0.00 [d]	2.17 ± 0.00 [c]	2.09 ± 0.04 [ab]	2.07 ± 0.01 [a]	2.06 ± 0.02 [a]
R (%)			86.10 ± 0.90 [ab]	93.77 ± 0.09 [c]	89.04 ± 0.11 [bc]	83.86 ± 1.63 [a]	83.44 ± 0.27 [a]	86.20 ± 0.96 [ab]
chlorogenic acid	21.96 ± 0.03 [A]	23.52 ± 0.04 [B]	17.38 ± 0.02 [a]	18.55 ± 0.08 [b]	16.93 ± 0.38 [a]	17.16 ± 0.23 [a]	17.14 ± 0.38 [a]	16.94 ± 0.07 [a]
R (%)			76.54 ± 0.08 [ab]	84.56 ± 0.35 [c]	77.00 ± 1.71 [b]	72.80 ± 0.97 [a]	73.01 ± 1.63 [a]	75.23 ± 0.30 [ab]
quercetin-3-*O*-glucoside	0.75 ± 0.01 [A]	0.98 ± 0.01 [B]	0.62 ± 0.03 [b]	0.63 ± 0.02 [bc]	0.53 ± 0.01 [a]	0.80 ± 0.03 [d]	0.66 ± 0.01 [bc]	0.71 ± 0.03 [c]
R (%)			78.94 ± 3.99 [bc]	81.90 ± 3.11 [c]	71.42 ± 1.18 [ab]	82.06 ± 2.64 [c]	66.44 ± 1.01 [a]	73.14 ± 2.85 [ab]
1,3-dicaffeoylquinic acid	1.41 ± 0.05 [A]	1.61 ± 0.00 [A]	1.06 ± 0.01 [a]	1.14 ± 0.01 [b]	1.06 ± 0.01 [a]	1.06 ± 0.02 [a]	1.03 ± 0.01 [a]	1.07 ± 0.02 [a]
R (%)			69.78 ± 0.90 [c]	78.15 ± 0.64 [d]	78.36 ± 0.48 [d]	66.16 ± 1.49 [ab]	63.58 ± 0.51 [a]	68.02 ± 1.33 [bc]
quercetin-3-*O*-rhamnoside	3.91 ± 0.01 [A]	4.81 ± 0.01 [B]	2.76 ± 0.05 [a]	3.00 ± 0.04 [b]	2.74 ± 0.10 [a]	2.99 ± 0.01 [b]	3.03 ± 0.01 [b]	3.02 ± 0.04 [b]
R (%)			68.46 ± 1.33 [c]	76.50 ± 0.94 [d]	70.23 ± 2.64 [c]	62.17 ± 0.15 [a]	62.95 ± 0.26 [a]	64.94 ± 0.77 [ab]
kaempferol-3-*O*-glucoside	1.04 ± 0.01 [A]	1.45 ± 0.02 [B]	0.74 ± 0.00 [a]	0.73 ± 0.00 [a]	0.73 ± 0.02 [a]	0.95 ± 0.03 [c]	0.89 ± 0.02 [b]	0.90 ± 0.01 [bc]
R (%)			73.11 ± 0.30 [b]	69.65 ± 0.39 [b]	70.60 ± 1.43 [b]	64.59 ± 1.78 [a]	62.00 ± 1.55 [a]	63.59 ± 0.97 [a]
1,5-dicaffeoylquinic acid	0.25 ± 0.01 [A]	0.27 ± 0.00 [A]	0.17 ± 0.00 [b]	0.17 ± 0.00 [b]	0.15 ± 0.01 [a]	0.16 ± 0.01 [ab]	0.14 ± 0.00 [a]	0.15 ± 0.00 [ab]
R (%)			59.07 ± 0.05 [a]	68.18 ± 0.71 [b]	57.83 ± 2.62 [a]	60.75 ± 2.63 [a]	54.18 ± 2.17 [a]	58.43 ± 1.22 [a]
3,4-dicaffeoylquinic acid	0.24 ± 0.00 [A]	0.33 ± 0.00 [B]	0.16 ± 0.00 [cd]	0.16 ± 0.00 [d]	0.14 ± 0.01 [abc]	0.15 ± 0.01 [bcd]	0.13 ± 0.00 [a]	0.14 ± 0.00 [ab]
R (%)			55.58 ± 0.81 [c]	67.49 ± 1.49 [d]	59.33 ± 3.01 [c]	46.54 ± 2.00 [b]	39.55 ± 1.12 [a]	43.32 ± 0.15 [ab]
kaempferol-3-*O*-rhamnoside	1.31 ± 0.00 [A]	1.89 ± 0.01 [B]	0.71 ± 0.01 [a]	0.71 ± 0.02 [a]	0.70 ± 0.05 [a]	1.00 ± 0.03 [b]	1.06 ± 0.01 [b]	0.98 ± 0.00 [b]
R (%)			53.56 ± 0.68 [a]	54.42 ± 1.29 [a]	53.52 ± 3.90 [a]	53.37 ± 1.69 [a]	55.94 ± 0.33 [a]	52.29 ± 0.03 [a]
quercetin-3-*O*-galactoside	0.52 ± 0.02 [A]	0.69 ± 0.01 [B]	0.21 ± 0.01 [a]	0.28 ± 0.01 [b]	0.21 ± 0.00 [a]	0.32 ± 0.01 [c]	0.28 ± 0.01 [b]	0.29 ± 0.01 [bc]
R (%)			40.68 ± 1.62 [a]	52.80 ± 0.90 [c]	42.55 ± 0.27 [ab]	46.08 ± 0.96 [b]	39.87 ± 1.46 [a]	42.45 ± 0.91 [ab]
caffeic acid	0.35 ± 0.01 [A]	0.34 ± 0.00 [A]	0.15 ± 0.00 [cd]	0.16 ± 0.00 [d]	0.14 ± 0.00 [c]	0.12 ± 0.01 [ab]	0.11 ± 0.00 [a]	0.12 ± 0.01 [b]
R (%)			41.92 ± 0.52 [cd]	44.54 ± 0.73 [d]	38.91 ± 0.51 [bc]	34.79 ± 2.07 [ab]	32.80 ± 0.34 [a]	39.74 ± 1.84 [bc]
4,5-dicaffeoylquinic acid	0.96 ± 0.00 [A]	1.40 ± 0.02 [B]	0.38 ± 0.02 [a]	0.39 ± 0.00 [a]	0.36 ± 0.01 [a]	0.51 ± 0.01 [b]	0.50 ± 0.01 [b]	0.50 ± 0.01 [b]
R (%)			36.22 ± 1.46 [a]	40.60 ± 0.00 [b]	37.78 ± 0.59 [a]	36.84 ± 0.60 [a]	35.54 ± 0.48 [a]	36.03 ± 0.58 [a]

Different uppercase letters ([A,B]) indicate that the phenolics content of aqueous and ethanol extracts is significantly different ($p < 0.05$). Different lowercase letters ([a,b,c,d]) in the same row indicates significant differences of phenolic compounds in power samples at the level of $p < 0.05$. WE, aqueous extract; EE, ethanol extract; MD, malto-dextrin (10–13 DE); GA, gum acacia. The numbers indicate the wall materials ratio; 1 = only MD, 2 = MD:GA of 1:1 ratio, 3 = MD:GA of 1:3 ratio.

To further visualize the distribution of individual phenolic compounds among powder samples, unsupervised PCA analysis was performed using SIMCA 16 software and the plot is shown as Figure 3. The first two principal components explained a total of 88.7% of variables (59.5% and 29.2%, respectively). The PCA Biplot showed a clear separation between powders derived from different *C. paliurus* extracts. For example, powders of the aqueous extract were located on the left side of the plot, whereas ethanol extract powders were on the right with high loadings of most flavonols such as kaempferol-3-*O*-rhamnoside, kaempferol-3-*O*-glucoside, quercetin-3-*O*-glucoside, quercetin-3-*O*-rhamnoside, etc. Noteworthy, the aqueous extract of *C. paliurus* encapsulated by MD-GA blends of 1:1 (i.e., the WE-MD/GA-2 sample) showed high correlations with one flavonol (myricetin-3-*O*-galactoside) and all phenolic acids except 4,5-dicaffeoylquinic acid.

The distribution of phenolic compounds among powder samples was related to their initial concentrations in the extracts as shown in Table 4. Zhou et al. found that the extraction solvent applied significantly influenced the bioactive profiles of *C. paliurus* extracts [11]. They reported that water was an efficient solvent for extracting phenolic acids, especially 3-*O*-caffeoyluinic acid and 4-*O*-caffeoyluinic acid, while the ethanol yielded high concentrations of triterpenoid and flavonoids [11]. We also observed solvent-specificity towards different phenolic compounds. From Table 4, a significantly higher number of flavonols were extracted by the ethanol than the aqueous solution, with the only exception being that for myricetin-3-*O*-galactoside (aqueous, 1.24 mg/g dw; ethanol, 0.96 mg/g dw). For some phenolic acids, the ethanol solution exhibited higher efficiency than the aqueous solution, such as chlorogenic (35.52 vs. 21.96 mg/g dw), 3,4-dicaffeoylquinic (0.33 vs. 0.24 mg/g dw), and 4,5-dicaffeoylquinic (1.40 vs. 0.96 mg/g) acids. On the other hand, other phenolic acids e.g., 5-*O*-caffeoylquinic, caffeic, 1,3-dicaffeoylquinic, and 1,5-dicaffeoylquinic acids, showed no significant quantitative differences in the aqueous and ethanol extracts. The discrepancy between our results and those reported by Zhou in phenolic acid extraction might be due to the concentration of ethanol used for extraction, i.e., 30% ethanol in the current study vs. 70% in the study of Zhou et al. [11], which affected the polarity of the solvent, and consequently the compound solubility and extraction efficiency.

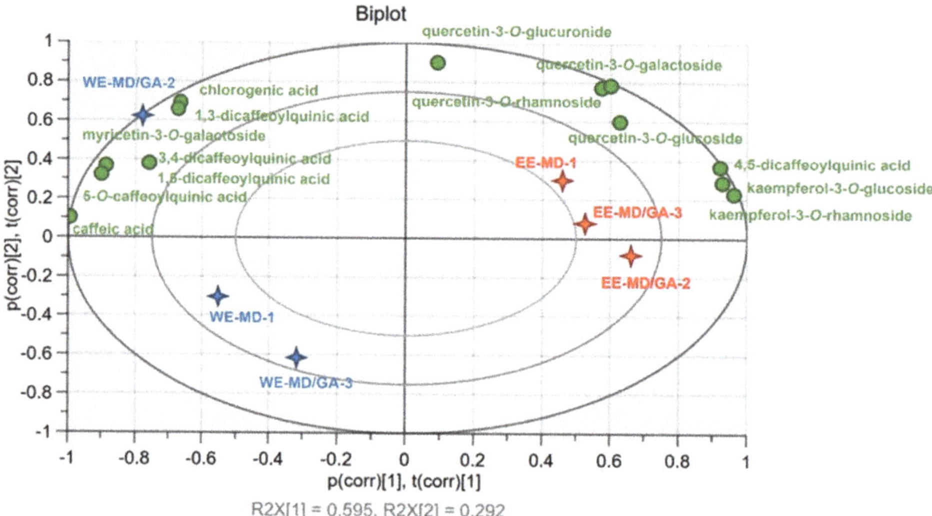

Figure 3. PCA biplot showing the distribution of phenolic compounds in different spray-dried *C. paliurus* powders; WE, aqueous extract; EE, ethanol extract; MD, maltodextrin (10–13 DE); GA, gum acacia. The numbers indicate the wall materials ratio; 1 = only MD, 2 = MD:GA of 1:1 ratio, 3 = MD:GA of 1:3 ratio.

3.5. Retention of Individual Phenolic Compounds after Spray Drying

The hot air applied during spray drying could induce changes to the phenolic compounds. Individual phenolics showed different rates of compound retention, as shown in Table 4. Two flavanols (i.e., quercetin-3-*O*-glucuronide and myricetin-3-*O*-galactoside) were retained at above 100% (115.63–152.6%) in all samples after drying. Similarly, in a previous study, we have found that the contents of myricetin-3-*O*-galactoside and quercetin-3-galactoside in the spray-dried powers were significantly higher than that of freshly prepared powers (302.0 vs. 74.3 µg/g and 479.5 vs. 205.5 µg/g, respectively) after 8 weeks' storage at 45 °C [20]. This indicated that some flavonols in the core material might change under high temperatures even after encapsulation. As for the spray-drying process, a shell was gradually formed at the droplet surface, during which the core material suffers thermal treatment up to 180 °C. This offers the possibility of increased contents of quercetin-3-*O*-glucuronide and myricetin-3-*O*-galactoside in the resulting powders. Zhang et al. [20] thought that the above compounds might be released from the phenolic polymers or transformed from other unstable glycosides. Quercetin arabino-furanoside and myricetin arabinoside are less stable than other glycosides and, thus, could be more susceptible to thermal degradation during processing. For example, only 24% and 22% of quercetin arabino-furanoside and myricetin arabinoside were retained in cranberry juices pasteurized at 90 °C for 10 min [47]. Similarly, an accelerated storage of apple juices at 70–100 °C has also confirmed the stability of different glycosides of quercetin. The order of quercetin glycosides stability was found as quercetin galactoside > quercetin glucoside > quercetin arabinoside [48]. Therefore, during the spray-drying process, flavonol arabinoside and arabino-furanoside could be degraded at high temperatures and transformed into more stable chemical forms, i.e., galactoside and glucuronide. These arguments offer explanations for the increased contents of myricetin-3-*O*-galactoside and quercetin-3-*O*-glucuronide after spray drying, as shown in Table 4. However, further research would be necessary to elucidate and confirm this phenomenon, and also investigate the effectiveness of wall materials to protect these unstable flavonol glycosides.

The maximum retention levels obtained for 5-*O*-caffeoylquinic acid, chlorogenic acid, quercetin-3-*O*-glucoside, and 1,3-dicaffeoylquinic acid were 93.77%, 84.56%, 82.06%, and 78.36%, respectively. Chlorogenic acid could be transformed into 5-*O*-caffeoylquinic acid when heated between 125 to 225 °C [49]. Since the *C. paliurus* powders were prepared at an inlet temperature of 180 °C, this conversion might be active in all the spray-dried powder samples, explaining the lower recovery value of chlorogenic acid (72.80–84.56%) than that of the 5-*O*-caffeoylquinic acid (83.86–93.77%). Relatively low retention rates were observed for quercetin-3-*O*-rhamnoside, 3,4-dicaffeoylqunic acid, kaempferol-3-*O*-glucoside, and 1,5-dicaffeoylqunic acid in the powders, with the highest attainment found in the WE-MD/GA-2 sample ($p < 0.05$). Other phenolic compounds have retentions of below 60%, with caffeic acid being the least recovered phenolic (32.80%) in the EE-MD/GA-2 sample.

Despite the fact that the individual phenolic content and recovery varied among samples, significantly higher retention of all the 14 phenolic compounds, especially those widely reported in *C. paliurus* (i.e., chlorogenic acid, quercetin-3-*O*-glucuronide, and 5-*O*-diffeolyquic acid), were significantly higher ($p < 0.05$) in the WE-MD/GA-2 sample. Overall, this sample also exhibited high retention of TPC. The results indicate that microencapsulation of aqueous extract of *C. paliurus* using a combination of MD and GA at 1:1 ratio could be a promising way to protect the phenolic compounds of *C. paliurus*.

4. Conclusions

The present study has provided insights into the microencapsulation of *C. paliurus* aqueous- and ethanolic-extracts using a mono-fluidic-jet spray drying technique. The powders produced were uniform in particle sizes with dented and marginally spherical-shaped microstructures. They were bright yellowish to brown colors, resembling the typical color of most tea-derived products. Their solubility in water was excellent, ranging from 88.81 and 99.12%. In addition, the powders have good storage stability, as reflected by their low moisture content (4.09 to 6.64%), water activity (0.11 to 0.19) and hygroscopicity. The phenolic compounds of *C. paliurus* were well-protected in the spray-dried powders, with the highest retention being > 94%. The individual phenolic profile, TPC and *AOC* results have further proven the preservation of phenolic compounds after spray drying. A combination of MD and GA at a ratio of 1:1 gave the best retention of phenolic compounds from the aqueous extract. Overall, the current study demonstrated a promising way to protect the phenolic and antioxidant capacities of *C. paliurus* extracts. The study could serve as a reference for future research and commercial scale processing to produce novel functional food powder with good physicochemical properties and health benefits from *C. paliurus*. Transforming the *C. paliurus* extracts into a versatile powder format could provide a window of opportunity for its application in various functional food products for a wider population.

Supplementary Materials: The following are available online at https://www.mdpi.com/article/10.3390/foods10122910/s1, Table S1: Calibration curves for the quantitation of individual phenolic compounds, Table S2: TPC, DPPH, and FRAP results of the optimized aqueous (WE) and ethanol extracts (EE).

Author Contributions: Conceptualization: S.Y.Q., J.X.; investigation, S.C., P.T., Y.P.; formal analysis, S.C., X.C.; writing-original draft, X.C.; writing - revising & editing, S.Y.Q.; supervision, S.Y.Q.; project administration, S.Y.Q. All authors have read and agreed to the published version of the manuscript.

Funding: The work was supported by a grant from the Enshi Tujia & Miao Autonomous Prefecture Academy of Agricultural Sciences, P.R. China (UniServices grant no. 36062.001) to S.Y.Q.

Acknowledgments: The authors wish to thank Yongchao Zhu and Sreeni Pathirana for the technical support. The authors also acknowledge the China Scholarship Council, China, for a Doctoral Scholarship (Ref. 201,806,790,005) awarded to the first author (X.C.).

Conflicts of Interest: The authors declare no conflict of interest.

References

1. Fang, S.; Yang, W.; Chu, X.; Shang, X.; She, C.; Fu, X. Provenance and temporal variations in selected flavonoids in leaves of *Cyclocarya paliurus*. *Food Chem.* **2011**, *124*, 1382–1386. [CrossRef]
2. Wang, K.Q.; Cao, Y. Research progress in the chemical constituents and pharmacologic activities of *Cyclocarya paliurus* (Batal.) Iljinshaja. *Heilongjiang Med. J.* **2007**, *31*, 577–579.
3. Wang, Q.; Jiang, C.; Fang, S.; Wang, J.; Ji, Y.; Shang, X.; Ni, Y.; Yin, Z.; Zhang, J. Antihyperglycemic, antihyperlipidemic and antioxidant effects of ethanol and aqueous extracts of *Cyclocarya paliurus* leaves in type 2 diabetic rats. *J. Ethnopharmacol.* **2013**, *150*, 1119–1127. [CrossRef]
4. Xie, J.-H.; Xie, M.-Y.; Nie, S.-P.; Shen, M.-Y.; Wang, Y.-X.; Li, C. Isolation, chemical composition and antioxidant activities of a water-soluble polysaccharide from *Cyclocarya paliurus* (Batal.) Iljinskaja. *Food Chem.* **2010**, *119*, 1626–1632. [CrossRef]
5. Lei, X.; Hu, W.-B.; Yang, Z.-W.; Hui, C.; Wang, N.; Liu, X.; Wang, W.-J. Enzymolysis-ultrasonic assisted extraction of flavanoid from *Cyclocarya paliurus* (Batal.) Iljinskaja:HPLC profile, antimicrobial and antioxidant activity. *Ind. Crop. Prod.* **2019**, *130*, 615–626.
6. Xie, J.H.; Shen, M.Y.; Xie, M.Y.; Nie, S.P.; Chen, Y.; Li, C.; Huang, D.F.; Wang, Y.X. Ultrasonic-assisted extraction, antimicrobial and antioxidant activities of *Cyclocarya paliurus* (Batal.) Iljinskaja polysaccharides. *Carbohydr. Polym.* **2012**, *89*, 177–184. [CrossRef]
7. Liu, X.; Xie, J.; Jia, S.; Huang, L.; Wang, Z.; Li, C.; Xie, M. Immunomodulatory effects of an acetylated *Cyclocarya paliurus* polysaccharide on murine macrophages RAW264. 7. *Int. J. Biol. Macromol.* **2017**, *98*, 576–581. [CrossRef]
8. Xie, M.Y.; Li, L.; Nie, S.P.; Wang, X.R.; Lee, F.S. Determination of speciation of elements related to blood sugar in bioactive extracts from *Cyclocarya paliurus* leaves by FIA-ICP-MS. *Eur. Food Res. Technol.* **2006**, *223*, 202–209. [CrossRef]
9. Zhang, J.; Shen, Q.; Lu, J.-C.; Li, J.-Y.; Liu, W.-Y.; Yang, J.-J.; Li, J.; Xiao, K. Phenolic compounds from the leaves of *Cyclocarya paliurus* (Batal.) Ijinskaja and their inhibitory activity against PTP1B. *Food Chem.* **2010**, *119*, 1491–1496. [CrossRef]
10. Hu, W.-B.; Zhao, J.; Chen, H.; Xiong, L.; Wang, W.-J. Polysaccharides from *Cyclocarya paliurus*: Chemical composition and lipid-lowering effect on rats challenged with high-fat diet. *J. Funct. Foods* **2017**, *36*, 262–273. [CrossRef]
11. Zhou, M.; Chen, P.; Lin, Y.; Fang, S.; Shang, X. A Comprehensive assessment of bioactive metabolites, antioxidant and antiproliferative activities of *Cyclocarya paliurus* (Batal.) Iljinskaja leaves. *Forests* **2019**, *10*, 625. [CrossRef]
12. Yang, Z.; Wang, J.; Li, J.; Xiong, L.; Chen, H.; Liu, X.; Wang, N.; Ouyang, K.; Wang, W. Antihyperlipidemic and hepatoprotective activities of polysaccharide fraction from *Cyclocarya paliurus* in high-fat emulsion-induced hyperlipidaemic mice. *Carbohydr. Polym.* **2018**, *183*, 11–20. [CrossRef]
13. Zhou, M.; Quek, S.Y.; Shang, X.; Fang, S. Geographical variations of triterpenoid contents in Cyclocarya paliurus leaves and their inhibitory effects on HeLa cells. *Ind. Crop. Prod.* **2021**, *162*, 113314. [CrossRef]

14. Minatel, I.O.; Borges, C.V.; Ferreira, M.I.; Gomez, H.A.G.; Chen, O.; Lima, G. Phenolic Compounds: Functional Properties, Impact of Processing and Bioavailability. In *Phenolic Compounds—Biological Activity*; Soto-Hernandez, M., Palma-Tenango, M., del Rosario Garcia Mateos, M., Eds.; InTech: Nappanee, IN, USA, 2017. [CrossRef]

15. Wu, W.D.; Amelia, R.; Hao, N.; Selomulya, C.; Zhao, D.; Chiu, Y.-L.; Chen, X.D. Assembly of uniform photoluminescent microcomposites using a novel micro-fluidic-jet-spray-dryer. *AIChE J.* **2010**, *57*, 2726–2737. [CrossRef]

16. Zhang, C.; Khoo, S.L.A.; Chen, X.D.; Quek, S.Y. Microencapsulation of fermented noni juice via micro-fluidic-jet spray drying: Evaluation of powder properties and functionalities. *Powder Technol.* **2019**, *361*, 995–1005. [CrossRef]

17. Wang, Y.; Liu, W.; Chen, X.D.; Selomulya, C. Micro-encapsulation and stabilization of DHA containing fish oil in protein-based emulsion through mono-disperse droplet spray dryer. *J. Food Eng.* **2016**, *175*, 74–84. [CrossRef]

18. Huang, E.; Quek, S.Y.; Fu, N.; Wu, W.D.; Chen, X.D. Co-encapsulation of coenzyme Q10 and vitamin E: A study of micro-capsule formation and its relation to structure and functionalities using single droplet drying and micro-fluidic-jet spray drying. *J. Food Eng.* **2019**, *247*, 45–55. [CrossRef]

19. Fu, N.; You, Y.-J.; Quek, S.Y.; Wu, W.D.; Chen, X.D. Interplaying Effects of Wall and Core Materials on the Property and Functionality of Microparticles for Co-Encapsulation of Vitamin E with Coenzyme Q10. *Food Bioprocess Technol.* **2020**, *13*, 705–721. [CrossRef]

20. Zhang, J.; Zhang, C.; Chen, X.; Quek, S.Y. Effect of spray drying on phenolic compounds of cranberry juice and their stability during storage. *J. Food Eng.* **2019**, *269*, 109744. [CrossRef]

21. Chen, X.; Ting, J.L.H.; Peng, Y.; Tangjaidee, P.; Zhu, Y.; Li, Q.; Shan, Y.; Quek, S.Y. Comparing Three Types of Mandarin Powders Prepared via Microfluidic-Jet Spray Drying: Physical Properties, Phenolic Retention and Volatile Profiling. *Foods* **2021**, *10*, 123. [CrossRef]

22. Pasrija, D.; Ezhilarasi, P.; Indrani, D.; Anandharamakrishnan, C. Microencapsulation of green tea polyphenols and its effect on incorporated bread quality. *LWT—Food Sci. Technol.* **2015**, *64*, 289–296. [CrossRef]

23. Ballesteros, L.F.; Ramirez, M.J.; Orrego, C.; Teixeira, J.; Mussatto, S.I. Encapsulation of antioxidant phenolic compounds extracted from spent coffee grounds by freeze-drying and spray-drying using different coating materials. *Food Chem.* **2017**, *237*, 623–631. [CrossRef]

24. Cian, R.E.; Campos-Soldini, A.; Chel-Guerrero, L.; Drago, S.R.; Betancur-Ancona, D. Bioactive Phaseolus lunatuspeptides release from maltodextrin/gum arabic microcapsules obtained by spray drying after simulated gastrointestinal digestion. *Int. J. Food Sci. Technol.* **2018**, *54*, 2002–2009. [CrossRef]

25. Zhang, C.; Chen, X.; Zhang, J.; Kilmartin, P.A.; Quek, S.Y. Exploring the effects of microencapsulation on odour retention of fermented noni juice. *J. Food Eng.* **2019**, *273*, 109892. [CrossRef]

26. Mazumder, M.A.R.; Ranganathan, T.V. Encapsulation of isoflavone with milk, maltodextrin and gum acacia improves its stability. *Curr. Res. Food Sci.* **2020**, *2*, 77–83. [CrossRef]

27. Tontul, I.; Topuz, A. Spray-drying of fruit and vegetable juices: Effect of drying conditions on the product yield and physical properties. *Trends Food Sci. Technol.* **2017**, *63*, 91–102. [CrossRef]

28. Wandrey, C.; Bartkowiak, A.; Harding, S.E. Materials for encapsulation. In *Encapsulation Technologies for Active Food Ingredients and Food Processing*; Springer: New York, NY, USA, 2010; pp. 31–100.

29. Rigon, R.T.; Norena, C.P.Z. Microencapsulation by spray-drying of bioactive compounds extracted from blackberry (rubus fruticosus). *J. Food Sci. Technol.* **2015**, *53*, 1515–1524. [CrossRef] [PubMed]

30. Nadeem, H.; Torun, M.; Özdemir, F. Spray drying of the mountain tea (*Sideritis stricta*) water extract by using different hydrocolloid carriers. *LWT—Food Sci. Technol.* **2011**, *44*, 1626–1635. [CrossRef]

31. Tonon, R.V.; Brabet, C.; Pallet, D.; Brat, P.; Hubinger, M.D. Physicochemical and morphological characterisation of açai (Euterpe oleraceae Mart.) powder produced with different carrier agents. *Int. J. Food Sci. Technol.* **2009**, *44*, 1950–1958. [CrossRef]

32. Dantas, D.; Pasquali, M.A.; Cavalcanti-Mata, M.; Duarte, M.E.; Lisboa, H.M. Influence of spray drying conditions on the properties of avocado powder drink. *Food Chem.* **2018**, *266*, 284–291. [CrossRef] [PubMed]

33. Quek, S.Y.; Chok, N.K.; Swedlund, P. The physicochemical properties of spray-dried watermelon powders. *Chem. Eng. Process. Process. Intensif.* **2007**, *46*, 386–392. [CrossRef]

34. Bhandari, B.; Senoussi, A.; Dumoulin, E.; Lebert, A. Spray drying of concentrated fruit juices. *Dry. Technol.* **1993**, *11*, 1081–1092. [CrossRef]

35. Tolun, A.; Altintas, Z.; Artik, N. Microencapsulation of grape polyphenols using maltodextrin and gum arabic as two alter-native coating materials: Development and characterization. *J. Biotechnol.* **2016**, *239*, 23–33. [CrossRef]

36. Silva, F.; Torres, L.; Silva, L.; Figueiredo, R.; Garruti, D.; Araujo, T.; Duarte, A.; Brito, D.; Ricardo, N. Cashew gum and mal-trodextrin particles for green tea (Camellia sinensis var Assamica) extract encapsulation. *Food Chem.* **2018**, *261*, 169–175. [CrossRef] [PubMed]

37. Suravanichnirachorn, W.; Haruthaithanasan, V.; Suwonsichon, S.; Sukatta, U.; Maneeboon, T.; Chantrapornchai, W. Effect of carrier type and concentration on the properties, anthocyanins and antioxidant activity of freeze-dried mao [*Antidesma bunius* (L.) Spreng] powders. *Agric. Nat. Resour.* **2018**, *52*, 354–360. [CrossRef]

38. Sun, S.; Du, G.J.; Qi, L.W.; Williams, S.; Wang, C.Z.; Yuan, C.S. Hydrophobic constituents and their potential anticancer activities from Devil's Club (*Oplopanax horridus* Miq.). *J. Ethnopharmacol.* **2010**, *132*, 280–285. [CrossRef]

39. Kwapinska, M.; Zbicinski, I. Prediction of Final Product Properties after Cocurrent Spray Drying. *Dry. Technol.* **2005**, *23*, 1653–1665. [CrossRef]

40. Solval, K.M.; Sundararajan, S.; Alfaro, L.; Sathivel, S. Development of cantaloupe (*Cucumis melo*) juice powders using spray drying technology. *LWT—Food Sci. Technol.* **2012**, *46*, 287–293. [CrossRef]

41. Zhang, C.; Fu, N.; Quek, S.Y.; Zhang, J.; Chen, X.D. Exploring the drying behaviors of microencapsulated noni juice using reaction engineering approach (REA) mathematical modelling. *J. Food Eng.* **2019**, *248*, 53–61. [CrossRef]

42. Idham, Z.; Muhamad, I.I.; Sarmidi, M.R. Degradation kinetics and color stability of spray-dried encapsulated anthocyanins from *Hibiscus sabdariffa* L. *J. Food Process. Eng.* **2011**, *35*, 522–542. [CrossRef]

43. Tsali, A.; Goula, A.M. Valorization of grape pomace: Encapsulation and storage stability of its phenolic extract. *Powder Technol.* **2018**, *340*, 194–207. [CrossRef]

44. Ferrari, C.C.; Germer, S.P.M.; Alvim, I.D.; De Aguirre, J.M. Storage Stability of Spray-Dried Blackberry Powder Produced with Maltodextrin or Gum Arabic. *Dry. Technol.* **2013**, *31*, 470–478. [CrossRef]

45. Sólyom, K.; Solá, R.; Cocero, M.J.; Mato, R.B. Thermal degradation of grape marc polyphenols. *Food Chem.* **2014**, *159*, 361–366. [CrossRef] [PubMed]

46. Kowalski, S. Changes of antioxidant activity and formation of 5-hydroxymethylfurfural in honey during thermal and mi-crowave processing. *Food Chem.* **2013**, *141*, 1378–1382. [CrossRef]

47. White, B.L.; Howard, L.R.; Prior, R.L. Impact of Different Stages of Juice Processing on the Anthocyanin, Flavonol, and Procyanidin Contents of Cranberries. *J. Agric. Food Chem.* **2011**, *59*, 4692–4698. [CrossRef] [PubMed]

48. van der Sluis, A.A.; Dekker, M.; van Boekel, M.A. Activity and concentration of polyphenolic antioxidants in apple juice. 3. Stability during storage. *J. Agric. Food Chem.* **2005**, *53*, 1073–1080. [CrossRef]

49. Dawidowicz, A.L.; Typek, R. Thermal Stability of 5-o-Caffeoylquinic Acid in Aqueous Solutions at Different Heating Conditions. *J. Agric. Food Chem.* **2010**, *58*, 12578–12584. [CrossRef] [PubMed]

 **foods**

Article

Effect of Water Temperature and Time during Heating on Mass Loss and Rheology of Cheese Curds

Ran Feng [1], Søren K. Lillevang [2] and Lilia Ahrné [1,*]

[1] Department of Food Science, University of Copenhagen, Rolighedsvej 26, 1958 Frederiksberg, Denmark; ran.feng@food.ku.dk

[2] Arla Foods Amba, Argo Food Park 19, 8200 Aarhus N, Denmark; sklv@arlafoods.com

* Correspondence: lilia@food.ku.dk

Abstract: During the manufacturing of mozzarella, cheese curds are heated to the desired stretching temperature traditionally by immersion in water, which influences the curd characteristics before stretching, and consequently the final cheese properties. In this study, cheese curds were immersed in hot water at 60, 70, 80 and 90 °C up to 16 min and the kinetics of mass loss and changes of rheological properties were investigated. The total mass of cooked curds increased up to 10% during the first minute, independent of the temperature, as a consequence of water retention. Fat was the main component lost into the cooking water (<3.5% w/w), while the concentration of protein increased up to 3.4% (w/w) compared to uncooked curds due to the loss of other components. Curds macrostructure during cooking showed that curds fully fuse at 70 °C/4 min; 80 °C/2 min and 90 °C/1 min, while after intensive cooking (>8 min) they lost the ability to fuse as a consequence of protein contraction and fat loss. Storage modulus, representing the curd strength, was dependent on cooking temperature and positively, and linearly, correlated with curd protein content (21.7–24.9%). This work shows the potential to modify curd composition and structure, which will have consequences for further processing and final product properties.

Keywords: cheese curd; cooking time and temperature; mass loss; rheology

Citation: Feng, R.; Lillevang, S.K.; Ahrné, L. Effect of Water Temperature and Time during Heating on Mass Loss and Rheology of Cheese Curds. *Foods* **2021**, *10*, 2881. https://doi.org/10.3390/foods10112881

Academic Editors: Harjinder Singh and Alejandra Acevedo-Fani

Received: 28 October 2021
Accepted: 16 November 2021
Published: 22 November 2021

Publisher's Note: MDPI stays neutral with regard to jurisdictional claims in published maps and institutional affiliations.

1. Introduction

Pasta-filata cheese processing involves a combined cooking-stretching step, during which the cheese curd is first cooked in hot water followed by shear stretching. The purpose of the cooking is to achieve the gel-sol transition temperature (i.e., the temperature where storage modulus G' equals loss modulus, G'') that is necessary to plasticize the cheese curd during stretching [1]. Typically, curd cooking is done in hot water, with a curd-to-water ratio of 1 to 1.4 [2–4], at temperatures that usually range from 60 °C to 85 °C and time ranging from 4 to 27 min [5,6]. Phase transition, molecular redistribution and compositional loss occur in cheese curd during this process: the curd mass transits from a viscoelastic solid to a viscoelastic liquid state during heating due to weakening of protein phase [7]; fat coalescence arises with increasing curd temperatures [8]; protein, fat, and calcium migrate from the curd into the water [6]. The extent of these transformations and mass loss depend on the equipment and processing parameters selected that determine the heat and mass transfer from the heating medium to the curd.

Combined thermal and mechanical effects on cheese composition have been reported in literature since cooking-stretching is usually applied in a single unit operation in the dairy industry. Bähler et al. (2013) reported that an increase in cooking temperature from 55 to 70 °C reduced the cheese yield from 0.88 to 0.59 g/g [4]. Banville et al. (2016) reported higher protein and lower fat contents of the cheeses after thermomechanical processing in 60–70 °C water, and an average weight loss of 10% was observed that was mainly caused by the fat loss [9]. Our previous study showed that increasing cooking residence time of cheese curd in cooker-stretcher induced increased water and decreased protein and fat contents of the cheese, by respectively 5.6%, 5.4% and 8.8% (*w/w*) [6]. However, due to the various equipment and processing parameters used, as well as the combination of thermal and mechanical effects, it is difficult to extract the effects of the cooking process alone on mass loss and cheese curd composition.

There is also limited information about the rheological properties of the curd during the cooking process, although the rheological properties of cheese have been extensively studied. Cheese is a viscoelastic material and its rheological properties have been considered more crucial than flavor attributes for consumer preferences [7]. The number, strength, and type of bonds between casein molecules constitute the basis of the cheese rheological properties [7], while these attributes are influenced by cheese composition and microstructure. Influence of fat on cheese rigidity (storage modulus, G') was found to be dependent on temperature: increased fat content induced increased G' at 10 °C, whereas the effect was not apparent at 25 °C because G' of fat approaches that of the protein gel [10]. Gel-sol transition temperature was found to increase with calcium-to-protein ratio [11], and the higher transition temperature indicates more difficult plasticization and melting due to higher proportion of colloidal Ca and strengthened protein phase [7,12]. Hence, changes of rheological properties of the curd during the cooking process are important to investigate as they will determine the structural arrangements of the curd at the starting point of further processing.

Although, the cooking step determines the curd yield, composition, structural properties and consequently influences the functionality of the final cheese product, knowledge about the mass loss during the cooking process at different temperatures, as well as consequences for physical–chemical and structural properties of the curd have not been assessed. This knowledge is relevant to understand the performance of curds during further processing (e.g., stretching) and develop innovative dairy products. Thus, the current study aims to determine the kinetics of mass loss from the curd during cooking at temperatures from 60 to 90 °C up to 16 min and consequences for structural and rheological properties of the curd.

2. Materials and Methods

2.1. Sample Preparation and Cooking Process

Frozen (−20 °C) 'Cagliata' mozzarella curd (24.0% protein, 27.0% fat, 43.7% water, 0.82% Ca, 0.69% NaCl) was provided by Arla Foods (Fredericia, Denmark). The curd was fermented by starter culture brined in NaCl and had a pH of 5.6. It was stored in a 4 °C refrigerator for three days before processing in order to fully defrost. The curd was then cut into 1 cm^3 cubes using a wire cutter and 30 g of curd was cooked with 2.5% (*w/v*) NaCl solution at 60, 70, 80 and 90 °C in a shaking water bath (Grant Instruments, Cambridge, UK) at 125 rpm. The curd-to-water ratio was 1:1 (*w/w*) (30 g curd and 30 g water) and each experiment was performed in duplicate. Based on pre-experiments, the liquid was separated from the curd using a sieve after 1-, 2-, 4-, 8-, 12- and 16-min cooking time, and stored at −20 °C until composition analysis. The cooked curd was weighted, photographed, transferred into a cylindrical cup and stored at 4 °C overnight for rheological measurements on the following day.

2.2. Composition Analysis of Cooking Water

Nitrogen, fat and water content were determined by the Kjeldahl [13], Gerber [14] and the oven-drying [15] methods, respectively. Protein was quantified by Kjeldahl using a nitrogen conversion factor of 6.38. The water content was determined at 105 °C. Calcium and sodium content was measured by Inductively Coupled Plasma Optical Emission Spectroscopy (ICP-OES; Agilent Technologies, Santa Clara, CA, USA). Before analysis, 1 g of sample was digested with 8 ml HNO_3 (65%) and 2 ml HCl (37%) using a Multiwave GO microwave (Anton Paar, Graz, Austria). Standards in a range of 0.04–20 mg L^{-1} were prepared from multi-element (Ag, Al, B, Ba, Bi, Ca, Cd, Co, Cr, Cu, Fe, Ga, In, K, Li, Mg, Mn, Na, Ni, Pb, Sr, Tl, Zn) standard solution IV from Merck KGaA. Standard curves were determined at the wavelengths of 396.847 nm and 588.995 nm for calcium and sodium, respectively. All analysis were performed in duplicate.

2.3. Rheological Properties of Cooked Curd

Rheological properties were determined by the methods of Feng et al. (2021) with some modifications [6]. The rheological properties of the cheeses were studied on a rheometer (Discovery HR-2, TA Instruments, New Castle, DE, USA) with 25 mm diameter serrated parallel plates. Disc-shaped cheese samples of 25 mm diameter and ~2 mm thickness were prepared and equilibrated for 5 min at the test temperature. A 1N normal force was used to define the measurement gap. A ring of paraffin oil was placed around the sample periphery to avoid moisture loss during rheological measurements.

Frequency sweeps were conducted by applying frequencies in descending order from 50 Hz to 0.1 Hz at 20 °C using 0.1% strain amplitude (within linear viscoelastic region). The data of storage modulus G' was fitted to the following equation:

$$G' = G'_1 \cdot \omega^n \tag{1}$$

where G'_1 is the storage modulus measured at 1 Hz, ω is the frequency in Hz, and n is the degree of frequency dependence.

In temperature sweeps, strain and frequency were 0.1% and 1 Hz respectively and the temperature was increased from 20 to 100 °C. The rate of the Peltier heating system was set at $3\,°C \cdot min^{-1}$. All rheological measurements were conducted at least in duplicate.

2.4. Thermal Analysis of Uncooked Curd by Differential Scanning Calorimetry

The thermal properties of the uncooked Cagliata curd were evaluated by differential scanning calorimetry (DSC 1, Star^e System, Mettler Toledo, Glostrup, Denmark). Samples of 10–15 mg were placed in aluminum pans and heated from 20 °C up to 90 °C with a heating rate of $5\,°C \cdot min^{-1}$. An empty sealed aluminum pan was used as reference in every test. Heat flow $(W \cdot g^{-1})$ versus temperature curves was obtained. Triplicate was measured for the curd sample.

2.5. Statistical Analysis

One-way analysis of variance (ANOVA) followed by a Duncan test was done to verify differences between means using IBM SPSS Statistics 28 (IBM Corporation, Somers, NY, USA). Differences were considered significant at the probability level $p < 0.05$.

3. Results and Discussion

3.1. Mass Transfer from Curd to the Cooking Water

The total mass loss (yield) (Table 1) shows an initial mass gain, independent of the cooking water temperature during the first 4 or 8 min of cooking up to 10% followed by a mass loss up to 7% for curds cooked for 16 min at 70 or 80 °C. These results can be explained by the hydration of the curds when immersed in water and by the changes in the individual components of the curd as will be discussed below.

Water, protein, fat, Ca in the curds after cooking at 60, 70, 80 and 90 °C for 1, 2, 4, 8, 12 and 16 min are shown in Figure 1. The figures on the left side (Figure 1a,c,e,g,i) show the mass in grams of each component remained in the 30 g original curd, while those on the right side (Figure 1b,d,f,h,j) show the final % (*w/w*) of each component in the cooked curd.

Table 1. Yield (g cooked curd/g initial uncooked curd), reported as means ± standard deviation.

Cooking Time (Min)	Yield (G Cooked Curd/G Initial Uncooked Curd)			
	Water Temperature During Cooking			
	60 °C	70 °C	80 °C	90 °C
1	1.08 ± 0.01 [Aa]	1.08 ± 0.01 [Aa]	1.10 ± 0.01 [Aa]	1.09 ± 0.01 [Aa]
2	1.05 ± 0.01 [Ba]	1.07 ± 0.01 [Aa]	1.06 ± 0.00 [Ba]	1.08 ± 0.01 [Aa]
4	1.03 ± 0.01 [Ca]	1.02 ± 0.01 [Ba]	1.04 ± 0.01 [Ba]	1.03 ± 0.02 [Ba]
8	1.01 ± 0.00 [Dab]	0.99 ± 0.00 [Ca]	1.00 ± 0.01 [Cab]	1.02 ± 0.01 [Bb]
12	0.98 ± 0.00 [Ea]	0.94 ± 0.02 [Da]	0.98 ± 0.02 [Ca]	0.96 ± 0.01 [Ba]
16	0.98 ± 0.00 [Ea]	0.93 ± 0.01 [Db]	0.93 ± 0.01 [Db]	0.96 ± 0.01 [Ba]

[A–E] Different letters in the same column indicate significantly difference ($p < 0.05$). [a–b] Different letters in the same row indicate significantly difference ($p < 0.05$).

A significant increase of water content from 13.1 g to 16.2–16.5 g ($p < 0.05$) in 30 g of curd is observed in the first minute of cooking (Figure 1a, because cooking water gets attached to the curd surface and entrapped between the individual curd cubes. With the longer cooking until 8 min, water content in the cooked curds was always above the water content level of the uncooked curd due to the presence of surface and entrapped cooking water. The amount of the entrapped water decreased from 16.2–16.5 g (1 min) to 12.6–13.7 g (16 min) in 30 g curd with increased cooking time ($p < 0.05$) since the cooked curd gradually lost the ability to retain the water. The cooking temperature did not play a significant role in affecting the water content until 12 min, likely because of the surface temperature. To identify the temperature effect after 12 min, Figure 2 shows the water content in 30 g curds cooked for 12–16 min in comparison with uncooked curds. No significant difference between the cooked and uncooked curds was observed, due to the large standard deviations, though the curds processed at 70 °C/12–16 min displayed significantly lower ($p < 0.05$) water content (12.5–12.6 g) compared to curds obtained at 80 °C/12 min and 90 °C/12–16 min (13.4–13.8 g). As previously reported, cheese heating at 60–100 °C can induce contraction and shrinkage of the protein network and simultaneous expulsion of water as a result of increased hydrophobic interactions between casein molecules [16]. Hence, the slightly lower water content at 70 °C might be attributed to the structural changes of the casein network that expelled water. However, curds cooked at 80–90 °C presented a higher water content probably because of the sudden shrinkage of the curd protein matrix and skin formation that hindered syneresis at such high temperatures [17,18]. These changes in the casein network might be a result of Ca precipitation due to heat treatment that affected Ca-protein interactions [19]. This is also visible in Figure 1b that shows that the water in the cooked curd reduced by 5% during cooking at 60–80 °C, whereas to a less extent (2%) at 90 °C due to the formed skin as mentioned.

The protein content of the curd was mainly affected by the cooking time and not by the temperature in the range studied (Figure 1c). The rate of protein release into the water slowed down after 2 min cooking and tends to stabilize for longer cooking times, suggesting that most protein losses occurred at the curd surface at initial cooking phase. The absolute protein loss in 30 g curd was only 0.17 g, which represent a 3.5% increase of protein content (Figure 1d) in the cooked curd as a result of loss of other components, i.e., fat, etc. The cooking temperature had little effect on protein content, only at 16 min cooking statistically significant differences were observed between 60/90 °C and 70/80 °C ($p < 0.05$). This could be an indication of a weakened protein matrix that requires further studies in the future.

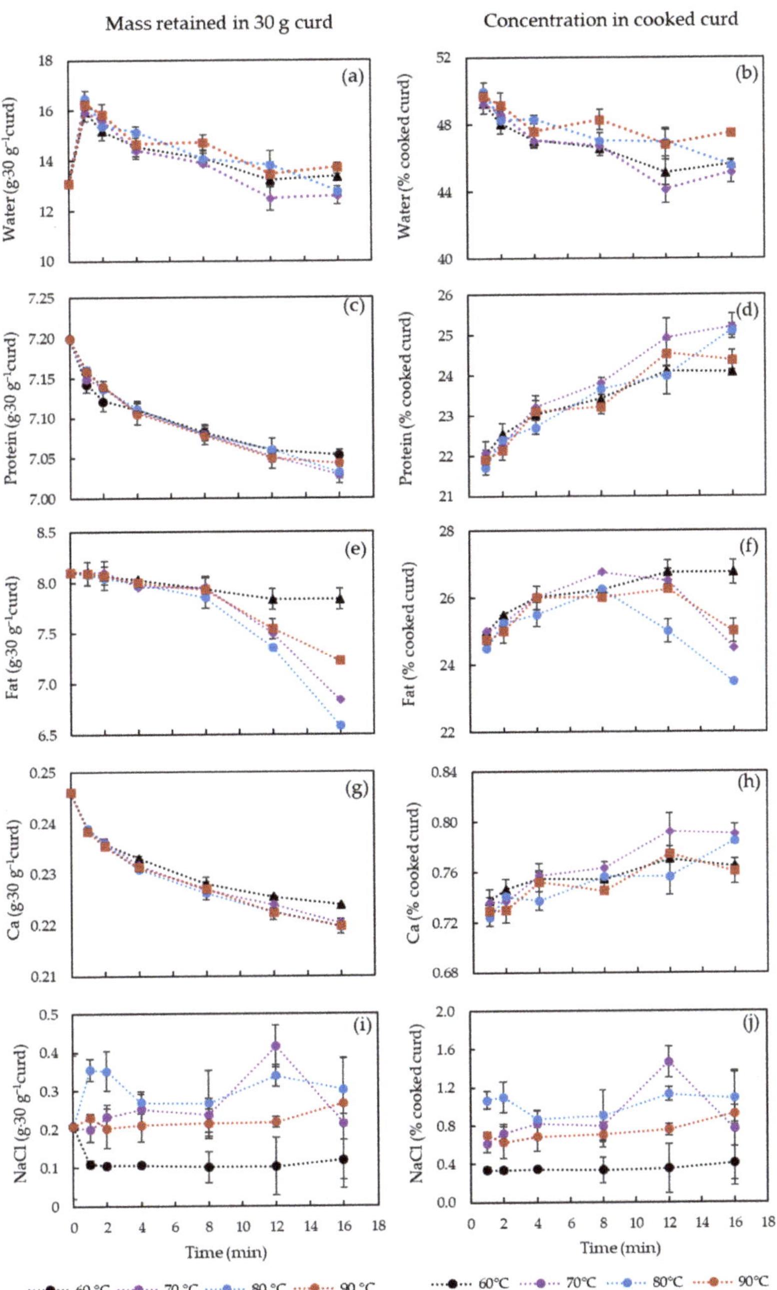

Figure 1. Content (%, *w/w*) of water, protein, fat, calcium and NaCl retained in curd at varied cooking conditions. (**a,c,e,g,i**) show the content in 30 g original curd while (**b,d,f,h,j**) show the % concentration in final curd. The error bars indicate standard deviations.

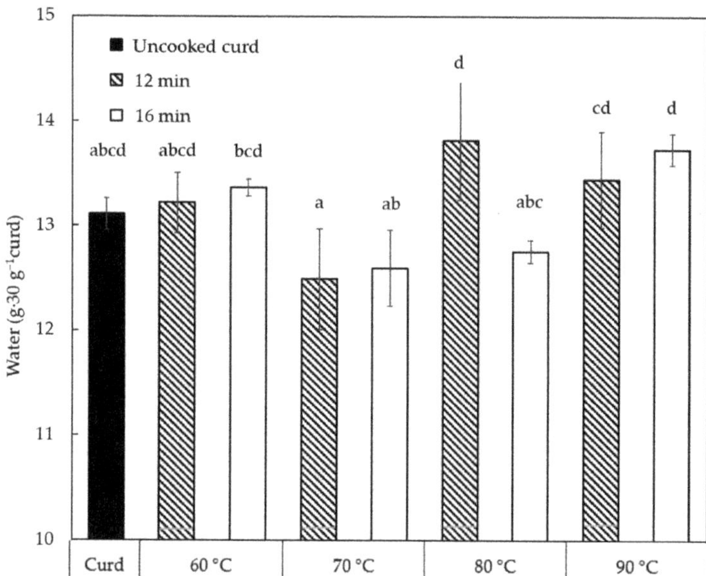

Figure 2. Water content retained in 30 g curd at varied cooking temperatures for 12–16 min. Values with different superscripts differ significantly ($p < 0.05$). The error bars indicate standard deviations.

Fat was the main component lost into the cooking water, which at maximum decreased from 8.1 g to 6.6 g in 30 g curd for 80 °C as cooking time increased up to 16 min (Figure 1e). Longer cooking resulted in increased curd temperature and thus enhanced fat mobility. The highly mobile fat can, therefore, lead to fat coalescence and free oil release [4]. Different from protein content, fat content exhibited a sudden decrease at 8 min cooking for all temperatures, except 60 °C, suggesting that the coalesced fat is released from the curd casein matrix. Similar to water and protein kinetics, fat content at 90 °C decreased to a less extent compared to that at 70 and 80 °C, which may be due to the fact that fat migration is limited by the toughened protein matrix and skin formation at surface of the curds [20]. Thus, at cooking temperature of 90 °C, the curd surface might contract, forming a skin that hindered the rate of fat release. Figure 1f shows that, in fact, the % of fat content in the cooked curds increased with cooking time up to 8 min and this was not strongly influenced by temperature, whereas after 8 min significant differences are observed. At 60 °C the % fat content is higher for curds cooked at 16 min compared to 1 min cooking, whereas at 70–90 °C a significant decrease of fat content is observed ($p < 0.05$). The increase in the initial times of cooking was simply due to the loss of other components, specially entrapped water, and relatively limited fat release. After 8–12 min, as all the curd mass reached the water temperature, significant fat loss occurred as observed in Figure 1e ($p < 0.05$).

The amount of Ca in the curd (Figure 1g) not only decreased with cooking time but also reached a lower value at temperatures of 70–90 °C, compared to 60 °C. Ca seems to have been continuously lost into the water together with protein, as similar trend is observed for Ca and protein (Figure 1d). Ca:protein ratio in the curd (Figure 3) was relatively less affected by the cooking temperature. It was reported that an increase from 15.8 to 31 mg Ca g⁻¹ protein induced decreased gel-sol temperature from 60 to 51.7 °C [11]. Thus, the slight reduction of Ca:protein from 34.2 to 31.2 mg Ca g⁻¹ protein is not expected to induce significant changes on the curd properties. However, the proportion of soluble and insoluble Ca might be different, which is highly affected by water and temperature, and has been reported to play an important role in melting properties of cheese [21].

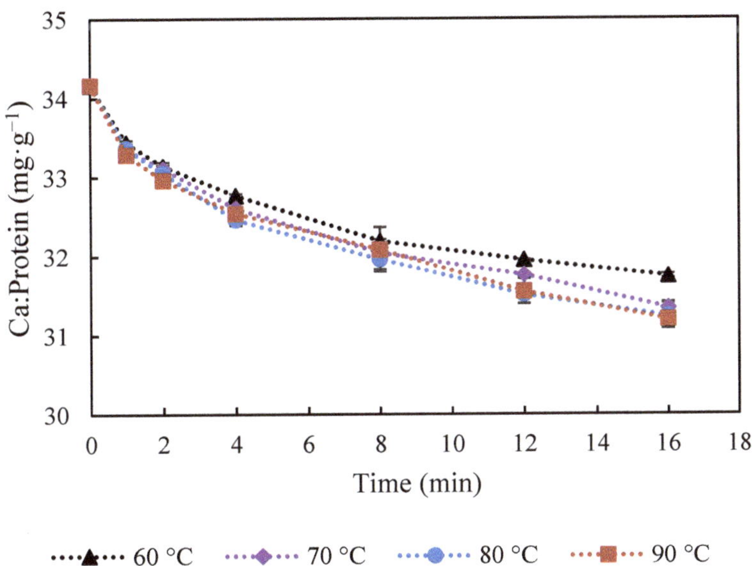

Figure 3. Calcium-to-protein ratio in mg calcium·g^{-1} protein at during cooking in water at 60, 70, 80 and 90 °C. The error bars indicate standard deviations.

Limited uptake or loss of NaCl from the curds to the water was observed during the cooking process, however, differences in salt content is observed at different temperatures (Figure 1i,j). After 16 min the cooked curds at 60 °C exhibited a lower NaCl content (0.4 ± 0.2% *w/w*) compared to cooked curds at 70, 80 and 90 °C curds that had respectively 0.8 ± 0.6%, 1.3 ± 0.3% and 0.9 ± 0.1%), which was also similar to the content after 1 min cooking. These differences are due, as previously discussed, to the surface water and entrapped cooking water that contained 2.5% NaCl. Therefore, at water temperature >70 °C acceleration of salting was observed.

3.2. Macrostructural and Rheological Properties of Curds during Water Cooking

The appearance of cooked curd after cooking at 60 to 90 °C/1–16 min is shown in Figure 4, and a summary of the time, at which the macrostructure transitions occurred are presented in Table 2. It was noticed that the curds were always kept separated in the cooking water during cooking. Independent of the temperature, a general observation is that, as cooking time proceeds, curd cubes were able to fuse into one block after being removed from the water. However, after a given cooking time, the curds lost the ability to fuse together and form a block, instead separated cube particles can be observed. At 60 °C, the curd kept the cube shape during the first 2 min, starting to melt together after 4 min. However, the curd at 60 °C did not fully fuse to the same extent as observed at other temperatures (70 °C/4 min and 80–90 °C/2 min). After 12 min cooking, the curd appeared as separated cube particles again with visible edges and the same occurred at earlier times for higher cooking temperatures, i.e., 8 min at 70–90 °C as displayed in Table 2. Furthermore, curds cooked at 90 °C for longer times (12–16 min) presented visually the coarsest particles with a rough surface, indicating extensive mass loss.

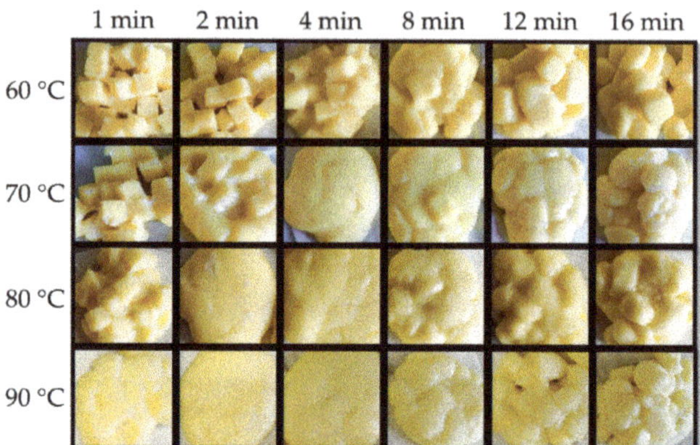

Figure 4. Appearance of curd after cooking at varied conditions.

Table 2. Cooking time (min) when curd started melting, fully melted and separated.

Phase Transition	Water Temperature During Cooking			
	60 °C	70 °C	80 °C	90 °C
Melting of curds				
Surface melting	4	2	1	1
Fully fused	-*	4	2	2
Separation of curds				
Separated	12	8	8	8

* Curd cooked at 60 °C did not fully fuse.

The melting of curds has been attributed to liquefaction of fat, and softening of protein matrix due to decreased number and strength of protein-protein bonds [7]. The thermo-rheological measurements of the uncooked curd (Figure 5) showed that G' and G'' decreased with the increase of temperature and loss tangent continuously increased during heating, reaching its maximum value (<1) at about 90 °C. A previous study of cheese has shown similar thermo-rheological behavior and two temperature ranges, one at 20–35 °C and another at 50–65 °C with a faster rate of decrease for moduli that were respectively attributed to melting of the fat phase and softening of the protein matrix [22]. The uncooked curd also showed two temperature ranges with faster decrease of moduli: the melting temperature of fat at 25–45 °C and the softening temperature of the protein matrix at 70–90 °C. DSC measurements of the uncooked curd confirmed that the fat melts (endothermic peaks) at 25.2–41.8 °C with a representative curve shown in Figure 6.

Different from the cheese, uncooked curd contains tiny native fat globules that are protected by protein layers [6], which may justify the higher melting temperature. In addition, the higher Ca content (0.82%) and Ca:protein (34.2 mg Ca g^{-1} protein) in the uncooked curd suggested more protein bound to Ca together with an increased insoluble Ca content, which may also justify the higher protein softening temperature. Loss tangent was lower than the reported value of 2.64 for cheese [22]. The maximum loss tangent value is often regarded as an indicator of melt functionality. However, for the uncooked curd not stretched, a lower maximum value indicates a viscoelastic solid that was not plasticized. Fully melted curd was not observed at 60 °C (Figure 4), but only for temperature of 70 °C or above, which was consistent with the observed 70–90 °C protein softening range (Figure 5). The separation stage could be linked to the contraction of the casein matrix as mentioned in Section 3.1, and the formed skin hindered further fusion of protein matrix with visually coarser curd surface. Furthermore, the markedly reduced fat content at 8 min cooking (Figure 1e,f) likely contributed to this separation phenomenon.

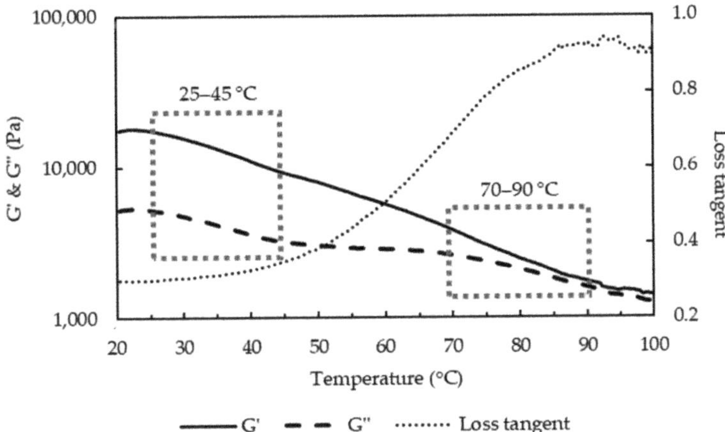

Figure 5. Storage and loss moduli during temperature sweep test on uncooked curd.

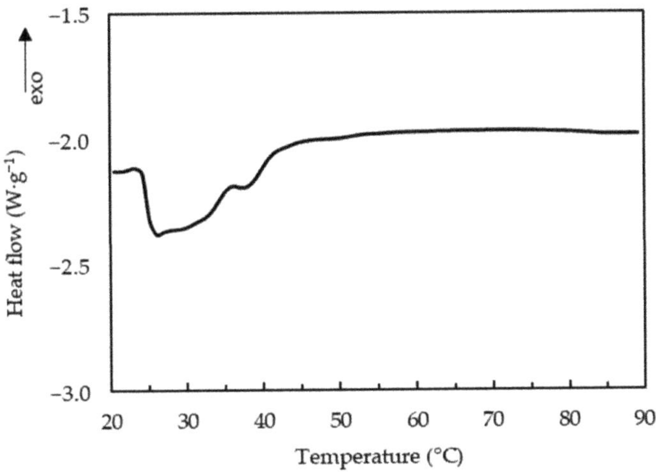

Figure 6. Differential scanning calorimetry curve of uncooked curd.

The rheological properties of the cooked and uncooked curds were evaluated and Figure 7 shows the storage modulus at 1 Hz, G'_1, of the cooked curd produced at 60, 70, 80 and 90 °C up to 16 min. Storage modulus reflects the total number and strength of protein-protein bonds in the protein matrix [7]. Curds cooked at 70 °C and 80 °C for 16 min and 90 °C for 12–16 min could not be measured because it was impossible to form a disk to be placed between the parallel plates of the rheometer. The reason could be the changes in calcium phosphate, inducing Ca-protein interactions, which caused a strongly contracted protein network with coarse curd surface that could not fuse to form a whole curd block anymore. Moreover, these conditions corresponded to extensive fat loss as previously discussed.

G'_1 significantly increased as cooking time and temperature increased, which were related to the loss of entrapped water and fat, and relatively increased protein concentration, as well as the proportion of insoluble Ca. Lower moisture content was usually linked to increased cheese hardness and G' due to a lower degree of casein hydration [23]. The observation was also in agreement with previous studies, which reported higher firmness of low-fat cheeses than that of full- or reduced-fat cheeses that indicated increases in the protein concentration and a more cross-linked structure [24,25]. Rheological properties of cheese at high temperatures are also associated with levels of insoluble Ca. The strength of Ca-protein binding was found to increase with increasing temperature [26], and G′ at 70 °C increased significantly with increasing concentrations of colloidal calcium phosphate [27]. The positive correlation between G'_1 and protein content (21.7 to 24.9% *w/w*, Figure 8) in the cooked curd is temperature dependent, showing that enhanced protein-protein interactions as protein content increased have contributed to the increased strength of the protein network. A rapid increase of G'_1 is observed at higher cooking temperatures 80–90 °C in comparison to 60–70 °C, as shown by the two trend lines for curds cooked at 60–70 °C and 80–90 °C instead of one for all the curds. The higher slope for 80–90 °C treated curds could be caused by the precipitation of the soluble Ca that led to Ca/phosphate-mediated protein-protein interactions between casein molecules [16].

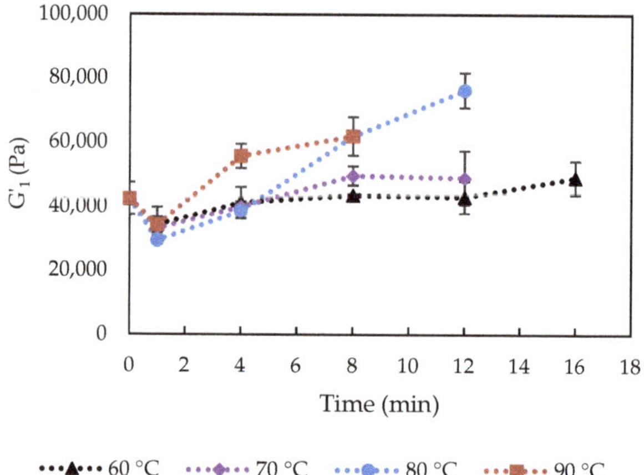

Figure 7. Storage modulus at 1 Hz (G'_1) of curd cooked at varied conditions. The error bars indicate standard deviations.

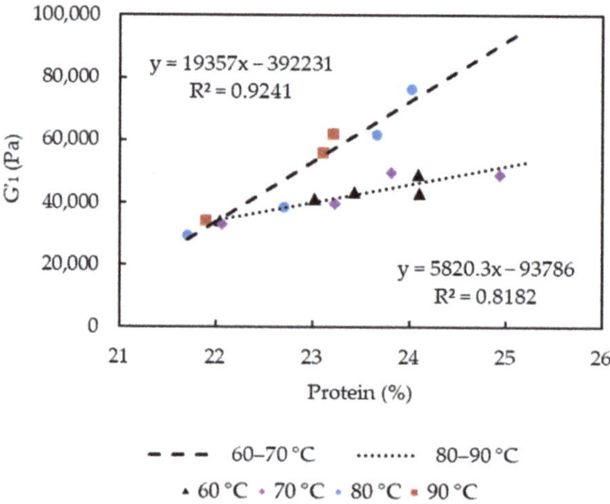

Figure 8. Plot of storage modulus versus curd protein content.

The crossover temperature of storage and loss moduli, T_c, (Figure 9) give an indication of the gel-sol transition point of cheese during the temperature sweep test, which can be associated with the point at which the cheese can rapidly melt and flow [28]. The uncooked curd did not show a crossover point (Figure 5), indicating a hard protein matrix structure with small native fat globules that impede melting. All the cooked curds showed values of T_c ranging from 67 to 82 °C, which indicates structural modifications caused by the cooking process. It is interesting to note that significantly higher T_c was observed after 1 min cooking at 60 and 70 compared to 80–90 °C ($p < 0.05$). Furthermore, T_c significantly decreased from 79 to 68 °C during cooking 60 °C, while at cooking at 70 °C, first a decrease in T_c from 76 to 67 °C was observed, but suddenly increased after 8 min cooking to 75 °C ($p < 0.05$). During cooking at 80 and 90 °C, an increase of T_c is observed along the cooking time from 68 to 82 °C and from 69 to 80 °C, respectively.

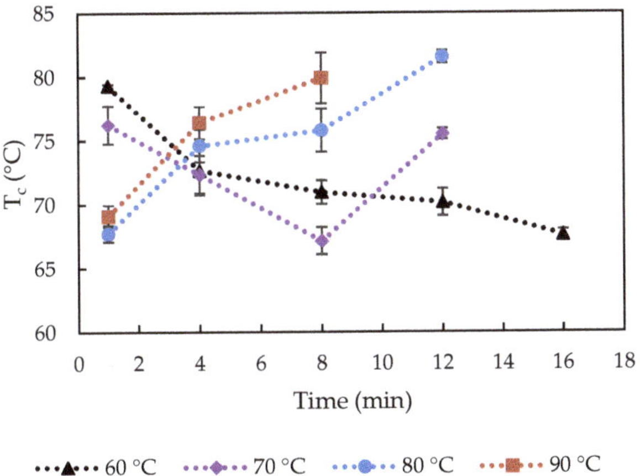

Figure 9. Crossover temperature (T_c) of curd cooked at varied cooking conditions. The error bars indicate standard deviations.

These differences were unexpected but might be linked to differences in fat content and entrapped/surface water (from cooking). Higher fat and water content were associated with higher meltability [29,30]. Initially, with cooking at low temperatures 60–70 °C, the T_c value likely depended on fat content. Increased % fat content with longer cooking at 60 °C/1–16 min and 70 °C/1–8 min (Figure 1f) could lead to earlier melting, i.e., lower T_c, and the sudden decrease of fat content from 8 min cooking at 70 °C delayed the crossover of G′ and G″. Ibáñez et al. (2020) also observed a lower T_c in low-fat cheese [25]. As the cooking temperature increased, the rheological properties of curd depend on both water and fat content. The effect of reduced amount of entrapped water in curds cooked at 80–90 °C/1–8 min (Figure 1b) preceded that of increased % fat content on melting, resulting in the higher T_c values. The longer holding times might also lead to the formation of new heat-induced calcium-phosphate structures, which might explain the increased T_c with extended temperature/time.

3.3. Schematic Description of Mass Loss and Structural Changes in Curds as a Consequence of Cooking

A schematic model is proposed in Figure 10 to summarize the changes in curd during the water cooking process. The cooking process is complex and depend on: (i) cooking process parameters (e.g., equipment design, curds/water ratio and water agitation); (ii) water properties (e.g., temperature and composition (salt content, Ca, pH)); and (iii) curd properties (size, composition, temperature).

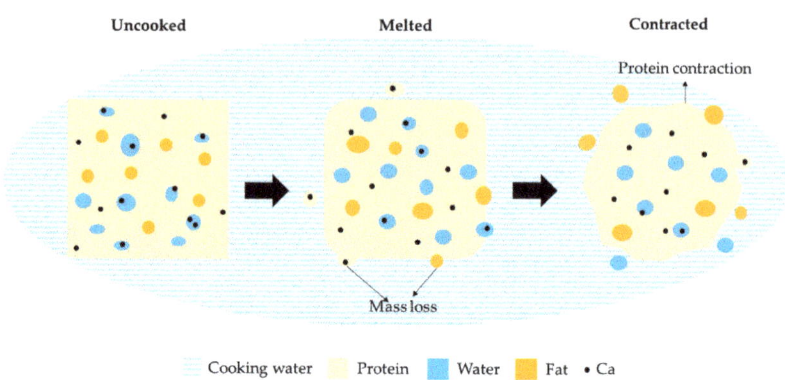

Figure 10. Schematic model proposed for kinetics of mass transfer during cooking of cheese curd.

In this study, three stages were identified to describe the molecular migration and microstructural changes taking place during water cooking. The original curd, consists of a protein (casein) matrix with homogeneously distributed water (serum) and fat globules. Ca is either soluble in the water phase or bound to the protein matrix. When the curd is immersed in water, heat is transferred from the water surrounding the curd cubes to the cube surface, which causes an increase of the water content in the curds. As the surface temperature of the cube increases, the curds surface softens after a few minutes of cooking and the cubes lose the edges due to the water agitation. As heat is transported from the cube surface to the interior of the cube, the curds can then fuse into a continuous curd block once removed from the water. During this stage, protein, fat and Ca are released into the cooking water, but the mass loss seems to occur mainly at surface. Ca becomes more insoluble and is lost together with protein. As cooking proceeds, temperature of the cubes increases, leading to contraction of the protein network due to an increase in hydrophobic interactions and Ca/phosphate-mediated interactions between the para-casein molecules, which causes the observed separated curd appearance. Protein and Ca losses continue at reduced rate compared to the melted stage, whereas the contraction of protein network seems to accelerate fat loss.

In summary, the conditions selected for cooking process can significantly alter cheese curd status and influence its structural and rheological properties. Further studies are needed to further understand changes at molecular and microstructural level.

4. Conclusions

Cooking time and temperature induce significant changes in cheese curd composition and rheological properties. Increase of water temperatures from 60 °C to 90 °C and long cooking times significantly accelerate these changes. A significant amount of water coats the curds surface after immersion in water, which has an important role both for mass and heat transfer. The total mass of cooked curds increased up to 10% during the first minutes of cooking, independent of the cooking temperature, although for longer cooking times the mass loss reached 7%. Fat was the main component lost from curd into the cooking water (<3.5% loss based on final curd mass), due to fat liquefaction and migration as the heating intensity increased. Although slight protein and Ca losses were observed, the final concentrations in the cooked curds increased up to 3.4% and 0.07%, respectively. The rate of protein and Ca losses decreased with the cooking time, while the rate of fat loss increased with longer cooking. Thus, fat content in the cooked curds exhibited a sudden decrease after 8 min cooking for all temperatures, which was coincided with the separation phenomenon observed, i.e., inability of curds to fuse. G' and T_c were highly influenced by cooking time and temperature, which were related to loss of water and fat, increase of protein content and increased insoluble Ca-phosphate, and G′ was linearly correlated with increased protein content 21.7–24.9%.

This cooking study provides initial knowledge about compositional and rheological changes of curd during cooking, and the influence of water temperature and cooking time. Further studies are needed at molecular and microstructural level to further understand the changes observed and to assess the effect of other relevant parameters.

Author Contributions: Conceptualization, R.F., S.K.L. and L.A.; methodology, R.F., S.K.L. and L.A.; formal analysis, R.F.; investigation, R.F.; writing—original draft preparation, R.F.; writing—review and editing, S.K.L. and L.A.; supervision, L.A.; project administration, L.A.; funding acquisition, L.A. All authors have read and agreed to the published version of the manuscript.

Funding: This project has received funding from the European Union's Horizon 2020 research and innovation programme under the Marie Skłodowska-Curie Grant Agreement No. 801199.

And the Danish Dairy Rationalisation Fund (DDRF) and Arla Foods as part of the platform for Novel Gentle Processing.

Data Availability Statement: Not applicable.

Acknowledgments: We authors acknowledge the funding bodies for supporting this work.

Conflicts of Interest: The authors declare no conflict of interest.

References

1. Bähler, B.; Hinrichs, J. Characterisation of mozzarella cheese curd by means of capillary rheometry. *Int. J. Dairy Technol.* **2013**, *66*, 231–235. [CrossRef]
2. Guinee, T.P.; Mulholland, E.O.; Mullins, C.; Corcoran, M.O. Effect of salting method on the composition, yield and functionality of low moisture Mozzarella cheese. *Milchwissenschaft* **2000**, *55*, 135–138.
3. Locci, F.; Ghiglietti, R.; Francolino, S.; Iezzi, R.; Mucchetti, G. Effect of stretching with brine on the composition and yield of high moisture Mozzarella cheese. *Milchwissenschaft* **2012**, *67*, 81–85.
4. Bähler, B.; Ruf, T.; Samudrala, R.; Schenkel, P.; Hinrichs, J. Systematic approach to study temperature and time effects on yield of pasta filata cheese. *Int. J. Dairy Technol.* **2015**, *69*, 184–190. [CrossRef]
5. Renda, A.; Barbano, D.M.; Yun, J.J.; Kindstedt, P.S.; Mulvaney, S.J. Influence of Screw Speeds of the Mixer at Low Temperature on Characteristics of Mozzarella Cheese. *J. Dairy Sci.* **1997**, *80*, 1901–1907. [CrossRef]
6. Feng, R.; Barjon, S.; Berg, F.W.V.D.; Lillevang, S.K.; Ahrné, L. Effect of residence time in the cooker-stretcher on mozzarella cheese composition, structure and functionality. *J. Food Eng.* **2021**, *309*, 110690. [CrossRef]
7. Lucey, J.; Johnson, M.; Horne, D. Invited Review: Perspectives on the Basis of the Rheology and Texture Properties of Cheese. *J. Dairy Sci.* **2003**, *86*, 2725–2743. [CrossRef]
8. Rowney, M.; Roupas, P.; Hickey, M.; Everett, D. The Effect of Compression, Stretching, and Cooking Temperature on Free Oil Formation in Mozzarella Curd. *J. Dairy Sci.* **2003**, *86*, 449–456. [CrossRef]
9. Banville, V.; Chabot, D.; Power, N.; Pouliot, Y.; Britten, M. Impact of thermo-mechanical treatments on composition, solids loss, microstructure, and rheological properties of pasta filata-type cheese. *Int. Dairy J.* **2016**, *61*, 155–165. [CrossRef]
10. Rogers, N.; McMahon, D.; Daubert, C.; Berry, T.; Foegeding, E. Rheological properties and microstructure of Cheddar cheese made with different fat contents. *J. Dairy Sci.* **2010**, *93*, 4565–4576. [CrossRef]
11. Kern, C.; Weiss, J.; Hinrichs, J. Additive layer manufacturing of semi-hard model cheese: Effect of calcium levels on thermo-rheological properties and shear behavior. *J. Food Eng.* **2018**, *235*, 89–97. [CrossRef]
12. Schenkel, P.; Samudrala, R.; Hinrichs, J. Thermo-physical properties of semi-hard cheese made with different fat fractions: Influence of melting point and fat globule size. *Int. Dairy J.* **2013**, *30*, 79–87. [CrossRef]
13. ISO. *ISO 8968-1: 2014 (IDF 20-1: 2014) Milk and Milk Products: Determination of Nitrogen Content-Part 1: Kjeldahl Principle and Crude Protein Calculation*; International Organization for Standardization: Geneva, Switzerland, 2014.
14. IDF. *Milk–Determination of Fat Content–Gerber Butyrometers*; International Dairy Federation: Brussels, Belgium, 1981; p. 105.
15. Horwitz, W. *Official Methods of Analysis of Association of Official Analytical Chemists International*; Association of Official Analytical Chemists International: Arlington, VA, USA, 2000.
16. Fox, P.F.; Guinee, T.P.; Cogan, T.M.; McSweeney, P.L.H. Processed Cheese and Substitute/Imitation Cheese Products. In *Fundamentals of Cheese Science*; Springer: Boston, MA, USA, 2017; pp. 589–627.
17. Huber, P.; Fertsch, B.; Schreiber, R.; Hinrichs, J. Dynamic model system to study the kinetics of thermally-induced syneresis of cheese curd grains. *Milchwissenschaft* **2001**, *56*, 549–552.

18. Walstra, P.; Wouters, T.M.; Geurts, T.J. *Dairy Science and Technology*, 2nd ed.; CRC Press; Taylor & Francis Group: Boca Raton, FL, USA, 2006.

19. Tamime, A.Y. *Processed Cheese and Analogues: An Overview*; John Wiley & Sons: Hoboken, NJ, USA, 2011; pp. 1–24. [CrossRef]

20. Sharma, P.; Munro, P.A.; Dessev, T.T.; Wiles, P.G.; Buwalda, R.J. Effect of shear work input on steady shear rheology and melt functionality of model Mozzarella cheeses. *Food Hydrocoll.* **2016**, *54*, 266–277. [CrossRef]

21. Dave, R.I.; McMahon, D.J.; Broadbent, J.R.; Oberg, C.J. Reversibility of the Temperature-Dependent Opacity of Nonfat Mozzarella Cheese. *J. Dairy Sci.* **2001**, *84*, 2364–2371. [CrossRef]

22. Sharma, P.; Munro, P.A.; Dessev, T.T.; Wiles, P.G. Shear work induced changes in the viscoelastic properties of model Mozzarella cheese. *Int. Dairy J.* **2016**, *56*, 108–118. [CrossRef]

23. Tunick, M.H.; Mackey, K.L.; Shieh, J.J.; Smith, P.W.; Cooke, P.; Malin, E.L. Rheology and microstructure of low-fat Mozzarella cheese. *Int. Dairy J.* **1993**, *3*, 649–662. [CrossRef]

24. Fenelon, A.M.; Guinee, T.P. Primary proteolysis and textural changes during ripening in Cheddar cheeses manufactured to different fat contents. *Int. Dairy J.* **2000**, *10*, 151–158. [CrossRef]

25. Ibáñez, R.; Govindasamy-Lucey, S.; Jaeggi, J.; Johnson, M.; McSweeney, P.; Lucey, J. Low- and reduced-fat milled curd, direct-salted Gouda cheese: Comparison of lactose standardization of cheesemilk and whey dilution techniques. *J. Dairy Sci.* **2020**, *103*, 1175–1192. [CrossRef] [PubMed]

26. Dalgleish, D.G.; Parker, T.G. Binding of calcium ions to bovine α sl-casein and precipitability of the protein–calcium ion complexes. *J. Dairy Res.* **1980**, *47*, 113–122. [CrossRef]

27. O'Mahony, J.; McSweeney, P.; Lucey, J. A Model System for Studying the Effects of Colloidal Calcium Phosphate Concentration on the Rheological Properties of Cheddar Cheese. *J. Dairy Sci.* **2006**, *89*, 892–904. [CrossRef]

28. Gunasekaran, S.; Chang-Hwan, H.; Ko, S. Cheese melt/flow measurement methods–recent developments. *Aust. J. Dairy Technol.* **2002**, *57*, 128.

29. Cais-Sokolińska, D.; Pikul, J. Cheese meltability as assessed by the Tube Test and Schreiber Test depending on fat contents and storage time, based on curd-ripened fried cheese. *Czech J. Food Sci.* **2009**, *27*, 301–308. [CrossRef]

30. McMahon, D.J.; Oberg, C. Influence of fat, moisture and salt on functional properties of Mozzarella cheese. *Aust. J. Dairy Technol.* **1998**, *53*, 98.

 *foods*

Article

Understanding In Vivo Mastication Behaviour and In Vitro Starch and Protein Digestibility of Pulsed Electric Field-Treated Black Beans after Cooking

Marbie Alpos [1,2], **Sze Ying Leong** [1,2], **Veronica Liesaputra** [3], **Candace E. Martin** [4] **and Indrawati Oey** [1,2,*]

1   Department of Food Science, University of Otago, Dunedin 9054, New Zealand;
    marbie.alpos@postgrad.otago.ac.nz (M.A.); sze.leong@otago.ac.nz (S.Y.L.)
2   Riddet Institute, Palmerston North 4442, New Zealand
3   Department of Computer Science, University of Otago, Dunedin 9054, New Zealand;
    veronica.liesaputra@otago.ac.nz
4   Department of Geology, University of Otago, Dunedin 9054, New Zealand; candace.martin@otago.ac.nz
*   Correspondence: indrawati.oey@otago.ac.nz; Tel.: +64-3-479-8735

Citation: Alpos, M.; Leong, S.Y.;
Liesaputra, V.; Martin, C.E.; Oey, I.
Understanding In Vivo Mastication
Behaviour and In Vitro Starch and
Protein Digestibility of Pulsed Electric
Field-Treated Black Beans after
Cooking. *Foods* **2021**, *10*, 2540.
https://doi.org/10.3390/
foods10112540

Academic Editors: Harjinder Singh
and Alejandra Acevedo-Fani

Received: 31 August 2021
Accepted: 18 October 2021
Published: 22 October 2021

Publisher's Note: MDPI stays neutral
with regard to jurisdictional claims in
published maps and institutional affiliations.

Abstract: The aim of this study was to understand (i) the in vivo mastication behaviour of cooked black beans (chewing duration, texture perception, oral bolus particle size, microstructure, and salivary α-amylase) and (ii) the in vitro digestibility of starch and protein of in vivo-generated black bean oral bolus under simulated gastrointestinal condition. The beans were pre-treated using pulsed electric field (PEF) with and without calcium chloride ($CaCl_2$) addition prior to cooking. The surface response model based on least square was used to optimise PEF processing condition in order to achieve the same texture properties of cooked legumes except for chewiness. In vivo mastication behaviour of the participants ($n = 17$) was characterized for the particle size of the resulting bolus, their salivary α-amylase activity, and the total chewing duration before the bolus was deemed ready for swallowing. In vitro starch and protein digestibility of the masticated bolus generated in vivo by each participant along the gastrointestinal phase were then studied. This study found two distinct groups of chewers—fast and slow chewers who masticated all black bean beans, on average, for <25 and >29 s, respectively, to achieve a bolus ready for swallowing. Longer durations of chewing resulted in boluses with small-sized particles (majorly composed of a higher number of broken-down cotyledons (2–5 mm^2 particle size), fewer seed coats (5–13 mm^2 particle size)), and higher activity of α-amylase. Therefore, slow chewers consistently exhibited a higher in vitro digestibility of both the starch and protein of processed black beans compared to fast chewers. Despite such distinct difference in the nutritional implication for both groups of chewers, the in vivo masticated oral bolus generated by fast chewers revealed that the processing conditions involving the PEF and addition of $CaCl_2$ of black beans appeared to significantly ($p < 0.05$) enhance the in vitro digestibility of protein (by two-fold compared to untreated samples) without stimulating a considerable increase in the starch digestibility. These findings clearly demonstrated that the food structure of cooked black beans created through PEF treatment combined with masticatory action has the potential to modulate a faster hydrolysis of protein during gastrointestinal digestion, thus offering an opportunity to upgrade the quality of legume protein intake in the daily diet.

Keywords: mastication; starch digestibility; protein digestibility; texture; particle size; α-amylase; black beans; calcium; pulsed electric field; legume; thermal processing; food oral processing

1. Introduction

Texture degradation of legumes due to thermal processing can be avoided by the addition of exogenous calcium to form crosslinks with demethoxylated pectins, due to activation of pectin methylesterase, to provide cell wall strengthening effect in the middle lamellae [1,2]. Recently, there is an interest to employ cell electroporation-based technology, namely pulsed electric field (PEF), to accelerate the uptake of exogenous calcium [3–5] and to modify the texture of plant materials [6,7]. So far, the application of PEF technology for legumes has not yet been studied, especially when this technology is used prior to hydrothermal processing. Additionally, it is not well understood how texture modification achieved using PEF technology in the presence of calcium prior to hydrothermal processing could impact the mastication behaviour of cooked legumes, and subsequently influencing the digestibility of macronutrients such as starch and protein at gastrointestinal phase.

Food digestion is a dynamic process. It starts in the mouth wherein food is broken down by the teeth through the process of chewing or mastication, which takes about 280 ms for most types of food [8]. A bolus is formed after the food is grinded to small-sized particles and lubricated by the saliva, which contains α-amylase that can hydrolyse starches. The bolus is then swallowed through the oesophagus into the stomach, where further size reduction occurs by the gastric acids, enzymatic action, and peristalsis. Food bolus is then hydrolysed in the small intestine, where most of the digestive enzymes, such as pancreatic α-amylase and proteases and brush border enzymes, are present. Bolus formation and its particle size, and the consequent digestibility of its starch and protein, can be influenced by numerous factors. This includes the properties of the food such as texture, specifically hardness, portion size, moisture and fat contents; gender, age, and the dental status of the individual chewing the food; and the mastication behaviour such as the cycle, force and duration, and saliva production [8]. Previous studies have found that food hardness greatly affects the particle size of food after mastication and softer food requires a shorter duration of mastication, resulting in larger bolus particle size [5,9]. However the effect of food chewiness is not yet much studied. Moreover, bolus particle size was reported to affect the nutrient digestibility of food such as starch, wherein larger particles, with smaller surface area, were less accessible to salivary amylase and would take a longer time to be digested at the gastrointestinal phase [10–13].

The aim of the present investigation was to understand (i) the in vivo mastication behaviour of cooked black beans (chewing duration, texture perception, oral bolus particle size, microstructure, and salivary α-amylase) and (ii) the in vitro digestibility of starch and protein of in vivo-generated black bean oral bolus under simulated gastrointestinal condition. The beans were treated using PEF processing with and without calcium chloride prior to cooking. The optimal PEF processing parameters to achieve the same texture properties of cooked legumes except for chewiness were firstly selected. One of the novelties of this study is the use of image analysis to estimate the particle size distribution of beans in the oral boluses after in vivo mastication prior to swallowing. Compared to conventional techniques, such as laser diffraction and sieving methods [14,15], image analysis is a more convenient and fast approach providing a more relevant and accurate information on the particle size of heterogenous food structure with multiple layers (e.g., seed coat), such as black beans, and irregularly shaped particles, such as masticated bolus. This technique is able to differentiate components of food when there are visible distinguishable features such as colour [13], which is highly applicable to black beans due to contrasting colours of its outer seed coat (black) and inner cotyledon (cream white). Moreover, there is very little information in the current literature on characterising the breakdown of these components during in vivo mastication of legumes.

2. Materials and Methods

2.1. Raw Material

One single batch of dried black beans (10 kg) procured from a local store was used. Upon arrival, they were sorted and those with physical damage and discoloration were excluded from the study.

2.2. Selection of PEF Treatment Parameters and Other Processing Variables

Different PEF processing parameters (electric field strength up to 2.3 kV/cm, specific energy up to 134 kJ/kg, pulse width of 20 µs, and frequency of 50 Hz) and calcium chloride ($CaCl_2$) solution (up to 300 ppm) were systematically screened in this study. Soaked beans were placed in a PEF treatment chamber and immersed either in distilled water or in calcium chloride ($CaCl_2$) solution with a bean-to-solution ratio of 1:2. The samples were PEF-treated (ELCRACK-HPV 5 PEF batch system, German Institute of Food Technologies, Quakenbrück, Germany) based on previous work [5]. The PEF chamber consisted of 2 parallel stainless-steel electrodes with an electrode distance of 80 mm. Different concentrations of calcium chloride (0 up to 300 ppm $CaCl_2$ solution) resulted in differences in electrical conductivity averaged between 61 and 507 µS/cm. Pulse shape (square wave bipolar) during PEF treatment was monitored using a digital oscilloscope (UTD2042C, Uni-Trend, Dongguan City, Guangdong, China). The change in conductivity and temperature before and after PEF treatment was measured with a conductivity meter (CyberScan CON 11, Eutech Instruments, Queenstown, Singapore). The specific energy input for each PEF treatment was estimated using Equation (1).

$$\text{Specific energy input (kJ/kg)} = \text{(Pulse energy of the PEF generator} \times \text{Pulse number)/Total weight of black beans and } CaCl_2 \text{ solution} \tag{1}$$

All PEF-treated beans were then cooked at 80 °C for 60 min and their texture properties were determined. The texture profile analysis (TPA) for each cooked black bean sample type (approximately 20 g to allow at least 8–10 independent texture analyses per sample) was conducted using a texture analyser (TA-HD plus, Stable Micro Systems, Surrey, UK) based on a double compression or 'two-bite' test, as described by Alpos et al. [16]. The samples were compressed at 50% strain (5 beans per compression) using a 50 mm diameter cylinder probe attached in a 250 kg load cell at a test speed of 1 mm/s. The textural properties of each sample including hardness (peak force during the first compression, Newton), cohesiveness, springiness, chewiness, and resilience was determined automatically from the TPA curve using an in-house macro, developed in the Exponent software (Stable Micro Systems, Surrey, UK).

For calcium content analysis, the outer seed coat and inner cotyledon of three cooked black bean samples were individually analysed (Supplementary Materials Figure S1) due to the considerable difference in distribution of calcium in typical legumes (70% in seed coat vs. 30% in cotyledon) [17]. They were manually separated, frozen using liquid nitrogen (N_2), freeze-dried (Labconco FreeZone freeze dryer, Kansas City, MO, USA), and then milled using a mortar and pestle to pass through a 425 µm mesh screen. The freeze-dried seed coat and cotyledon (0.1 g) were firstly digested by adding 2 mL of concentrated nitric acid (70%, v/v) in a 50 mL plastic tube. The tubes were then placed in a water bath at 95 °C for 10 min to allow complete digestion. After that, the samples were diluted to 50 mL with milli-Q water and filtered using polytetrafluoroethylene (PTFE) membrane filter (0.45 µm). Calcium concentration in the samples were determined using an inductively coupled plasma mass spectrometer (Agilent 7900 ICP-MS, Santa Clara, CA, USA). The auto tune function and helium gas mode of the instrument was utilised to minimise interferences and instrumental drift while maximising sensitivity. All samples, blanks (70% (v/v) nitric acid), and calibration standards (NIST traceable Agilent multi element standards) were introduced to the instrument in 2% (v/v) nitric acid. Scandium was used as the internal standard for calcium analysis which was added online to correct for matrix effects and instrumental drift. Calcium content measurements were performed in triplicate for each sample treatment.

A least squares model (minimizing the sum of the squares of the residuals) based on input variables (field strength, energy input, and calcium concentration) and responses (hardness, cohesiveness, springiness, chewiness, and resilience) was constructed to find the optimal settings of the input variables leading to change in chewiness, whilst preserving the other black beans texture parameters (i.e., hardness, cohesiveness, springiness, and resilience) during cooking by considering main effects, two-way interactions, and quadratic terms of the input variables. Estimates, standard error of the estimates, and *p*-value of the estimate for each model term were obtained (JMP Pro v.14, Cary, NC, USA). The "prediction profiler" feature in JMP was also employed to provide cross-sectional views of the fitted surface response and to observe how the prediction model changes across an individual input variable while holding the other inputs at fixed values.

2.3. Preparation of Black Bean Samples for In Vivo Mastication Study

The in vivo mastication study was conducted on three types of cooked black bean samples (i.e., shared similar texture properties except for chewiness) based on the result obtained from Section 2.2: PEF-treated in the absence of calcium chloride solution (thereafter referred as "sample A"), PEF-treated in the presence of 300 ppm $CaCl_2$ solution (thereafter referred as "sample B"), and non-PEF treated black beans (thereafter referred as "sample C"). Sample A treatment involved 2000 pulses and resulted in specific energy input of 8 kJ/kg, which took 40 ms of treatment time (pulse number × pulse width). Sample B (when using 300 ppm $CaCl_2$ as electrical conducting medium) underwent PEF treatment for 18 ms with 900 pulses and resulted in 13 kJ/kg specific energy input. Sample C referred to soaked bean samples that were not treated with PEF but were thermally processed in 300 ppm $CaCl_2$ solution right after overnight soaking.

The experimental setup of the in vivo mastication study, which comprised of sample preparation, sample collection, and subsequent analysis, is summarised in Figure 1. A day before the in vivo mastication study, black beans were soaked in distilled water (1:3 (w/v) seed-to-water ratio) for 24 h at 20 °C. The soaking water was discarded the next day. The soaked beans were divided into three lots, where the first two lots were allocated as samples A and B and proceed to PEF-treatment as described in Section 2.2. The remaining soaked beans were used as sample C. After that, all sample types were immediately thermally processed in a temperature-controlled water bath at 80 °C for 60 min. Then, samples were cooled in an ice water bath (<4 °C) for at least 20 min to remove the thermal load. The sequential PEF and thermal processing for each sample was performed in 100 g lots and then only the cooked beans (i.e., solution was removed) were pooled to obtain enough samples for the in vivo mastication study. All sample preparations were performed under food-grade conditions, ensuring adherence to food hygiene and safety protocols. Samples for each treatment were used immediately on the day for the in vivo mastication (see Section 2.4) and a small aliquot (20 g per sample) was used for texture measurement and calcium quantification (see Section 2.2) to ensure the expected texture parameters and calcium content were obtained prior to the in vivo mastication study.

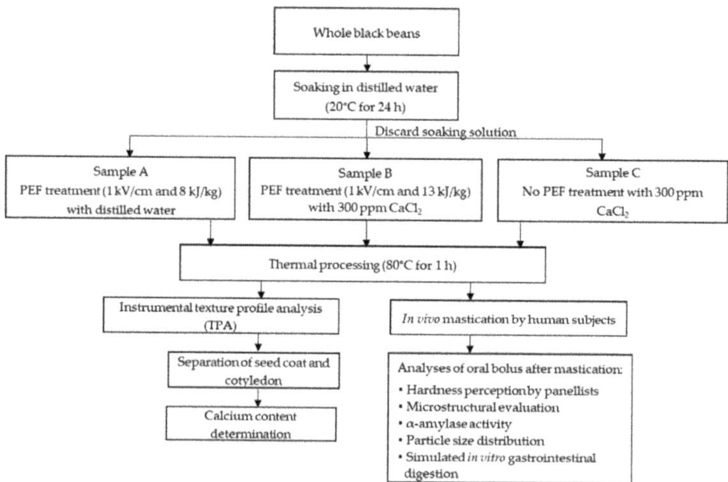

Figure 1. Experimental setup of this study. (PEF: Pulsed electric field).

2.4. In Vivo Mastication Study

Participants were recruited based on the following inclusion criteria: Complete dentition (without removable dentures and no dental treatments for the past 3 months), no difficulty in masticating, no pre-existing medical conditions, and no legume allergy. This study was approved by the University of Otago human ethics committee (approval reference code: 20/038) and participants were informed in detail about the objectives and methodology of the study before signing a written consent form. The results collected from 17 participants (comprising of 9 males and 8 females with a mean age of 27.2 ± 9.5 years) were analysed and reported.

The study was conducted in individual sensory booths in a well-ventilated room with fluorescent lighting. Each booth was equipped with 1 sample tray, 1 plastic spoon, a paper questionnaire, a pen, a digital timer, and a glass of drinking water (Supplementary Materials Figure S2). Every participant was provided with four separate portions of 5 g (a total of 20 g) of each sample to perform the in vivo mastication. Each participant was sequentially served (in a balanced random sample presentation order to limit first order and carry-over effects) with three separate trays for the three samples containing a total of 12 closed containers each tray (Supplementary Materials Figure S3): Four for each 5 g samples, 4 for the mouth rinsing water, and 4 for collection of the oral boluses. All containers were pre-labelled with random 3-digit codes corresponding to each sample type to prevent biased testing and it was made sure that the spitting container had the same code as the sample container.

For each sample tray, the participants were asked to randomly select one of the four sample containers and chew all the beans inside (5 g) in one mouthful until their swallowing threshold. They pressed the start button of the timer once they began chewing the sample and stopped when the sample was ready to be swallowed. The participants were not required to consume any of the sample, but to expectorate the oral boluses into the plastic spitting containers pre-labelled with the 3-digit code that matched with the sample code. The participants were then asked to rinse their mouth with the rinsing water provided (30 mL) to make sure that no food material remained in the mouth and were expectorated into the same plastic container containing the bolus. The participants were asked to repeat the same step for each of the sample containers remaining on the tray. The participants were instructed to write down, on the questionnaire, the total duration that it took for them to masticate each of the 5 g sample. After consuming all cooked beans (coming from the same process treatment) from the 4 plastic containers in the tray, the participants were asked to rate their texture perception of the sample on the questionnaire using a five-point hedonic scale [18]: very soft, soft, neither too soft nor too hard, hard, and very hard. The participants were allowed at least 30 s breaks before being presenting with the next tray of sample. The steps were repeated for the cooked beans from the second and third sample (either A, B, or C).

A total of 12 boluses were collected from each participant, 4 for each sample A, B, and C. The collected boluses, in four separate containers for each sample, were allocated for the determination of particle size distribution of bolus (see Section 2.5), α-amylase activity in the oral bolus (see Section 2.6), microstructural evaluation (see Section 2.7) and the amount of D-glucose released and L-serine in the digest when subjected to a subsequent 6 h long in vitro simulated gastrointestinal digestion (see Section 2.8). For those samples intended for the in vitro digestion, 1 mL of 1 M hydrochloric acid (HCl) was immediately added after mastication to each of the oral boluses to stop the activity of salivary amylase. All the collected boluses were frozen using liquid nitrogen (N_2) and stored at $-18\ ^\circ C$ until analysis.

2.5. Particle Size Distribution of Oral Boluses Using Image Analysis

2.5.1. Image Capturing and Processing of the Oral Bolus

The particle size distribution of the oral boluses across three sample types from every participant after masticating a portion of 5 g cooked black beans was determined using the image analysis technique according to the work of Bornhorst, Kostlan and Singh [13], with modifications. Each bolus was prepared by washing them on a 1 mm sieve in running tap water for 1 min to prevent the aggregation of the particles. Particles bigger than 1 mm that remained on the sieve were weighed (0.5 g) into plates (17.7 cm width) with blue-coloured background. Fifty millilitres of water were added to disperse the particles and gently stirred to prevent particles overlapping (Supplementary Materials Figure S4, left). The particles inside the entire blue plate were photographed alongside with a geometrical reference (American Board of Forensic Odontology no. 2 photomacrographic standard reference scale) for spatial calibration to convert pixels to millimetres during image processing. The camera used to capture the image was a Canon 6D Mark II (26.2 mp, Full Frame CMOS sensor, Ota City, Tokyo, Japan) with a Canon EF 50 mm f/1.7 STM lens, positioned 40 cm above the base of the blue plate. The camera was set as followed: no flash, aperture F22, ISO 3200, and shutter speed 1/200 s. Four to ten images were taken for each bolus, depending on the amounts of particles remaining on the sieve.

Captured images were analysed using the algorithm in the OpenCV software library (Open Source Computer Vision Library v.4.4.0, Palo Alto, CA, USA). The blue plate with the particles inside was first isolated using colour thresholding. Then, heuristics image analysis was performed to remove unwanted light reflection from the plate. Canny edge detection was used to identify "all" the particles in the plate and the OpenCV's counting pixels function was used to count the total amount of pixels or the area of each identified bean particle, according to its size and shape (whether it was complex or irregular shaped). Further segmentation was conducted wherein the "white" and "black" particles, corresponding to the inner cotyledon and outer seed coat, respectively, were distinguished using k-means clustering method (Supplementary Materials Figure S4, right). The measured pixels of "white", "black", and "all" particles were converted to mm^2 using the reference scale.

2.5.2. Modelling the Particle Area Distribution of Oral Bolus

The calculated area for each particle was used to determine the cumulative particle area distribution. The cumulative area percentage was then fitted to the Rosin-Rammler model (Equation (2)) to describe the particle size distribution of the in vivo masticated black bean bolus. This model was previously used to characterise the particle size of almond and rice during oral mastication as determined by image analysis [13,19].

$$C_{area} = 1 - \exp\left(-\left(\frac{x}{x_{50}}\right)^b \ln(2)\right) \tag{2}$$

where C_{area} is the cumulative area percentage of each particle from 0 to 100%, x_{50} is the median particle area in mm^2, and b represents the dimensionless distribution breadth constant (higher b value corresponds to a narrower distribution spread). Nonlinear regression (function 'nls') in R software (v.4.0.4 2021) as commanded by R Studio (v.1.4.1103 2021, Boston, MA, USA), was used to estimate the model parameters x_{50} and b for each bolus.

2.6. Determination of α-Amylase Activity in Oral Boluses

The α-amylase activity in the oral boluses was determined using the Ceralpha method of the Megazyme alpha-amylase assay kit (K-CERA 06/18, Wicklow, Ireland). The procedure made use of the oligosaccharide of "non-reducing-end blocked *p*-nitrophenyl malto-heptaoside" (BPNPG7) with excess levels of thermostable α-glucosidase as substrate [20]. When α-amylase reacted with the substrate, the former cleaved a bond within the latter. Then, the excess α-glucosidase further hydrolysed the reaction product *p*-nitrophenyl maltosaccharide into glucose and free *p*-nitrophenol. The reaction was stopped by the addition of tri-sodium phosphate which changed the colour of the sample to yellow [20].

The sample was prepared by centrifugation of the oral boluses at $1000 \times g$ for 10 min (IEC Micromax, Thermo Electron Corp., Milford, MA, USA). The supernatant (0.05 mL) was then suitably diluted with the buffer (0.95 mL, 0.1 M sodium malate, 0.1 M sodium chloride, 4 mM calcium chloride, pH 5.4) provided in the kit. The diluted samples (0.1 mL) were transferred into bottom of 15 mL plastic tubes and pre-incubated at 40 °C for 5 min. The substrate blocked *p*-nitrophenyl maltoheptaoside (BPNPG7) was also pre-incubated (40 °C for 5 min) at the same time as the samples. After 5 min of equilibration, substrate (0.1 mL) was added directly to each sample tube, vortexed and incubated at 40 °C for another 10 min. Immediately after, 3 mL of stopping reagent (20% (*w/v*) tri-sodium phosphate solution, pH 11) was added and vortex mixed. The absorbance of the sample solutions and reaction blank (0.1 mL of simulated saliva juice (2 mM sodium chloride, 2 mM potassium chloride, and 25 mM sodium bicarbonate) reacted with 3 mL stopping reagent followed by addition of 0.1 mL of substrate) was measured using a spectrophotometer (Specord 250 Plus, Analytik Jena, Jena, Germany) at 400 nm against distilled water. The result was expressed as α-amylase (Ceralpha Unit, CU) in the oral boluses needed to free one micromole of *p*-nitrophenol from BPNPG7 in one minute. α-amylase measurements were performed in triplicate for each participant.

2.7. Microstructural Evaluation of Oral Boluses

The effect of oral mastication on the microstructure of the black beans was visualized under light microscope as described previously by Gwala et al. [21]. Briefly, boluses were lyophilised (Labconco FreeZone freeze dryer, Kansas City, MO, USA) and milled to pass through a 425 μm mesh screen. A small amount of the sample powder was dispersed in distilled water on a microscope slide, covered, and then observed under a light microscope (Ceti, Auckland, New Zealand). Micrographs were viewed at 10× magnification and captured using a camera (Medline Scientific, Oxfordshire, UK) attached to the microscope controlled by a camera control software (ToupTek ToupView, Hangzhou, Zhejiang, China).

2.8. Simulated In Vitro Human Gastric Intestinal Digestion Assay and Determination of Starch and Protein Digestibility

The availability of starch and protein for hydrolysis in black beans after in vivo oral mastication was determined using the harmonised static in vitro digestion method developed by Infogest [22]. The digestion solutions were freshly prepared on the day of the assay according to the work of Abduh et al. [23]. Simulated saliva juice and α-amylase solution was not added to the oral bolus sample due to prior structure breakdown and hydrolysis occurred in vivo in the mouth. Therefore, only the gastric and small intestinal digestion phases were simulated in vitro.

Twenty millilitres of simulated gastric solution (4% (*w/v*) porcine stomach pepsin (AppliChem A4289, 0.7 FIP-U/mg, Barcelona, Spain) in 1 mM hydrochloric acid (pH 3) containing 151 mM sodium chloride and 28 mM potassium chloride) was firstly added to the oral bolus and incubated at 37 °C (Contherm Scientific Ltd., Hutt City, Wellington, New Zealand) for 120 min with shaking (55 strokes/min, rocking motion tilt angle of 7°, DLAB, SK-R1807-S, New Territories, Hong Kong). After 2 h, pepsin was deactivated by adjusting the pH to 7 with 1 M NaOH. Then, 40 mL of simulated small intestinal solution (1% (*w/v*) porcine pancreas pancreatin (Sigma P1750, 4 × USP, Sigma-Aldrich, St. Louis, MO, USA) and 0.85% (*w/v*) g porcine bile extract (ChemCruz SC-214601, Santa Cruz Biotechnology, Dallas, TX, USA) in 0.1 M sodium bicarbonate (pH 7) was added and incubated at 37 °C for the next 240 min with shaking.

2.8.1. Starch and Protein Digesta Collection and Measurement

Starch digesta (0.5 mL) were collected at 0 and 120 min of the gastric phase (after addition of simulated gastric solution) and at 0, 20, 30, 40, 60, 90, 120, 180 and 240 min of the small intestinal phase (after addition of simulated small intestinal solution). To inactivate the digestive enzymes after each sampling, the digesta were immediately heat-shocked in a boiling water bath for at least 10 min [24]. Then, to the collected digesta, 2.5 mL of 100 mL sodium acetate buffer at pH 5 were added and centrifuged at 2056× *g* (Beckman GPR Centrifuge, Brea, CA, USA) for 20 min. The amount of hydrolysed starch in the supernatant was measured using the D-glucose assay kit (Megazyme, Bray, Wicklow, Ireland) according to previous work [25–27]. Result expressed as amount of D-glucose in digest (mg).

In this study, different starch fractions from the in vivo-generated oral boluses of cooked black beans as digested in the small intestine were determined, which includes readily digestible starch (RDS), digested after 20 min; slowly digestible starch (SDS), digested between 20 to 120 min (slow but complete digestion); and resistant starch (RS) which passed through the small intestine undigested [28,29], using Equations (3)–(5), respectively, as follows:

$$\text{RDS } (\%) = (\text{G20} \times 0.9 \times 100)/\text{Total starch} \tag{3}$$

$$\text{SDS } (\%) = ((\text{G120} - \text{G20}) \times 0.9 \times 100)/\text{Total starch} \tag{4}$$

$$\text{RS } (\%) = (\text{TS} - (\text{RDS} + \text{SDS}) \times 100)/\text{Total starch} \tag{5}$$

where G20 and G120 represent the D-glucose released after 20 and 120 min of small intestinal digestion, respectively, and 0.9 is the factor to convert the measured glucose value to polysaccharide based on the molecular mass ratio of starch to glucose (162/180) [28]. Total starch of the oral bolus was determined according to previous work [16] using the total starch assay kit (Megazyme, Bray, Wicklow, Ireland).

Protein digesta (0.5 mL) were collected at 0, 30, 60, and 120 min of the gastric phase (after addition of simulated gastric solution) and at 0, 20, 30, 40, 60, 90, 120, 180 and 240 min of the small intestinal phase (after addition of simulated small intestinal solution). They were placed in Eppendorf tubes already containing 0.5 mL of 20 (*v/v*%) TCA [30] and centrifuged at 13,000× *g* (IEC Micromax, Thermo Electron Corp., Milford, MA, USA) for 5 min. The *o*-phthaldialdehyde (OPA) assay, as described by Liu, Oey, Bremer, Silcock and Carne [30], was used to measure the hydrolysed protein in the oral bolus by determining the free α-amino groups of the peptide fractions, using L-serine standard curve. Results were expressed as the amount of L-serine equivalents in digest (mg).

2.8.2. Kinetic Modelling of In Vitro Starch and Protein Digestibility at the Small Intestinal Phase

To estimate the extent and rate of starch hydrolysis of oral boluses, starch digestion kinetics during the small intestinal phase were modelled by a fractional conversion model (Equation (6)). The use of fractional conversion model to describe the starch digestibility behaviour of legume has been evidenced in other previous studies [11,31].

$$S_{(t)} = S_f + \left(S_0 - S_f\right) \times \exp\left(k_s \cdot t\right) \tag{6}$$

where $S_{(t)}$ is the D-glucose released at digestion time t, S_0 is the amount of D-glucose at the start of small intestinal phase (0 min), S_f is the D-glucose released at the end of the small intestinal digestion, and k_s is the rate constant of starch digestion. The model fitting and estimation of kinetic parameters k_s, S_0, and S_f were estimated using nonlinear regression function 'nls' in R software (v.4.0.4 2021) and R Studio (v.1.4.1103 2021, Boston, MA, USA).

Zero order kinetics model (Equation (7)) was used to estimate the extent and rate of protein hydrolysis of oral boluses during in vitro small intestinal phase.

$$P_{(t)} = P_0 + k_p \cdot t \tag{7}$$

where $P_{(t)}$ is the amount of L-serine at digestion time t, P_0 is the amount of L-serine at the start of small intestinal phase (0 min), and k_p is the rate constant of protein digestion. The model fitting and estimation of kinetic parameter (k_p) was achieved using linear regression function 'lm' in R software (v.4.0.4 2021) and R Studio (v.1.4.1103 2021, Boston, MA, USA).

To evaluate the goodness of fit of the kinetic models to the experimental data (starch and protein digestibility) obtained in this study, adjusted R^2 was calculated, and residual (random distribution of error) and parity plots were assessed [21].

2.9. Statistical Data Analysis

Statistical analyses on the result from in vivo mastication study were performed using Statistical Package for the Social Sciences (SPSS) version 25 (IBM Corp., Armonk, NY, USA) to determine the statistical significances among sample type and participants. Data collected such as texture parameters, estimated Rosin-Rammler model parameters x_{50} and b for particle size, starch and protein digestibility, and α-amylase activity were assessed for homogeneity using Levene's test. Then, student's *t*-test was conducted for single comparison and analysis of variance (ANOVA) with Tukey as post hoc multiple comparison test at 0.05 level of significance. Pearson's correlation coefficient (r) was also used to determine the linear relationship between the black beans texture, participants' mastication duration, bolus particle size, salivary α-amylase activity, and starch and protein hydrolysis.

3. Results and Discussion

3.1. Selection of PEF Treatment Parameters and Other Processing Variables

From the process optimization study, pre-treating the black beans at varying levels of electric field strength and energy input in the presence of calcium did not influence their hardness, cohesiveness, springiness, and resilience after thermal processing (Supplementary Material Table S1). A similarity in the hardness result for any PEF-treated black bean samples was unexpected, since an application of PEF pre-treatment on other plant matrices has been reported to cause significant texture softening effect due the ability of PEF in disrupting the cell structure integrity [6,7,32]. It could be that the addition of calcium chloride during the PEF treatment of the black bean samples facilitated formation of crosslinks with demethoxylated pectins in the middle lamella during thermal processing, preserving their texture from thermal degradation [2].

It is interesting to observe, from the process optimization study, that the chewiness of cooked black beans was the only texture parameter affected significantly by all the input variables applied to the black beans prior to thermal processing (Supplementary Material Table S1). In other words, the chewiness of the cooked black beans can be modulated by varying the intensity of electric field strength and energy input, and concentration of calcium during PEF and thermal treatment, without compromising the other four texture parameters. The result from the response surface model also showed the potential of creating cooked black beans with similar predicted cohesiveness with increasing energy input at low field strength or high field strength at low energy input in the absence of calcium (Supplementary Material Figure S5). However, increasing the energy input further at high field strength would negatively reduce the chewiness of cooked beans without affecting the other four texture parameters. On the contrary, cooked black beans are likely to share similar cohesiveness across any levels of energy input and electric field strength in the presence of calcium during PEF (Supplementary Material Figure S6).

Considering these results, the process parameters of PEF treatments used for the in vivo mastication study were set at an electric field strength of 1 kV/cm, commonly used for plant matrix [6,33], to cause effective cell permeabilization and energy input of approximately 10 kJ/kg in the presence of 300 ppm $CaCl_2$, which enabled the creation of black beans with the same hardness, cohesiveness, springiness, and resilience but not chewiness after thermal processing.

3.2. Texture Profile and Calcium Content of Differently Processed Black Beans Used for the In Vivo Mastication Study

From the process optimization study and response surface model, three types of black bean samples were used for the in vivo mastication study (sample A: PEF-treated and then cooked without $CaCl_2$ addition, sample B: PEF-treated and then cooked in the presence of $CaCl_2$, sample C: non-PEF-treated but cooked in the presence of $CaCl_2$).

The only texture parameter that was significantly different ($p < 0.05$) between the samples was chewiness, wherein sample B was slightly chewier than sample C (Supplementary Material Figure S7), thus requiring more energy to masticate the black beans.

This can be attributed to the firming effect of calcium with improved infusion by PEF [5]. Aligning with the calcium content result of the cooked black beans (Figure 2), the addition of CaCl$_2$ during PEF and thermal processing (sample B) significantly increased the calcium content in the seed coat of black beans as compared to the sample thermally processed with CaCl$_2$ without PEF treatment (sample C). This indicated that PEF processing played an important role in facilitating the uptake of calcium, which is likely to occur in the testa rather than the cotyledon. Yi et al. [34] found that pectin in the cell wall of the seed coat of common beans was capable of binding to exogenous calcium ions (Ca^{2+}). The increase in calcium concentration in sample B can also be explained by the ion channels present in the cell membranes of plants such as legumes which can be controlled by certain stimuli. It was known that the activation of Ca^{2+} channel, which contains calcium-selective pores, is voltage-dependent, such that application of electric pulses by PEF would open calcium channels allowing the calcium ions to pass through [35]. Another study by Zhou et al. [36] reported that the addition of CaCl$_2$ (about 600 ppm) enhanced the influx and accumulation of Ca^{2+} in mung bean cells. Moreover, diffusion could have happened due to the concentration gradient. In the present study, raw black beans were soaked in water overnight and absorbed a volume twice its weight. The soaking solution was then replaced during PEF treatment with 300 ppm CaCl$_2$. It can be inferred that a net movement of substance from the calcium-containing medium (higher concentration) into the beans (lower concentration) took place, as facilitated by the supposed cell electroporation effect by PEF.

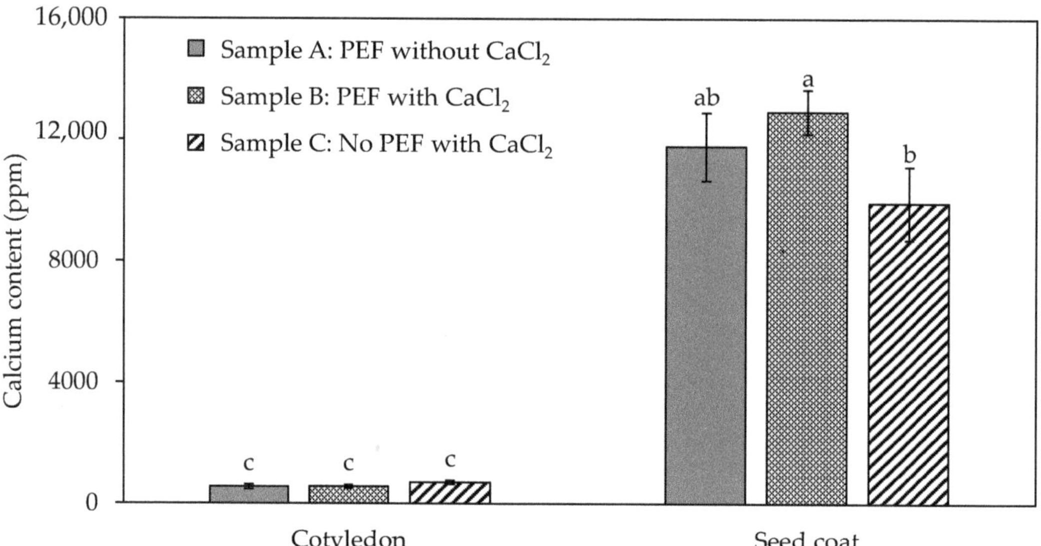

Figure 2. Calcium content of the cotyledon and seed coat of black bean samples A (PEF and thermally processed without CaCl$_2$ addition), B (PEF and thermally processed with CaCl$_2$ addition), and C (No PEF, thermally processed with CaCl$_2$). Data presented as mean $\pm$ standard deviation of three independent batches of cooked black beans (n = 3). Values with different letters between sample type for each seed component are significantly different ($p < 0.05$).

Overall, results revealed that the three cooked bean samples used for the in vivo mastication study shared similarity in most of the texture properties (except for chewiness) when evaluated using the texture analyser instrument. While the application of PEF treatment in the presence of calcium enhanced the uptake of calcium in the black beans, the implications of these on the bolus formation during in vivo mastication by human participants are detailed in the following sections.

3.3. Characterisation of Mastication Behaviour of Participants for PEF and Calcium Pre-Treated Cooked Black Beans

3.3.1. Texture Perception of Cooked Beans Rated by the Participants

The participant's perception of hardness when masticating the three different cooked black bean samples was evaluated. Most of the participants (8–12 of them out of 17 participants) perceived all three black bean samples as "hard" on a five-point hedonic scale, followed by "neither too soft nor too hard" (Supplementary Material Figure S8). Two participants rated Sample C to be "soft" while another two participants rated both PEF-treated samples (Samples A and B) to be "very hard" (Supplementary Material Figure S8). This result suggested that instrumental measurement and consumer perception of texture could differ, because, despite the similarity in hardness of these black beans measured in the texture analyser (Supplementary Material Figure S7), participants may rate their hardness differently during consumption. Clearly, a minority of participants was rather sensitive in perceiving differences in the hardness between the cooked black bean samples due to the different pre-treatments applied. The results also suggested that it is unlikely that the chewing duration of the participants to masticate each sample to its preferable size/form prior to swallowing influences how the hardness of samples being perceived. Taking participants 10 and 12 as examples, who chewed, on average, the three types of black bean samples for 24 and 74 s, respectively—they both rated the three samples the same ("hard"), but their chewing duration varied greatly.

3.3.2. Chewing Duration of Cooked Beans before Ready for Swallowing

The total durations taken for every participant to fully masticate each sample (in one mouthful of 5 g beans), before they are about to swallow, are illustrated in Figure 3. Of all the 17 participants, the chewing duration ranged from 11.25 to 78.5 s for the three types of samples, clearly illustrating the large variation in the chewing durations between participants. The median chewing duration across the participants was 28.5, 29, and 24.75 s for samples A, B, and C, respectively (as represented by the long, short and dotted lines in Figure 3). Six participants were found to consistently take a longer chewing duration than the median duration to masticate all the samples. They were categorised as "slow chewers" in this study. On the contrary, nine participants were categorised as "fast chewers" in this study, who masticated the beans faster than the median duration. Two out of the 17 participants were considered "inconsistent chewers" due to their varying chewing pattern for each sample, specifically partly above and below the median duration. For instance, participant 18 took a shorter duration when masticating samples A and C but a longer chewing duration when masticating sample B. This was opposite for participant 1, who chewed sample A for a longer duration but samples B and C for shorter times. These results showed a great variability of chewing behaviour among individuals in this study.

Due to inter-individual variability, the average chewing duration was not significantly different ($p > 0.05$) between the three samples (Supplementary Materials Figure S9). It is recognised that humans masticate the same type of food differently [37] and individuals chew food using different oral strategies to obtain a bolus ready for swallowing. However, it cannot be ruled out that the participants misjudging the readiness of the bolus could have also contributed to the large chewing time variation. Chewing and subsequent swallowing is an act of volition; hence, it differs with individual's preference and judgement [9].

While it is acknowledged that the chewing duration over the same sample varied greatly between participants, it was interesting to observe that sample C was chewed the fastest (averaged at 28 s) out of the three samples (averaged at 31 and 30 s for sample A and B, respectively) (Supplementary Materials Figure S9). The chewing duration positively correlated ($r = 0.88$) with the chewiness result in the texture analyser (Supplementary Material Figure S7) wherein sample C was significantly ($p < 0.05$) less chewy than sample B, which signified that the latter required more energy to masticate to be ready for swallowing, taking more time to chew. Other studies have also found a direct relationship between chewing behaviour and food texture [5,8,9,11]. Chewing time also positively correlated

(r = 0.83) with calcium content in the seed coat (Figure 2). This indicates that, due to the lower amount of calcium found in the seed coat of sample C (Figure 2), the formation of pectin–calcium crosslinks to strengthen the cell wall was lesser than in samples A and B, making sample C easier to be chewed.

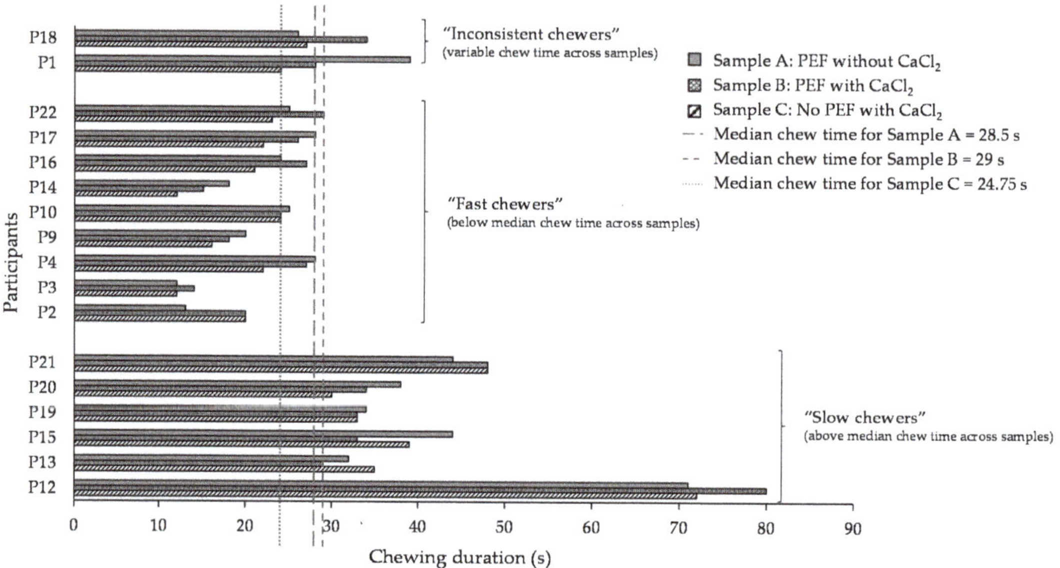

Figure 3. Variations in the chewing duration of participants (n = 17) for three different sample types (A: PEF and thermally processed without $CaCl_2$ addition; B: PEF and thermally processed with $CaCl_2$ addition; and C: No PEF, thermally processed with $CaCl_2$). Long, short, and dotted lines correspond to the median chewing duration for each sample type. Participants with variable chewing duration across three samples are defined as "inconsistent chewers", participants with chewing duration below the median line for all three samples are defined as "fast chewers", and participant with chewing duration above the median line for all three samples are defined as "fast chewers".

3.3.3. Particle Size Distribution after In Vivo Mastication of Differently Processed Black Beans

In this study, image analysis was found to be suitable in successfully quantifying the particle size distribution of the black beans in the oral bolus. The estimated Rosin-Rammler model parameters (x_{50} and b, Equation (2)) to describe the distribution of "all" the black bean particles, which encompassed "black" (seed coat) and "white" (cotyledon) particles (Figure 4) for each sample, are summarized in Table 1.

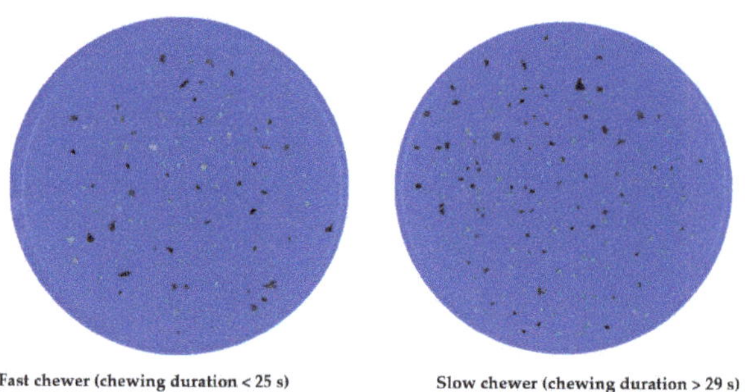

Figure 4. Example of processed image of black beans after mastication by a fast chewer and a slow chewer.

Table 1. Estimated Rosin-Rammler parameters (Equation (2)) of "all" black bean particles from samples A (PEF and thermally processed without CaCl$_2$ addition), B (PEF and thermally processed with CaCl$_2$ addition), and C (No PEF, thermally processed with CaCl$_2$) after in vivo human mastication (average of 17 participants).

Sample	Oral Bolus [†]	
	x_{50} (mm^2)	b
A	5.0 ± 1.3 [a]	1.63 ± 0.19 [a]
B	5.2 ± 1.2 [a]	1.60 ± 0.18 [a]
C	5.4 ± 0.9 [a]	1.53 ± 0.14 [a]

[†] Data are presented as the average ± standard deviation of the estimated value from 17 participants (n = 17) reporting according to the significant figures. Values with the same lowercase letters in superscript for the three samples within each parameter (per column) are not significantly different ($p > 0.05$).

The median sizes of the particles (x_{50}, indicate the 50% cumulative weight of the total food particles) for all participants ranged from 4 to 9 mm^2 for sample A, 4 to 9 mm^2 for sample B, and 4 to 7 mm^2 for sample C. On average, for the 17 participants, the estimated x_{50} for "all" particles derived from sample A was the lowest and the highest for masticated bolus of sample C (Table 1). Likewise, the masticated bolus of sample A from all 17 participants exhibited the highest average value of estimated b, suggesting a narrow spread of particle distribution, compared to masticated bolus from sample C (i.e., lowest average value of estimated b suggesting a broad spread of particle distribution) (Table 1). However, no significant difference ($p > 0.05$) in the median particle size (x_{50}) and distribution spread (b) between the samples after oral mastication was observed. The bolus particle size result was not in agreement with the texture perception by the participants, wherein sample B was majorly considered harder than the two other samples (Supplementary Material Figure S8). This shows that despite the differences in texture perception, humans are likely to masticate the samples until an acceptable bolus particle size ready to be swallowed is achieved.

In this study, one of the features utilised during the image analysis was to segregate and measure the particle size of the different components of the masticated beans, namely seed coat ("black") and cotyledon ("white"), to better describe each of their breakdowns during mastication (Supplementary Materials Figure S4). When the particle size result was analysed based on segregating the seed coat and cotyledon of the masticated beans, it was found that a higher median particle size and a wider distribution spread (lower b) were observed in "black" particles than in "white" (Table 2). This can also be observed in Figure 4 (both fast and slow chewers), wherein a higher number of cotyledon ("white") particles in bolus was evident. This was due to its larger area in the whole bean while the seed coat ("black") particles were much bigger than cotyledon after mastication. Based on the median particle size (x_{50}, Table 2) result and the image (Figure 4), it appeared that the participants may have difficulty breaking down the seed coat into smaller particle size to achieve a consistency that will flow smoothly during mastication as compared to the cotyledon.

With respect to the particle size differences of the oral boluses from 17 participants for three different sample types, it was found that the median particle size (x_{50}) of "white" particles ranged from 2 to 6 mm^2 for sample A, 2 to 5 mm^2 for sample B, and 3 to 5 mm^2 for sample C (Table 2). For the "black" particles, the size ranged from 5 to 11 mm^2 for sample A, 5 to 13 mm^2 for sample B, and 5 to 11 mm^2 for sample C (Table 2). Although significant difference between the three samples for the x_{50} and b averaged from all 17 participants was not detected in this study, it cannot be ruled out that different processing conditions including calcium addition, PEF, and thermal processing applied to the black beans might have impact on the breakdown of particles during in vivo mastication. This is because sample C consistently showed the highest estimated x_{50} and the lowest estimated b values compared to the in vivo masticated bolus from sample A and B based on the Rosin-

Rammler parameters considering "all" (Table 1), "black", and "white" (Table 2) particles, from all 17 participants.

Table 2. Estimated Rosin-Rammler parameters (Equation (2)) of the particles of black bean cotyledon ("white") and seed coat ("black") from samples A (PEF and thermally processed without $CaCl_2$ addition), B (PEF and thermally processed with $CaCl_2$ addition), and C (No PEF, thermally processed with $CaCl_2$) after in vivo human mastication (average of 17 participants).

Sample	Cotyledon ("White")		Seed Coat ("Black")	
	x_{50} (mm^2)	b	x_{50} (mm^2)	b
A	3.5 ± 0.9 [a]	1.9 ± 0.3 [a]	6.7 ± 1.8 [a]	1.7 ± 0.2 [a]
B	3.5 ± 0.8 [a]	1.9 ± 0.3 [a]	7.1 ± 2.0 [a]	1.7 ± 0.3 [a]
C	3.7 ± 0.6 [a]	1.8 ± 0.3 [a]	7.6 ± 1.4 [a]	1.58 ± 0.16 [a]

Data are presented as the average ± standard deviation of the estimated value from 17 participants ($n = 17$) reporting according to the significant figures. Values with the same lowercase letters in superscript for the three samples within each parameter (per column) are not significantly different ($p > 0.05$).

Figure 5 visualises the particle size distribution of a slow chewer (participant 12, average of 74 s chewing duration) and a fast chewer (participant 2, average of 18 s) for each sample. The number of particles (both cotyledon and seed coat) was observed to be mostly higher in a slow chewer, who had chewed the samples longer, and thus disintegrated the particles to many smaller sizes than a fast chewer. It was also interesting to observe from Figure 5 that the histograms of particle size distribution differed for one sample to another for both representative slow and fast chewers. For example, a higher number of larger "black" particles (i.e., seed coat) of samples B and C was produced by slow chewer compared to the "white particle" of similar size from the same sample. Likewise, a similar result was exhibited by fast chewer for samples B and C. Such differences in the particle size distribution between the samples suggests that the pre-treatment applied to the black beans prior to thermal processing could play some role in influencing the way beans are disintegrating when masticated by both fast and slow chewers.

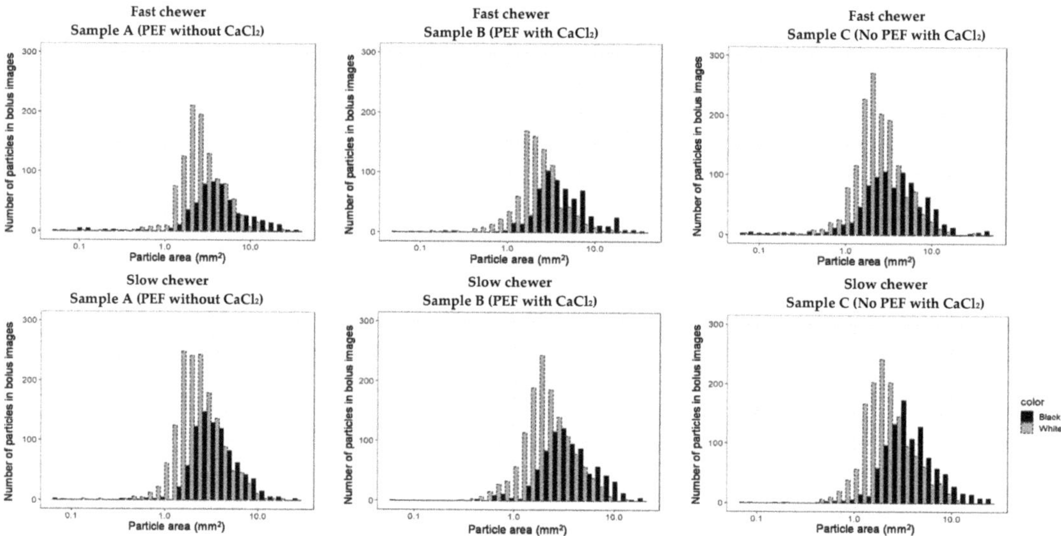

Figure 5. Representative examples of particle area distribution of separate components of black beans (black: seed coat, white: cotyledon) from the three samples (A: PEF and thermally processed without $CaCl_2$ addition; B: PEF and thermally processed with $CaCl_2$ addition; and C: No PEF, thermally processed with $CaCl_2$) as masticated by a fast chewer (participant 2) and a slow chewer (participant 12).

Histograms showing the distribution of the average Rosin-Rammler parameters (x_{50} and b) of "all" particles of the three black bean samples (A–C) after in vivo mastication among the participants were plotted (Supplementary Materials Figure S10) to obtain a comprehensive insight of the particle size distribution of different chewers in this study. Of the 17 participants, 53% masticated the black beans up to a median particle size of 6 mm^2 or smaller (Supplementary Materials Figure S10). Most of them are slow chewers who chewed the three samples for longer times (>29 s), resulting in an x_{50} between 4.18 and 4.41 mm^2. On the other hand, participants who chewed for <25 s (fast chewers) comminuted the beans to a larger particle size (>5 mm^2), who resulted in oral boluses with median particle sizes of between 5.25 and 6.70 mm^2. Moreover, the particle size distribution spread b was observed to be skewed to the right (Supplementary Materials Figure S10) which was attributed mostly to slow chewers. These results demonstrated that Rosin-Rammler parameters were clearly influenced by the participants' chewing duration where x_{50} was found to be smaller with a higher b in slow chewers than in fast chewers. An increase in b indicates that the distribution of the particles became narrower because the participants who masticated the samples for longer duration created a bolus with uniform small particle sizes.

Although the effect of PEF processing and calcium chloride addition on the particle size distribution of in vivo masticated cooked black beans was not statistically significant, an obvious difference in the estimated Rosin-Rammler model parameters and particle size distribution between the three samples was observed (Tables 1 and 2). Moreover, large variability in particle size between individuals was clearly exhibited in this study (Figure 5). This could be because every participant has their personal patterns of mastication and different perceptions of boluses that are deemed ready for swallowing [37]. Overall, this study revealed that median particle size (x_{50}) and distribution spread (b) were found to be negatively ($r = -0.64$) and positively ($r = 0.71$) correlated, respectively, with chewing time according to Pearson's correlation analysis. The correlation result is in agreement to the result of Olthoff et al. [38], wherein median particle size decreased as chewing strokes or duration increased.

3.3.4. The Activity of α-Amylase in Oral Boluses

Apart from the food properties and chewing behaviour, bolus formation could be influenced by the amount of saliva, which contains α-amylase, an enzyme that hydrolyses starch, secreted and incorporated in the food [8]. For this reason, the activity of α-amylase in the oral bolus after mastication and expectorated by each participant was measured using the Ceralpha method of Megazyme (Section 2.6). Previous studies [39,40] have looked at the effect of food properties on saliva production wherein dry and hard foods increased the secretion of saliva to lubricate them. Values obtained in this study cannot be directly compared to other studies due to differences in assay and units used, but it would be expected that the values reported here are lower, as the spitting water was included in the bolus which could have underestimated the α-amylase activity in the saliva.

It is not unexpected that a very wide inter-individual variation in α-amylase activity of the saliva was exhibited by the participants. For this reason, on average for 17 participants, salivary α-amylase activity in samples A, B, and C did not differ among each other ($p > 0.05$). In a way, the result suggests that cooked black beans pre-treated with PEF in the presence or absence of calcium did not affect the saliva production during human mastication. As shown in the histogram (Figure 6), the distribution of participants' salivary α-amylase activities is skewed on the left which indicates that most of the participants have lower salivary α-amylase values. Of the 17 participants, 41% had an average α-amylase activity of ≤ 0.1 CU/mL. This is mostly comprised of fast chewers with α-amylase activity ranged from 0.02 to 0.01 CU/mL. Clearly, α-amylase activity in the expectorated bolus was higher in slow chewers than in fast chewers.

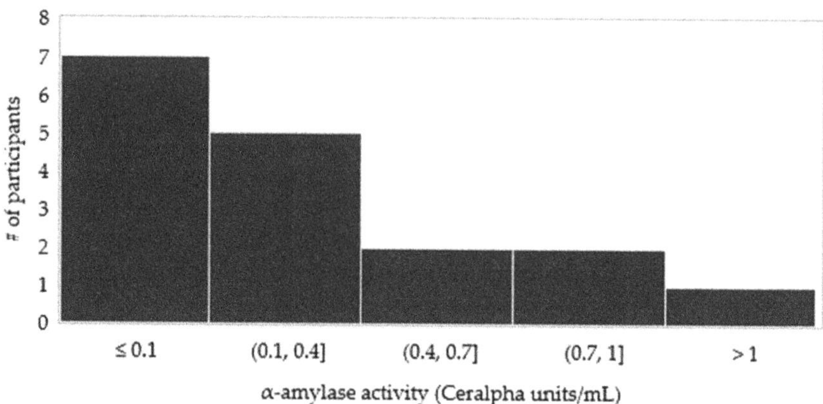

Figure 6. A histogram of the activity of α-amylase averaged from the oral bolus of the three samples (A: PEF and thermally processed without $CaCl_2$ addition; B: PEF and thermally processed with $CaCl_2$ addition; and C: No PEF, thermally processed with $CaCl_2$) masticated by 17 participants.

A positive correlation ($r = 0.88$) between chewing duration and α-amylase activity was found wherein prolonged chewing increased the amount of saliva and the activity of the enzyme. This is supported by the finding of a previous study wherein the longer the food is masticated in the mouth, the more saliva is incorporated in the bolus [9]. However, it cannot be ruled out that saliva production can vary between individuals in terms of flow rate and composition since saliva can be released in different salivary glands in the oral cavity wherein amylase-rich saliva (25% of whole saliva) is released from the parotid [8]. Overall, the results from the present study suggested that α-amylase activity increased with longer chewing duration possibly due to the increased requirement for saliva production.

3.3.5. Microstructural Changes of Black Bean Oral Bolus

Figure 7 shows the microstructural images, viewed under a light microscope, of the oral boluses of the three samples masticated by the participants within the average shortest (12 s) and longest (74 s) chewing duration. There is no remarkable difference in microstructure between samples A, B, and C as a result of varying the processing conditions. However, a more prominent change can be seen between the two chewing groups of participants. More intact cotyledon cells encapsulating the starch granules (Figure 7a–c) can be observed in the oral bolus of the fastest chewer. On the other hand, cotyledon cells that appeared to be ruptured were found in the oral bolus of the participant with the longest chewing duration with lesser amount of starch granules enclosed by the cell wall (Figure 7e,f). The presence of free starch granules was observed for both chewers (Figure 7). However, it was interesting to observe a few disrupted starch granules in the oral bolus of the slowest chewer (Supplementary Materials Figure S11). This could be a result of more shear bite forces applied to the beans with a longer duration of chewing.

The participants in this study have masticated the beans until they were ready for swallowing, hence the collected bolus will ideally be brought to the stomach for digestion. Therefore, the difference in the microstructures of the oral boluses as a result of extreme mastication duration might have an implication on the availability of starch to digestive enzymes.

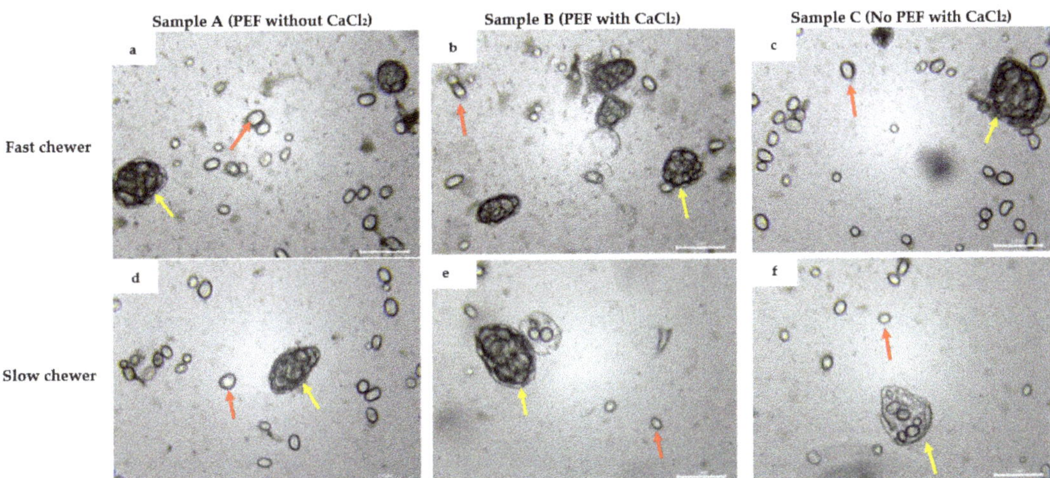

Figure 7. Selected light microscopic images of the starch granules from the oral bolus of three cooked black bean samples (A: PEF and thermally processed without CaCl$_2$ addition; B: PEF and thermally processed with CaCl$_2$ addition; and C: No PEF, thermally processed with CaCl$_2$) after in vivo mastication of the fastest (**a–c**) and slowest (**d–f**) chewers. Yellow arrows show the black bean cotyledon cells and red arrows point to free starch granules. Images were viewed under 10× magnification. Scale bar = 100 μm.

3.4. Characterisation of the Extent of In Vitro Starch Digestibility of In Vivo Masticated Black Beans by Different Participants

3.4.1. The Extent of Starch Digestibility during the In Vitro Gastric Phase

Figure 8 presents the amount of D-glucose digested at the start of gastric phase (0 min, end of oral mastication) and at the completion of gastric digestion at 120 min of selected participants with different mastication behaviour, specifically fast and slow chewers.

On average, for the three samples, the amount of D-glucose released from the bolus masticated by slow chewers at the start of the gastric phase was significantly ($p < 0.05$) higher than fast chewers (Figure 8). The amount of D-glucose released at 0 min gastric phase signified the starch that have been hydrolysed in the mouth by salivary α-amylase. The increase was also consistent at the end of the gastric phase (120 min) wherein all samples masticated by slow chewers were significantly higher in digested starch. Small sized particles found in the slow chewers (Figure 4) may have provided a greater surface area for salivary fluid to lubricate and increases the susceptibility of starch granules of black beans for enzyme hydrolysis. Moreover, a higher salivary α-amylase activity found mostly in slow chewers (Figure 6) could have facilitated a greater starch hydrolysis in the small sized particles and increased the amount of available glucose even before gastric digestion has started. It is argued that the role of salivary α-amylase is considered minimal in starch hydrolysis, as compared to pancreatic α-amylase, because of its short contact time with the food in the mouth and it is readily inactivated in the gastric phase [41]. Even though salivary amylase is typically inactivated at low pH (3.3–3.8) during gastric digestion, studies have found its significant contribution in gastric digestion up to 120 min and may even remain in the small intestine [42]. Moreover, it is possible that salivary amylase can be protected from inactivation by its substrate and hydrolysis product. Earlier study of Rosenblum et al. [43] revealed that starch (0.1%), partially hydrolysed starch (5%), maltose (5%), and maltotriose (5%) provided a protection to salivary amylase from inactivation, as evidenced by the retained activity up to 90% after 120 min of gastric digestion at pH 3, which was attributed to their interaction at the active site of the enzyme.

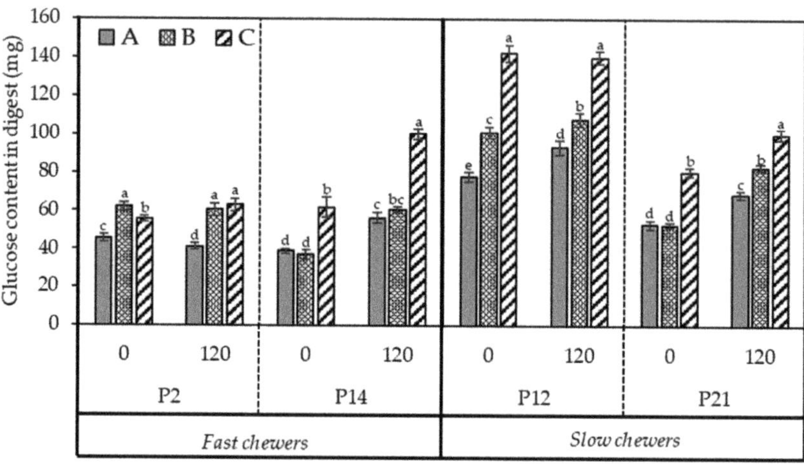

Gastric phase digestion duration (min)

Figure 8. In vitro gastric starch digestion of in vivo masticated black bean samples (A: PEF and thermally processed without $CaCl_2$ addition; B: PEF and thermally processed with $CaCl_2$ addition; and C: No PEF, thermally processed with $CaCl_2$) from selected fast and slow chewers. Data presented as mean ± standard deviation (n = 8 measurements). Values with different letters within each participant are significantly different ($p < 0.05$) with increasing digestion time.

Based on the results of individual participants, sample C appeared to be the most digestible during gastric phase compared to samples A and B due to a higher amount of D-glucose detected in the digest. However, the trend was not consistent for all the 17 participants so a direct conclusion cannot be made. When the statistical analysis of all the participants (n = 17) were considered, there was no significant difference ($p > 0.05$) in the amount of glucose detected in the digest between the different black bean samples A, B, and C. In other words, black beans processed using PEF, with or without calcium addition, then thermally processed, had the same starch digestibility at gastric phase as samples thermally processed alone without PEF pre-treatment. Therefore, PEF processing, which was previously determined to affect texture (Supplementary Material Figure S7) and enhance the uptake of calcium of cooked black beans (Figure 2), did not cause a considerable change in their starch digestibility at gastric phase. This indicates that PEF can potentially be used to modify the texture of legumes without posing implication on its starch hydrolysis.

3.4.2. The Proportion of Different Starch Fractions (RDS, SDS and RS) Digested during the In Vitro Small Intestinal Phase

The different categories of starch in the oral boluses of three cooked black bean samples as digested in the small intestine were determined in this study. Although significant differences ($p < 0.05$) in RDS, SDS, and RS can be observed between samples (A–C) for each individual participant (Table 3 presents a few representative examples), no significant differences ($p > 0.05$) was found between samples (A–C) for all the starch fractions when the statistical analysis of all the participants (n = 17) was considered. This supports the result in Section 3.4.1 wherein PEF processing, even with calcium addition, does not necessarily impact the nutritional implication of legumes, specifically starch.

The mastication behaviour of the participants remained a key factor driving the proportion of starch fractions digested along the small intestinal phase. In particular, chewing duration was found to be positively correlated with RDS and SDS (r = 0.54 and 0.66, respectively). From Table 3, slow chewers were characterized with a significantly ($p < 0.05$) higher RDS and SDS (up to 52 and 47%, respectively) than fast chewers. On the other hand, chewing time was negatively correlated with RS ($r = -0.61$). Results showed that fast chewers who masticated the beans for shorter times, resulting in larger particle sizes, were characterized with a significantly ($p < 0.05$) higher RS (up to 67%) than slow chewers (Table 3).

Table 3. Starch fractions of in vivo masticated boluses of the black bean samples (A: PEF and thermally processed without $CaCl_2$ addition; B: PEF and thermally processed with $CaCl_2$ addition; and C: No PEF, thermally processed with $CaCl_2$) from selected slow and fast chewers after in vitro small intestinal digestion.

Participants	RDS (%)			SDS (%)			RS (%)		
	Sample A	Sample B	Sample C	Sample A	Sample B	Sample C	Sample A	Sample B	Sample C
Fast chewers (chewed < 25 s)									
P2	$32.8 \pm 0.7\,^d{}_C$	$42.91 \pm 1.70\,^d{}_A$	$38.6 \pm 0.9\,^d{}_B$	$19.5 \pm 2.4\,^c{}_A$	$17.6 \pm 1.8\,^b{}_A$	$18.6 \pm 7.0\,^{cd}{}_A$	$47.8 \pm 1.9\,^b{}_A$	$39.52 \pm 1.80\,^c{}_B$	$42.9 \pm 6.8\,^c{}_{AB}$
P14	$37.7 \pm 1.9\,^c{}_A$	$37.0 \pm 2.0\,^e{}_A$	$27.7 \pm 1.1\,^f{}_B$	$21.8 \pm 0.9\,^c{}_A$	$19.3 \pm 0.9\,^b{}_B$	$10.9 \pm 1.7\,^e{}_C$	$40.5 \pm 1.4\,^c{}_C$	$43.71 \pm 2.50\,^b{}_B$	$61.32 \pm 1.10\,^a{}_A$
P16	$20.81 \pm 0.70\,^e{}_C$	$29.6 \pm 0.6\,^f{}_B$	$34.0 \pm 1.3\,^e{}_A$	$13.6 \pm 1.3\,^d{}_B$	$18.6 \pm 1.8\,^b{}_A$	$13.2 \pm 1.7\,^{de}{}_B$	$65.6 \pm 1.1\,^a{}_A$	$51.8 \pm 1.7\,^a{}_B$	$52.8 \pm 1.2\,^b{}_B$
Slow chewers (chewed > 29 s)									
P12	$39.0 \pm 1.5\,^c{}_C$	$47.8 \pm 1.3\,^c{}_B$	$57.4 \pm 2.3\,^b{}_A$	$26.2 \pm 3.5\,^b{}_A$	$24.5 \pm 0.9\,^a{}_A$	$19.6 \pm 1.9\,^{bc}{}_B$	$33.4 \pm 3.3\,^d{}_A$	$27.7 \pm 0.9\,^d{}_B$	$23.0 \pm 2.5\,^d{}_C$
P19	$56.4 \pm 1.0\,^a{}_B$	$69.1 \pm 1.1\,^a{}_A$	$51.6 \pm 0.9\,^c{}_C$	$26.4 \pm 1.4\,^b{}_A$	$19.1 \pm 2.2\,^b{}_B$	$24.3 \pm 2.3\,^{ab}{}_A$	$17.2 \pm 1.0\,^f{}_B$	$13.6 \pm 1.8\,^f{}_C$	$24.2 \pm 1.7\,^d{}_A$
P21	$46.2 \pm 1.4\,^b{}_C$	$54.9 \pm 2.2\,^b{}_B$	$61.8 \pm 1.5\,^a{}_A$	$29.9 \pm 3.1\,^a{}_A$	$27.4 \pm 3.2\,^a{}_A$	$28.3 \pm 3.9\,^a{}_A$	$24.0 \pm 1.9\,^e{}_A$	$17.7 \pm 2.1\,^e{}_B$	$11.7 \pm 1.5\,^e{}_C$

Data presented as mean $\pm$ standard deviation ($n = 8$ measurements) reporting according to the significant figures. Values with different lowercase letters in superscript are significantly different ($p < 0.05$) between participants for each starch fraction (RDS: readily digestible starch, SDS: slowly digestible starch, and RS: resistant starch). Values with different uppercase letters in subscript are significantly different ($p < 0.05$) between black bean samples (A, B, and C) for each participant within the same starch fraction.

These results imply that higher amounts of RDS and SDS, which are absorbed in the bloodstream, can cause an increase in blood glucose response and trigger metabolic disorder in diabetic patients [44]. On the other hand, RS which escapes the small intestine unhydrolyzed and is fermented in the large intestine by colonic microflora [45] can be beneficial in preventing type II diabetes, due to low levels of blood glucose and insulin requirement in the body [46]. Hence, it can be inferred that chewing duration and particle size can influence a foods' nutritional functionality.

3.4.3. The Kinetics of Starch Digestibility during the In Vitro Small Intestinal Phase

After the gastric phase, partially hydrolysed starch enters the small intestine. Most of the starch is then hydrolysed in the small intestine due to the presence of pancreatic amylase. The digested starch kinetics (amount of D-glucose in digest plotted against small intestinal digestion time) of selected participants are shown in Figure 9. The curves exhibited an increase in D-glucose released with small intestinal digestion time followed by a plateau for all the participants, regardless of the sample type. This clearly showed that the kinetics of starch digestibility during the in vitro small intestinal phase for PEF-treated and untreated cooked black beans in the presence or absence of calcium can be described using a fractional conversion model (Equation (6)).

Three starch digestion kinetic parameters (S_0, S_f, and k_s) were estimated using a fractional conversion model (Equation (6)) and the values are shown in Table 4 for selected participants categorised based on their chewing duration for each sample type. The estimated kinetic parameters describing the in vitro starch digestion were markedly varied for participants with different mastication duration. In agreement to the starch digestion kinetic curves visualized in Figure 9, slow chewers who masticated the beans for more than 29 s showed higher (up to 3-fold) estimated amounts of D-glucose released at the start (S_0: 103–168 mg) and the end (S_f: 216–343 mg) of small intestinal digestion, regardless of the sample type, than fast chewers (S_0: 51–113 mg, S_f: 134–240 mg) (Table 4). The result was expected as smaller starch granules with larger surface area from slow chewers are likely to be digested faster by amylases than larger particles with a smaller surface area from fast chewers [10]. This can be further supported by the microscopy images in Figure 7 and Supplementary Material Figure S11, wherein oral bolus of a slow chewer, predominantly, have ruptured cells and free starch granules. Free starch granules are more susceptible to amylase for digestion due to the absence of a physical barrier (cell wall) as similarly observed by Rovalino-Córdova et al. [47] in kidney beans.

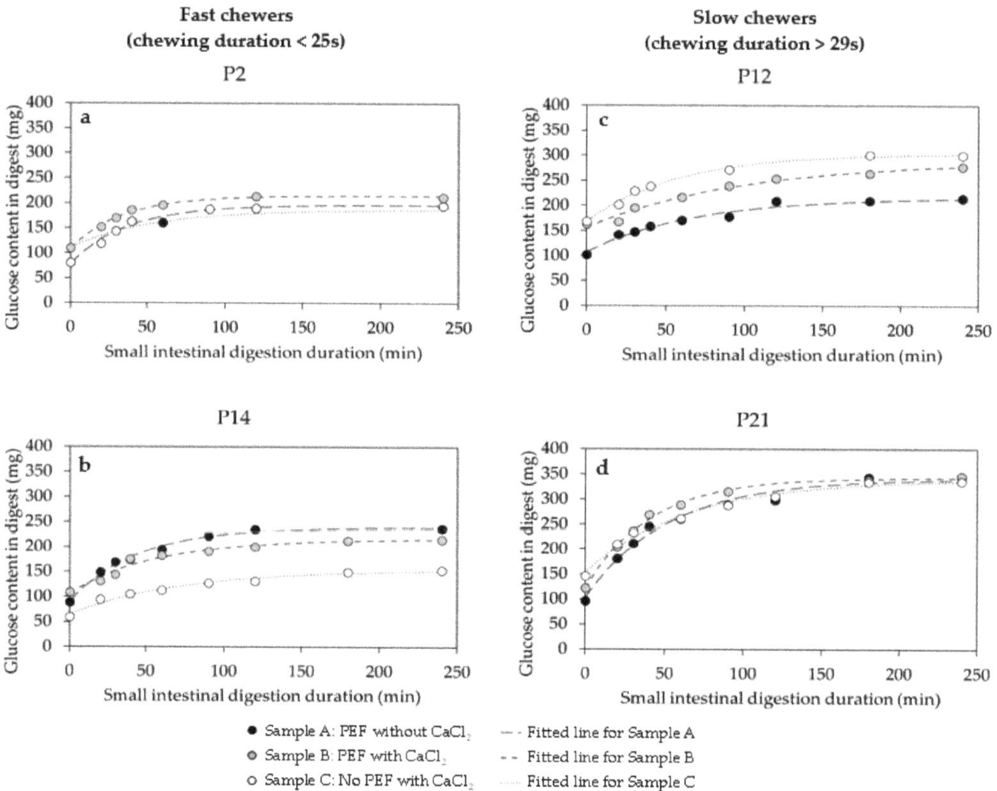

Figure 9. Kinetics of in vitro starch digestion (small intestinal phase) of in vivo masticated oral boluses of three cooked black bean samples (A: PEF and thermally processed without $CaCl_2$ addition; B: PEF and thermally processed with $CaCl_2$ addition; and C: No PEF, thermally processed with $CaCl_2$) from selected fast (**a**,**b**) and slow (**c**,**d**) chewers. Experimental data are represented by circle markers and the predicted values using fractional conversion model (Equation (6)) are shown in lines.

It is worthy to note that the in vitro small intestinal starch kinetic curves for some participants differ for different sample type (e.g., P12 and P14, Figure 9), but there was no consistent trend showing the starch in a particular sample being digested faster or slower than other sample (Table 4). The estimated starch digestion rate constant (k_s) did not vary much between the fast ($1.4–3.0 \times 10^{-2}$ min^{-1}) and slow ($1.0–2.5 \times 10^{-2}$ min^{-1}) chewers for the three sample types. In addition, statistical analysis taking into account all 17 participants revealed no significant difference ($p > 0.05$) between the k_s of samples A–C in terms of in vitro small intestinal starch digestion. Similar with the result in the in vitro gastric phase, calcium addition combined with PEF treatment did not seem to influence starch digestion of cooked black beans at small intestinal phase. This agrees with the findings of Abduh, Leong, Agyei and Oey [23] using the same level of PEF (up to 1.1 kV/cm electric field strength) wherein in vitro gastrointestinal starch digestion of potato was not affected considerably. However, using higher intensities of PEF treatment might have a different effect. For instance, Li et al. [48] found a significant increase in in vitro digestibility of wheat, potato, and pea starches when PEF electric field strength was applied up to 8.57 kV/cm.

Table 4. Kinetic parameters of in vitro small intestinal starch digestion of in vivo masticated oral boluses of the black bean samples (A: PEF and thermally processed without $CaCl_2$ addition; B: PEF and thermally processed with $CaCl_2$ addition; and C: No PEF, thermally processed with $CaCl_2$) from selected fast and slow chewers as estimated using fractional conversion model (Equation (6)).

Participants	S_0 (mg)			S_f (mg)			k_s ($\times 10^{-2}$ min^{-1})		
	Sample A	Sample B	Sample C	Sample A	Sample B	Sample C	Sample A	Sample B	Sample C
Fast chewers (chewed < 25 s)									
P2	78.4 ± 5.6	108.5 ± 3.3	110.6 ± 7.2	196.4 ± 4.4	214.6 ± 2.7	186.9 ± 6.7	2.7 ± 0.4	3.0 ± 0.2	1.9 ± 0.6
P14	90.5 ± 5.2	105.9 ± 6.6	64.4 ± 4.9	240.0 ± 4.7	216.0 ± 6.1	154.6 ± 6.1	2.3 ± 0.2	1.8 ± 0.3	1.4 ± 0.3
P16	50.6 ± 4.4	112.5 ± 4.2	103.4 ± 5.0	133.8 ± 4.5	209.5 ± 4.8	200.9 ± 4.6	1.6 ± 0.3	1.4 ± 0.2	1.8 ± 0.3
Average	73.18 ± 5.06 [a]	109.0 ± 4.7 [a]	92.8 ± 5.7 [a]	190.1 ± 4.6 [a]	213.4 ± 4.5 [a]	180.8 ± 5.8 [a]	2.2 ± 0.3 [a]	2.1 ± 0.3 [a]	1.7 ± 0.4 [a]
Slow chewers (chewed > 29 s)									
P12	107.3 ± 5.4	156.2 ± 6.1	168.0 ± 4.7	216.0 ± 6.2	289.2 ± 12.3	303.8 ± 4.5	1.5 ± 0.2	1.0 ± 0.2	1.8 ± 0.2
P19	152.3 ± 7.8	145.1 ± 11.9	128.6 ± 11.8	324.8 ± 5.7	306.40 ± 9.80	306.5 ± 9.3	2.3 ± 0.3	2.1 ± 0.4	2.1 ± 0.4
P21	102.6 ± 11.6	121.6 ± 5.2	152.9 ± 7.1	340.7 ± 9.9	343.0 ± 3.9	338.8 ± 7.7	1.9 ± 0.3	2.5 ± 0.1	1.6 ± 0.2
Average	120.7 ± 8.3 [a]	141.0 ± 7.7 [a]	149.8 ± 6.3 [a]	293.8 ± 7.3 [a]	312.9 ± 6.5 [a]	316.4 ± 7.2 [a]	1.9 ± 0.3 [a]	1.9 ± 0.3 [a]	1.8 ± 0.3 [a]

Data presented as estimated value ± standard error of the model (Equation (6)) reporting according to the significant figures. Average values with different lowercase letters in superscript are significantly different ($p < 0.05$) between samples (A–C) for each group of chewers within the same kinetic parameter. S_0: D-glucose released at the start of small intestinal digestion, S_f: D-glucose released at the end of small intestinal digestion, and k_s: estimated starch digestion rate constant. P2, P14, P16, P12, P19, and P21 corresponds to participants' numbers.

3.5. Characterisation of the Extent of In Vitro Protein Digestibility of In Vivo Masticated Black Beans by Different Participants

3.5.1. The Extent of Protein Digestibility during the In Vitro Gastric Phase

Figure 10 shows the amounts of peptides or α-amino groups (calculated as L-serine equivalents) released during protein digestion in the gastric phase of oral boluses masticated by selected participants that are fast and slow chewers. Clearly, a greater increase in the protein hydrolysis was observed in slow chewers than in fast chewers along the gastric phase. At 0 min, which corresponded to the amount of available amino acids before the action of digestive enzymes in the stomach, L-serine equivalents released was up to 45% higher (on average for three samples) in slow chewers in comparison to fast chewers. This was consistent up to the end of the gastric phase (120 min), wherein slow chewers hydrolysed up to 51% more proteins than fast chewers. It should be considered that the black bean samples had been thermally processed (80 °C for 1 h) prior to in vivo mastication, which could have possibly denatured the protein and reduced antinutrients such as trypsin inhibitors owing to the free amino acids at the start of the gastric phase. Protein is further denatured in the stomach due to the acidic environment and then hydrolysed by pepsin into large polypeptides and a small number of amino acids [49]. As shown earlier in Figures 4 and 7, the longer chewing time applied to the beans by slow chewers resulted in smaller particle sizes and possible ruptured cells and starch granules. This could have increased the accessibility of protein, which is bioencapsulated in the same cell compartment as starch, to proteases. A similar result in soybeans was observed wherein the degree of protein hydrolysis was increased as particle size was decreased [50]. The same study also reported a 41% increase in protein digestion when there was no physical barrier to digestive enzymes contributed by the cell wall. Nevertheless, protein digestion can still proceed in intact cells but at a slower rate. Protein digestibility was also increased in milled cowpea due to the smaller size, hence bigger surface area, for proteolytic attack [51].

Although significant differences between samples (A–C) were observed for each individual masticating the beans at each digestion time point, it was not consistent across all the participants (Figure 10). When statistical analysis considering 17 participants was conducted, the differences in protein digestibility in the gastric phase of the three samples (A–C) masticated by the participants were not significant ($p > 0.05$). Therefore, the cooked black beans pre-treated with PEF in presence or absence of calcium addition, resulted in the same degree of in vitro protein hydrolysis in the gastric phase as those processed conventionally by thermal processing. There is no known information to date on the effect of PEF on the protein digestibility of plant tissues—especially legumes—thus results cannot be validated with other studies.

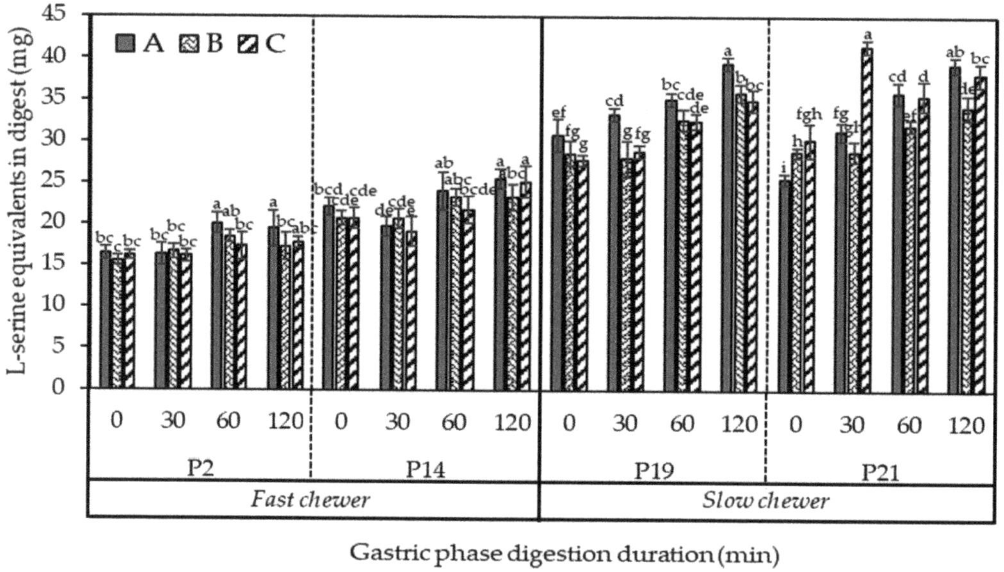

Figure 10. In vitro gastric protein digestion of in vivo masticated black bean samples (A: PEF and thermally processed without CaCl$_2$ addition; B: PEF and thermally processed with CaCl$_2$ addition; and C: No PEF, thermally processed with CaCl$_2$) from selected fast and slow chewers. Data presented as mean $\pm$ standard deviation (n = 8 measurements). Values with different letters within each participant are significantly different ($p < 0.05$) with increasing digestion time.

3.5.2. The Kinetic Behaviour of Protein Digestibility during the In Vitro Small Intestinal Phase

Figure 11 illustrates the time-course protein hydrolysis at small intestinal phase of in vivo masticated bolus from selected fast and slow chewers fitted well with the zero-order kinetic model (Equation (7)). The amount of hydrolysed protein at the end of in vitro small intestinal phase (240 min) was up to four-fold higher than the amount digested at the end of the gastric phase (0 min), indicating that protein is mostly hydrolysed in the small intestine, by pancreatic protease and trypsin, into free amino acids and small peptide chains of two to six amino acid residues [49]. Moreover, Figure 11 shows that slow chewers have higher free amino acids or L-serine equivalents (> 100 mg) released at the start of the small intestinal phase (0 min), and even higher than at the end of the small intestinal phase (240 min) in fast chewers.

The estimated kinetic rate (k_p) of protein digestion for selected participants categorised based on their chewing duration for each sample type is presented in Table 5. The rate of protein digestion was also generally higher in slow chewers (k_p = 6.7–16.1 $\times$ 10^{-2} min^{-1}) than in fast chewers (4.1–10.4 $\times$ 10^{-2} min^{-1}). The results in this study are consistent with the starch hydrolysis result wherein nutrients in small-sized particles, as observed in the oral bolus of slow chewers due to longer chewing time (Figure 4), were more prone to enzymatic attack—this was similarly observed by Paz-Yépez et al. [52] in the protein digestibility of walnuts and peanuts. Overall, result from the current study revealed that protein digestion at small intestinal phase can be influenced by the participants' chewing duration (r = 0.92) and particle size (x_{50}) ($r = -0.74$).

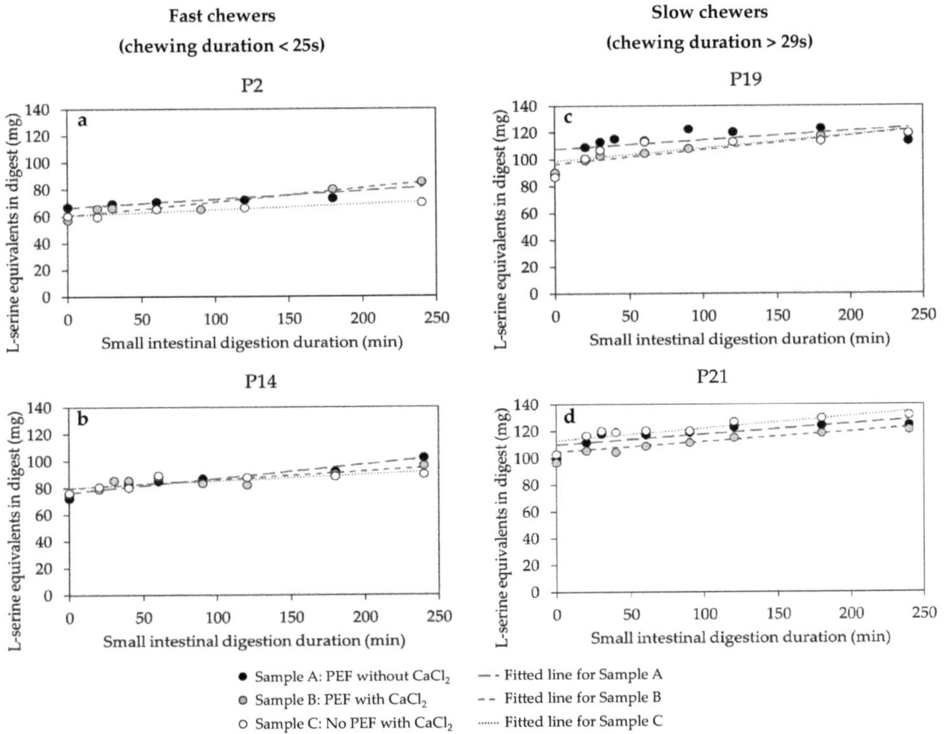

Figure 11. Kinetics of in vitro protein digestion (small intestinal phase) of in vivo masticated oral boluses of three cooked black bean samples (A: PEF and thermally processed without $CaCl_2$ addition; B: PEF and thermally processed with $CaCl_2$ addition; and C: No PEF, thermally processed with $CaCl_2$) from selected fast (**a,b**) and slow (**c,d**) chewers. Experimental data are represented by circle markers and predicted values are shown in lines. All protein digestion data did not fit into a fractional conversion model and was fitted with a zero-order kinetic model (Equation (7)) instead.

Table 5. Estimated rate of in vitro small intestinal protein digestion k_p ($\times 10^{-2}$ min^{-1}) of in vivo masticated oral boluses of the black bean samples (A: PEF and thermally processed without $CaCl_2$ addition; B: PEF and thermally processed with $CaCl_2$ addition; and C: No PEF, thermally processed with $CaCl_2$) from selected fast and slow chewers.

Participants	Sample A	Sample B	Sample C
Fast chewers (chewed < 25 s)			
P2	6.2 ± 1.5	10.4 ± 1.5	4.1 ± 0.9
P4	5.5 ± 1.6	9.2 ± 1.0	4.3 ± 1.9
P16	4.7 ± 1.5	8.4 ± 1.3	5.4 ± 1.8
Average	5.5 ± 1.5 [b]	9.3 ± 1.3 [a]	4.6 ± 1.5 [b]
Slow chewers (chewed > 29 s)			
P12	10.2 ± 1.5	16.1 ± 1.2	11.5 ± 2.7
P19	6.7 ± 4.0	10.5 ± 1.7	9.6 ± 3.3
P21	7.7 ± 2.5	7.5 ± 2.6	9.0 ± 2.1
Average	8.2 ± 2.7 [a]	11.3 ± 1.8 [a]	10.0 ± 2.7 [a]

Data presented as estimated value $\pm$ standard error of the model (Equation (7)) reporting according to the significant figures. Average values with different lowercase letters in superscript are significantly different ($p < 0.05$) between samples (A–C) for each group of chewers. P2, P4, P16, P12, P19, and P21 corresponds to participants' numbers.

Unlike the result of starch digestibility, statistical analysis revealed that protein in sample B (i.e., PEF treated with calcium then thermally processed) was digested significantly ($p = 0.001$) faster (up to 2-fold) than samples A (i.e., PEF treated without calcium then thermally processed) and C (i.e., thermally processed only in presence of calcium) in fast chewers (Table 5). In other words, PEF alone did not significantly affect the protein digestibility of black beans unless calcium was added. The cell electroporation induced by PEF of black beans potentially played a role in enhancing the digestion of protein even when calcium was added. Previous studies [30,53] found an increase in protein digestibility when egg whites and cooked beef were PEF treated (electric field strength of 0.60 kV/cm and 1.8, respectively) which was attributed to the electroporation effect and structural alteration of protein by PEF, enhancing the susceptibility of protein to digestive enzymes in the small intestine. However, calcium could also potentially contribute to the increased digestibility of protein. A recent study has found that addition of calcium (~600 ppm) enhanced the activity of phytase and acid phosphatase, which are enzymes that catalyse the hydrolysis or degradation of phytic acid [36]. Reportedly, phytic acid can lower protein digestibility by forming phytate-protein complexes with the positively charged basic amino acid of proteins [54], altering its structure and further affecting its solubility, digestibility and accessibility of protein to digestive enzymes [55,56]. However, the role of calcium in improving protein digestibility and its interaction with protein still needs further study. The results of this study are physiologically relevant because the use of PEF, a novel emerging processing, can improve protein digestion even when consumers only chew for shorter duration. This can be achieved without necessarily increasing starch digestibility which could have negative implication especially for diabetic people.

4. Conclusions

The current study has adopted several novel experimental and analysis approaches to understand the implication of using emerging PEF technology to process legumes on their starch and protein digestibility. The use of image analysis technique to quantify particle sizes of black beans composed of separate components (outer seed coat and inner cotyledon) has provided novel information on the breakdown of each component of untreated and PEF-treated black beans during in vivo mastication prior to swallowing. It is worthy to note that the use of emerging PEF technology on black beans was shown effective in facilitating the uptake of exogenous calcium (especially to the outer seed coat), help preventing the texture from softening during the subsequent thermal processing and able to modulate the chewiness of black beans without changing other texture parameters (i.e., hardness, cohesiveness, springiness, and resilience). Despite this, the result showed that cooked black beans with PEF pre-treatment in calcium posed no influence on the in vivo mastication behaviour among the participants that could negatively impact on the in vitro starch and protein digestibility. In fact, the observed differences in in vivo mastication behaviour between the participants, in terms of particle size, microstructure, and α-amylase activity of the resulting black beans oral bolus, was highly correlated ($r = 0.64–0.88$) with the chewing duration to masticate a mouthful of the cooked beans before it was deemed ready for swallowing. Two different groups of participants were successfully defined in this study on the basis of their chewing duration: slow (>29 s) and fast (<25 s) chewers. While the amount of protein and starch in masticated cooked black beans being digested during in vitro gastrointestinal phase was always higher in slow chewers compared to fast chewers, the combined PEF processing with $CaCl_2$ addition has significantly ($p < 0.05$) increased the rate of in vitro small intestinal protein digestion of cooked black beans especially for fast chewers by two-fold. For the first time in the literature, this finding indicates an opportunity of using PEF technology on legumes to improve the digestion of protein for a certain consumer group, such as fast chewers without triggering an increase in the starch digestibility. Clearly, PEF treatment application on black beans is shown promising to hasten the release and hydrolysis of protein in the gastrointestinal tract, but it is of future interest to investigate whether the hydrolysed proteins become more bioaccessible and then available for absorption in the blood stream to carry the relevant metabolic function in the body. Forthcoming studies should also address whether PEF treatment can be applied on other legume (types and physical structures, e.g., without outer seed coat) to facilitate release of nutritious high-quality plant proteins.

Supplementary Materials: The following are available online at https://www.mdpi.com/article/10.3390/foods10112540/s1, Figure S1: Black beans outer seed coat (blue arrow) and inner cotyledon (orange arrow) separated for calcium content analysis, Figure S2: Sensory booth set up for each participant during the in vivo mastication study, Figure S3: Preparation of separate trays for the three samples presented to the participants containing the sample containers (5 g sample per container, blue arrow), mouth rinsing water containers (30 mL per container, black arrow), and spitting containers (green arrow), Figure S4: Sample image processing of black beans oral bolus. Left: black beans dispersed in the plate and positioned for picture taking with the sample label and reference scale. Right: after image processing showing the segmentation of the "white" (marked green) and "black" (marked yellow) particles, Figure S5: Traces of prediction profile (from JMP Pro v.14) showing the predicted texture parameters of cooked black beans as either PEF electric field strength or energy input is changed, while the calcium concentration is held constant at 0 ppm, Figure S6: Traces of prediction profile (from JMP Pro v.14) showing the predicted texture parameters of cooked black beans as either PEF electric field strength or energy input is changed while the calcium concentration is held constant at 300 ppm, Figure S7: Texture parameters of the selected three black bean samples (A: PEF and thermally processed without $CaCl_2$ addition, B: PEF and thermally processed with $CaCl_2$ addition, and C: No PEF, thermally processed with $CaCl_2$) used for in vivo oral mastication study. Data presented as mean $\pm$ standard deviation of forty-eight independent measurements (n = 48). Values with different lowercase letters between sample type for each texture parameter are significantly different ($p < 0.05$), Figure S8: Participants ($n = 17$) perception of the hardness of three black bean samples (A: PEF and thermally processed without $CaCl_2$ addition; B: PEF and thermally processed with $CaCl_2$ addition; and C: No PEF, thermally processed with $CaCl_2$) rated on a five-point hedonic scale, Figure S9: Boxplot of the chewing duration of the three samples (A: PEF and thermally processed without $CaCl_2$ addition; B: PEF and thermally processed with $CaCl_2$ addition; and C: No PEF, thermally processed with $CaCl_2$) by 17 participants. Upper and lower lines of the box are the upper and lower quartiles, the lines inside the box represent the median values, and the lines extending outside of the box are the minimum and maximum values. No significant difference ($p > 0.05$) was observed between samples, Figure S10: Histograms showing the distribution of the average Rosin-Rammler parameters (x_{50} and b) of "all" particles of the three black bean samples (A: PEF and thermally processed without $CaCl_2$ addition; B: PEF and thermally processed with $CaCl_2$ addition; and C: No PEF, thermally processed with $CaCl_2$) after in vivo mastication ($n = 17$), Figure S11: Light microscopic image showing broken free starch granules from the oral bolus of a slow chewer. Images were viewed under $50\times$ magnification. Scale bar = 100 μm, Table S1: p-value for the estimates of the model parameters examining the response surface effect of input variables (PEF electric field strength, PEF energy input, and calcium concentration during PEF and thermal processing) against texture parameters of cooked black beans (hardness, cohesiveness, springiness, chewiness, and resilience).

Author Contributions: Conceptualization, I.O. and S.Y.L.; methodology, M.A., S.Y.L., V.L., C.E.M. and I.O.; validation, M.A., S.Y.L., V.L., C.E.M. and I.O.; formal analysis, M.A., V.L. and C.E.M.; investigation, M.A., S.Y.L. and I.O.; resources, V.L., C.E.M. and I.O.; data curation, M.A.; writing—original draft preparation, M.A.; writing—review and editing, M.A., V.L., C.E.M., S.Y.L. and I.O.; visualization, M.A., S.Y.L. and I.O.; supervision, I.O. and S.Y.L.; project administration, I.O.; funding acquisition, I.O. All authors have read and agreed to the published version of the manuscript.

Funding: This research was supported by the Riddet Institute, a New Zealand Centre of Research Excellence, funded by the Tertiary Education Commission.

Institutional Review Board Statement: This study was approved by the University of Otago human ethics committee (Approval reference code: 20/038, Date of approval: 28 September 2020).

Informed Consent Statement: Informed consent was obtained from all subjects involved in the study.

Acknowledgments: Alpos is a recipient of the New Zealand ASEAN Scholarship by the Ministry of Foreign Affairs and Trade making it possible for her to complete her Master's degree at the University of Otago. Technical assistance of Stella Green and Sarah Johnson is very much appreciated. The authors also thank Fiona Nyhof for her constructive feedback when preparing the human ethics approval application. Roman Karki is acknowledged for his valuable insights on the least squares model and prediction profile using the JMP Pro software.

Conflicts of Interest: The authors declare no conflict of interest. The funders had no role in the design of the study; in the collection, analyses, or interpretation of data; in the writing of the manuscript, or in the decision to publish the results.

References

1. Van Buggenhout, S.; Sila, D.; Duvetter, T.; Van Loey, A.; Hendrickx, M. Pectins in processed fruits and vegetables: Part III—Texture engineering. *Compr. Rev. Food Sci. Food Saf.* **2009**, *8*, 105–117. [CrossRef]
2. Van Buren, J.P. The chemistry of texture in fruits and vegetables. *J. Texture Stud.* **1979**, *10*, 1–23. [CrossRef]
3. Van der Sman, R. Impact of processing factors on quality of frozen vegetables and fruits. *Food Eng. Rev.* **2020**, *12*, 399–420. [CrossRef]
4. Moens, L.G.; Huang, W.; Van Loey, A.M.; Hendrickx, M.E.G. Effect of pulsed electric field and mild thermal processing on texture-related pectin properties to better understand carrot (*Daucus carota*) texture changes during subsequent cooking. *Innov. Food Sci. Emerg. Technol.* **2021**, *70*, 102700. [CrossRef]
5. Leong, S.Y.; Du, D.; Oey, I. Pulsed Electric Fields enhances calcium infusion for improving the hardness of blanched carrots. *Innov. Food Sci. Emerg. Technol.* **2018**, *47*, 46–55. [CrossRef]
6. Lebovka, N.I.; Praporscic, I.; Vorobiev, E. Effect of moderate thermal and pulsed electric field treatments on textural properties of carrots, potatoes and apples. *Innov. Food Sci. Emerg. Technol.* **2004**, *5*, 9–16. [CrossRef]
7. Toepfl, S.; Heinz, V.; Knorr, D. Applications of pulsed electric fields technology for the food industry. In *Pulsed Electric Fields Technology for the Food Industry: Fundamentals and Applications*; Raso, J., Heinz, V., Eds.; Springer: New York, NY, USA, 2006; pp. 197–221.
8. Bornhorst, G.M.; Singh, R.P. Bolus formation and disintegration during digestion of food carbohydrates. *Compr. Rev. Food Sci. Food Saf.* **2012**, *11*, 101–118. [CrossRef]
9. Chen, J.; Khandelwal, N.; Liu, Z.; Funami, T. Influences of food hardness on the particle size distribution of food boluses. *Arch. Oral Biol.* **2013**, *58*, 293–298. [CrossRef]
10. Tharanathan, R.N.; Mahadevamma, S. Grain legumes—A boon to human nutrition. *Trends Food Sci. Technol.* **2003**, *14*, 507–518. [CrossRef]
11. Pallares, A.P.; Loosveldt, B.; Karimi, S.N.; Hendrickx, M.; Grauwet, T. Effect of process-induced common bean hardness on structural properties of in vivo generated boluses and consequences for in vitro starch digestion kinetics. *Br. J. Nutr.* **2019**, *122*, 388–399. [CrossRef]
12. Pangborn, R.M.; Lundgren, B. Salivary secretion in response to mastication of crisp bread. *J. Texture Stud.* **1977**, *8*, 463–472. [CrossRef]
13. Bornhorst, G.M.; Kostlan, K.; Singh, R.P. Particle size distribution of brown and white rice during gastric digestion measured by image analysis. *J. Food Sci.* **2013**, *78*, E1383–E1391. [CrossRef] [PubMed]
14. Etzler, F.M.; Sanderson, M.S. Particle size analysis: A comparative study of various methods. *Part. Part. Syst. Charact.* **1995**, *12*, 217–224. [CrossRef]
15. Mora, C.; Kwan, A.; Chan, H. Particle size distribution analysis of coarse aggregate using digital image processing. *Cem. Concr. Res.* **1998**, *28*, 921–932. [CrossRef]
16. Alpos, M.; Leong, S.Y.; Oey, I. Combined effects of calcium addition and thermal processing on the texture and in vitro digestibility of starch and protein of black beans (*Phaseolus vulgaris*). *Foods* **2021**, *10*, 1368. [CrossRef]
17. Lombardi-Boccia, G.; Lucarini, M.; Di Lullo, G.; Del Puppo, E.; Ferrari, A.; Carnovale, E. Dialysable, soluble and fermentable calcium from beans (*Phaseolus vulgaris* L.) as model for in vitro assessment of the potential calcium availability. *Food Chem.* **1998**, *61*, 167–171. [CrossRef]
18. Wesley, R.; Rousselle, J.; Schwan, D.; Stadelman, W. Improvement in quality of scrambled egg products served from steam table display. *Poult. Sci.* **1982**, *61*, 457–462. [CrossRef]
19. Swackhamer, C.; Zhang, Z.; Taha, A.Y.; Bornhorst, G.M. Fatty acid bioaccessibility and structural breakdown from in vitro digestion of almond particles. *Food Funct.* **2019**, *10*, 5174–5187. [CrossRef]
20. McCleary, B.V.; McNally, M.; Monaghan, D.; Mugford, D.C. Measurement of α-amylase activity in white wheat flour, milled malt, and microbial enzyme preparations, using the Ceralpha assay: Collaborative study. *J. AOAC Int.* **2002**, *85*, 1096–1102. [CrossRef]
21. Gwala, S.; Wainana, I.; Pallares Pallares, A.; Kyomugasho, C.; Hendrickx, M.; Grauwet, T. Texture and interlinked post-process microstructures determine the in vitro starch digestibility of Bambara groundnuts with distinct hard-to-cook levels. *Food Res. Int.* **2019**, *120*, 1–11. [CrossRef]

22. Minekus, M.; Alminger, M.; Alvito, P.; Ballance, S.; Bohn, T.; Bourlieu, C.; Carrire, F.; Boutrou, R.; Corredig, M.; Dupont, D.; et al. A standardised static in vitro digestion method suitable for food an international consensus. *Food Funct.* **2014**, *5*, 1113–1124. [CrossRef]
23. Abduh, S.; Leong, S.Y.; Agyei, D.; Oey, I. Understanding the properties of starch in potatoes (*Solanum tuberosum* var. Agria) after being treated with pulsed electric field processing. *Foods* **2019**, *8*, 159. [CrossRef] [PubMed]
24. Pallares, A.P.; Miranda, B.A.; Truong, N.Q.A.; Kyomugasho, C.; Chigwedere, C.M.; Hendrickx, M.; Grauwet, T. Process-induced cell wall permeability modulates the in vitro starch digestion kinetics of common bean cotyledon cells. *Food Funct.* **2018**, *9*, 6544–6554. [CrossRef]
25. Piecyk, M.; Wołosiak, R.; Drużynska, B.; Worobiej, E. Chemical composition and starch digestibility in flours from Polish processed legume seeds. *Food Chem.* **2012**, *135*, 1057–1064. [CrossRef]
26. Wani, I.A.; Sogi, D.S.; Wani, A.A.; Gill, B.S. Physical and cooking characteristics of some Indian kidney bean (*Phaseolus vulgaris* L.) cultivars. *J. Saudi Soc. Agric. Sci.* **2017**, *16*, 7–15. [CrossRef]
27. Guiné, R.P.F.; Roque, A.R.F.; Seiça, F.F.A.; Batista, C.E.O. Effect of chemical pretreatments on the physical properties of kiwi. *Int. J. Food Eng.* **2016**, *2*, 90–95. [CrossRef]
28. Liu, X.; Lu, K.; Yu, J.; Copeland, L.; Wang, S.; Wang, S. Effect of purple yam flour substitution for wheat flour on in vitro starch digestibility of wheat bread. *Food Chem.* **2019**, *284*, 118–124. [CrossRef]
29. Englyst, H.N.; Veenstra, J.; Hudson, G.J. Measurement of rapidly available glucose (RAG) in plant foods: A potential in vitro predictor of the glycaemic response. *Br. J. Nutr.* **1996**, *75*, 327–337. [CrossRef]
30. Liu, Y.-F.; Oey, I.; Bremer, P.; Silcock, P.; Carne, A. In vitro peptic digestion of ovomucin-depleted egg white affected by pH, temperature and pulsed electric fields. *Food Chem.* **2017**, *231*, 165–174. [CrossRef]
31. Gwala, S.; Pallares, A.P.; Pälchen, K.; Hendrickx, M.; Grauwet, T. In vitro starch and protein digestion kinetics of cooked Bambara groundnuts depend on processing intensity and hardness sorting. *Food Res. Int.* **2020**, *137*, 109512. [CrossRef] [PubMed]
32. Barbosa-Cánovas, G.V.; Pothakamury, U.R.; Gongora-Nieto, M.M.; Swanson, B.G. *Preservation of Foods with Pulsed Electric Fields*; Academic Press: Cambridge, UK, 1999.
33. Parniakov, O.; Bals, O.; Lebovka, N.; Vorobiev, E. Effects of pulsed electric fields assisted osmotic dehydration on freezing-thawing and texture of apple tissue. *J. Food Eng.* **2016**, *183*, 32–38. [CrossRef]
34. Yi, J.; Njoroge, D.M.; Sila, D.N.; Kinyanjui, P.K.; Christiaens, S.; Bi, J.; Hendrickx, M.E. Detailed analysis of seed coat and cotyledon reveals molecular understanding of the hard-to-cook defect of common beans (*Phaseolus vulgaris* L.). *Food Chem.* **2016**, *210*, 481–490. [CrossRef] [PubMed]
35. Pereira, R.N.; Galindo, F.G.; Vicente, A.A.; Dejmek, P. Effects of pulsed electric field on the viscoelastic properties of potato tissue. *Food Biophys.* **2009**, *4*, 229–239. [CrossRef]
36. Zhou, T.; Wang, P.; Yang, R.; Wang, X.; Gu, Z. Ca^{2+} influxes and transmembrane transport are essential for phytic acid degradation in mung bean sprouts: Mechanism of the enhanced phytic acid degradation in mung bean sprouts under $CaCl_2$ treatment. *J. Sci. Food Agric.* **2017**, *98*, 1968–1976. [CrossRef]
37. Mishellany, A.; Woda, A.; Labas, R.; Peyron, M.-A. The challenge of mastication: Preparing a bolus suitable for deglutition. *Dysphagia* **2006**, *21*, 87–94. [CrossRef]
38. Olthoff, L.; Van der Bilt, A.; Bosman, F.; Kleizen, H. Distribution of particle sizes in food comminuted by human mastication. *Arch. Oral Biol.* **1984**, *29*, 899–903. [CrossRef]
39. Gavião, M.B.D.; Engelen, L.; Van Der Bilt, A. Chewing behavior and salivary secretion. *Eur. J. Oral Sci.* **2004**, *112*, 19–24. [CrossRef]
40. Engelen, L.; Fontijn-Tekamp, A.; van der Bilt, A. The influence of product and oral characteristics on swallowing. *Arch. Oral Biol.* **2005**, *50*, 739–746. [CrossRef]
41. Pedersen, A.; Bardow, A.; Jensen, S.B.; Nauntofte, B. Saliva and gastrointestinal functions of taste, mastication, swallowing and digestion. *Oral Dis.* **2002**, *8*, 117–129. [CrossRef]
42. Fried, M.; Abramson, S.; Meyer, J. Passage of salivary amylase through the stomach in humans. *Dig. Dis. Sci.* **1987**, *32*, 1097–1103. [CrossRef]
43. Rosenblum, J.L.; Irwin, C.L.; Alpers, D.H. Starch and glucose oligosaccharides protect salivary-type amylase activity at acid pH. *Am. J. Physiol. Gastrointest. Liver Physiol.* **1988**, *254*, G775–G780. [CrossRef]
44. Leong, S.Y.; Duque, S.M.; Abduh, S.B.M.; Oey, I. Carbohydrates. In *Innovative Thermal and Non-Thermal Processing, Bioaccessibility and Bioavailability of Nutrients and Bioactive Compounds*; Barba, F.J., Saraiva, J.M.A., Cravotto, G., Lorenzo, J.M., Eds.; Woodhead Publishing: Cambridge, UK, 2019; pp. 171–206.
45. Osorio-Díaz, P.; Bello-Pérez, L.A.; Sáyago-Ayerdi, S.G.; Benítez-Reyes, M.D.P.; Tovar, J.; Paredes-López, O. Effect of processing and storage time on in vitro digestibility and resistant starch content of two bean (*Phaseolus vulgaris* L.) varieties. *J. Sci. Food Agric.* **2003**, *83*, 1283–1288. [CrossRef]
46. Jenkins, A.L.; Marchie, A.; Kendall, C.W.C.; Augustin, L.S.A.; Hamidi, M.; Axelsen, M.; Franceschi, S.; Jenkins, D.J.A. Glycemic index: Overview of implications in health and disease. *Am. J. Clin. Nutr.* **2002**, *76*, 266S–273S. [CrossRef]
47. Rovalino-Córdova, A.M.; Fogliano, V.; Capuano, E. The effect of cell wall encapsulation on macronutrients digestion: A case study in kidney beans. *Food Chem.* **2019**, *286*, 557–566. [CrossRef] [PubMed]

48. Li, Q.; Wu, Q.-Y.; Jiang, W.; Qian, J.-Y.; Zhang, L.; Wu, M.; Rao, S.-Q.; Wu, C.-S. Effect of pulsed electric field on structural properties and digestibility of starches with different crystalline type in solid state. *Carbohydr. Polym.* **2019**, *207*, 362–370. [CrossRef] [PubMed]

49. Silk, D.; Grimble, G.; Rees, R.G. Protein digestion and amino acid and peptide absorption. *Proc. Nutr. Soc.* **1985**, *44*, 63–72. [CrossRef] [PubMed]

50. Zahir, M.; Fogliano, V.; Capuano, E. Food matrix and processing modulate in vitro protein digestibility in soybeans. *Food Funct.* **2018**, *9*, 6326–6336. [CrossRef] [PubMed]

51. Tinus, T.; Damour, M.; van Riel, V.; Sopade, P.A. Particle size-starch–protein digestibility relationships in cowpea (*Vigna unguiculata*). *J. Food Eng.* **2012**, *113*, 254–264. [CrossRef]

52. Paz-Yépez, C.; Peinado, I.; Heredia, A.; Andrés, A. Influence of particle size and intestinal conditions on in vitro lipid and protein digestibility of walnuts and peanuts. *Food Res. Int.* **2019**, *119*, 951–959. [CrossRef]

53. Bhat, Z.F.; Morton, J.D.; Mason, S.L.; Jayawardena, S.R.; Bekhit, A.E.-D.A. Pulsed electric field: A new way to improve digestibility of cooked beef. *Meat Sci.* **2019**, *155*, 79–84. [CrossRef] [PubMed]

54. Kumar, V.; Sinha, A.K.; Makkar, H.P.S.; Becker, K. Dietary roles of phytate and phytase in human nutrition: A review. *Food Chem.* **2010**, *120*, 945–959. [CrossRef]

55. Deshpande, S.; Damodaran, S. Effect of phytate on solubility, activity and conformation of trypsin and chymotrypsin. *J. Food Sci.* **1989**, *54*, 695–699. [CrossRef]

56. Knuckles, B.; Kuzmicky, D.; Betschart, A.A. Effect of phytate and partially hydrolyzed phytate on in vitro protein digestibility. *J. Food Sci.* **1985**, *50*, 1080–1082. [CrossRef]

 *foods*

Article

Antioxidants Bioaccessibility and *Lactobacillus salivarius* (CECT 4063) Survival Following the In Vitro Digestion of Vacuum Impregnated Apple Slices: Effect of the Drying Technique, the Addition of Trehalose, and High-Pressure Homogenization

Cristina Gabriela Burca-Busaga [1], Noelia Betoret [1], Lucía Seguí [1], Jorge García-Hernández [2], Manuel Hernández [2] and Cristina Barrera [1,*]

1   Instituto Universitario de Ingeniería de Alimentos para el Desarrollo de la Universitat Politècnica de València, 46022 Valencia, Spain; cribur@posgrado.upv.es (C.G.B.-B.); noebeval@tal.upv.es (N.B.); lusegil@upvnet.upv.es (L.S.)
2   Centro Avanzado de Microbiología de Alimentos de la Universitat Politècnica de València, 46022 Valencia, Spain; jorgarhe@btc.upv.es (J.G.-H.); mhernand@btc.upv.es (M.H.)
*   Correspondence: mcbarpu@tal.upv.es; Tel.: +34-629-987-104

Citation: Burca-Busaga, C.G.; Betoret, N.; Seguí, L.; García-Hernández, J.; Hernández, M.; Barrera, C. Antioxidants Bioaccessibility and *Lactobacillus salivarius* (CECT 4063) Survival Following the In Vitro Digestion of Vacuum Impregnated Apple Slices: Effect of the Drying Technique, the Addition of Trehalose, and High-Pressure Homogenization. *Foods* 2021, 10, 2155. https://doi.org/10.3390/foods10092155

Academic Editors: Harjinder Singh and Alejandra Acevedo-Fani

Received: 30 July 2021
Accepted: 9 September 2021
Published: 12 September 2021

Publisher's Note: MDPI stays neutral with regard to jurisdictional claims in published maps and institutional affiliations.

Abstract: To benefit the health of consumers, bioactive compounds must reach an adequate concentration at the end of the digestive process. This involves both an effective release from the food matrix where they are contained and a high resistance to exposure to gastrointestinal conditions. Accordingly, this study evaluates the impact of trehalose addition (10% *w/w*) and homogenization (100 MPa), together with the structural changes induced in vacuum impregnated apple slices (VI) by air-drying (AD) and freeze-drying (FD), on *Lactobacillus salivarius* spp. *salivarius* (CECT 4063) survival and the bioaccessibility of antioxidants during in vitro digestion. Vacuum impregnated apple slices conferred maximum protection to the lactobacillus strain during its passage through the gastrointestinal tract, whereas drying with air reduced the final content of the living cells to values below 10 cfu/g. The bioaccessibility of antioxidants also reached the highest values in the VI samples, in which the release of both the total phenols and total flavonoids to the liquid phase increased with in vitro digestion. The addition of trehalose and homogenization at 100 MPa increased the total bioaccessibility of antioxidants in FD and AD apples and the total bioaccessibility of flavonoids in the VI samples. Homogenizing at 100 MPa also increased the survival of *L. salivarius* during in vitro digestion in FD samples.

Keywords: apple structure; *Lactobacillus salivarius* spp. *salivarius* (CECT 4063); trehalose; high pressure homogenization; in vitro digestion; antioxidants bioaccessibility; air-drying; freeze-drying

1. Introduction

Health and wellness are trends that have conditioned the development of the food and beverage industry in the last decades [1]. Concern about boosting the immune system as a way to prevent non-communicable diseases and provide protection from pathogenic viral infections has grown considerably since the COVID-19 pandemic hit the world. As a result of the growing inclination of consumers towards preventive healthcare, the global functional food and beverage market size was valued at USD 258.80 billion in 2020 and is projected to double by 2028 [2]. In relation to ingredients, probiotics held the major share, which is mainly attributed to an increasing awareness regarding their prophylactic and therapeutic potential [3].

The functionality of probiotics is mainly based on bacterial interaction with host gut microbiota through a number of actions including increasing the production of vitamins, antioxidants, and short-chain fatty acids, modifying the intestinal microbiota, reducing the intestinal pH, or improving the intestinal barrier selectivity through higher mucin, immunoglobulin A, and defensins production [4]. Probiotics not only protect humans against gastrointestinal pathogens [5], but they may also alleviate gastrointestinal dysbiosis, lower serum cholesterol, ameliorate cancer and lactose intolerance, and prevent allergic and autoimmune disorders [6]. Due to their power to remedy both systematic metabolic diseases and genetic neurodegenerative disorders, probiotics are considered as the twenty-first century panpharmacon [3]. However, in order to achieve these health benefits, probiotics must be consumed regularly and reach the intestine as a viable strain in appropriate quantities (10^8–10^9 cfu/day) [7]. Therefore, probiotics used in food formulations must not only survive to the processing and storage stages, but they must also survive the harsh conditions of the gastrointestinal tract, such as the low pH of the stomach, and the digestive enzymes and bile salts of the small intestine [8]. Therefore, various methods have been proposed to enhance the viability of probiotic bacteria, such as strain and food carrier selection, the addition of prebiotics, cell immobilization, and microencapsulation or the induction of cellular stress-tolerance pathways [9].

With regard to the food matrix, research has mainly focused on the role played by its composition, which is often modified by adding ingredients that act as probiotic growth promoters (e.g., sugars, vitamins, minerals, prebiotics) or protectants (e.g., skim milk powder, whey protein, glycerol, lactose, fat, trehalose) [10]. Although dairy foods are the foods with the greatest potential as probiotic carriers, apples have also been demonstrated to be a suitable alternative [11,12] since they contain polyphenols (dihydrochalcones, flavanols, hydroxycinnamates, and flavanol), vitamins, minerals, lipids, peptides, and carbohydrates [13], in addition to fibers (cellulose, hemicellulose, and pectin) [14], that act as prebiotics. Phenolic compounds in apples are also responsible for the health-protecting effects related to apple consumption [15] due to their antioxidant properties and their capacity to neutralize free radicals and to protect cells against oxidative stress [16,17]. Furthermore, as it was recently reported by the authors of [18] for raw and fried tomato puree inoculated with *L. reuteri*, antioxidants may enhance the viability of probiotics during digestion. This makes some unit operations, such as high-pressure homogenization (HPH), particularly relevant to enhance probiotic survival under adverse conditions either in response to induced stress [19,20] or to the higher release of certain bioactives [21].

Denser, more viscous, or complex matrices have been reported to better protect probiotics during digestion, although they also limit their release [10]. However, studies which focus particularly on food structures are scarce and are based on cheese networks, in which it is easier to modify the structure without significantly varying the composition. Changes in the food structure are in most cases promoted by different processing techniques (e.g., boiling, roasting, frying, freezing, air-drying, freeze-drying). In particular, freeze-drying results in high-value products with advanced rehydration properties, a high porosity, and limited shrinkage which, due to the low processing temperatures, preserves their nutrients, color, aroma, and flavors [22]. On the contrary, convective dried products tend to have a low porosity, a high apparent density, poor rehydration properties and, due to high temperatures and long processing times, they lose a large amount of nutrients, flavor and aroma compounds.

For all the aforementioned reasons, this study aims to evaluate the effect that the stabilization of vacuum impregnated apple slices by convective drying and freeze-drying has on *Lactobacillus salivarius* spp. *salivarius* (CECT 4063) survival and on the bioaccessibility of antioxidants (total phenols, total flavonoids, and total ABTS and DPPH scavenging compounds) during an in vitro simulation of gastrointestinal digestion. Simultaneously, information about the impact of trehalose additions to the snack formulation and of the homogenization at 100 MPa of the liquid containing the probiotic are also gathered.

2. Materials and Methods

2.1. Microbial Strain and Raw Materials

The selection of both the microbial strain and the raw materials was based on our previous research of probiotic apple snacks with the potential ability to treat and prevent

Helicobacter pylori infection [23–26]. Therefore, the *Lactobacillus salivarius* spp. *salivarius* CECT 4063 strain was used as a bacterial culture. The lyophilized strain was supplied by the Spanish Type Culture Collection of Paterna (Valencia, Spain). Following the manufacturer's instructions, the microbial strain was first reactivated in sterile MRS broth (Scharlau Chemie®, Barcelona, Spain) at 37 °C for 24 h and then transferred to commercial clementine juice (Hacendado brand), which was used as a probiotic carrier. Finally, apples (var. Granny Smith) cut into 5 mm slices (20 mm internal diameter and 65 mm outer diameter) were used to host the probiotic. The selection of this solid matrix was based on its high homogeneity and porosity compared to other fruits [27].

2.2. Sample Preparation

Different food matrices used in this study were manufactured according to the procedure described in Burca-Busaga et al. [26]. On the one hand, vacuum impregnated samples (VI apples) were obtained in a VT 6130M Heraeus Vacutherm Oven (Thermo Scientific) connected to a LVS 210T laboratory vacuum system (Welch Ilmvac™, Fisher Scientific, Madrid, Spain) by the immersion of the apple slices in the corresponding impregnation liquid in a 1:5 (*w/v*) ratio, the application of a vacuum pressure of 50 mbar for 10 min, and the restoration of the atmospheric pressure, which was maintained for another 10 min. On the other hand, air-dried samples (AD apples) were obtained by drying VI apples in a CLW 750 TOP+ tray dryer (Pol-Eko-Aparatura SP.J.) with a cross flow of air at 2 m/s and 40 °C up to a water activity value of 0.35. Finally, freeze-dried samples (FD apples) were obtained by keeping VI apples at −40 °C for 24 h in a Matek CVN-40/105 ultra-freezer and the further sublimation of the frozen water at −45 °C and 0.1 mbar for 24 h in a Telstar Lioalfa-6 freeze-drier.

As the impregnation liquid, commercial clementine juice (Hacendado brand) containing 9.8 g/L of $NaHCO_3$ (Sigma-Aldrich, Madrid, Spain) and 5 g/L of yeast extract (Scharlau Chemie®, Barcelona, Spain), inoculated with $1.4 \pm 0.3 \times 10^6$ cfu/mL of *L. salivarius* and then incubated at 37 °C for 24 h was used (liquid 0%_0 MPa). In certain experiments, 100 g/kg of food-grade trehalose from tapioca starch (TREHATM, Cargill Ibérica) was added to the juice prior to inoculation (liquid 10%_0 MPa). In other experiments, the fermented liquid was homogenized at 100 MPa in a laboratory scale high pressure homogenizer (Panda Plus 2000, GEA-Niro Soavi) before it was used as an impregnation liquid (liquid 0%_100 MPa). Based on previous findings [28], homogenization at 100 MPa was not applied to the liquid containing 10% (*w/w*) of trehalose since it significantly reduced (*p*-value < 0.05) the survival of *L. salivarius* following the in vitro digestion of the inoculated clementine juice. Final counts in any of the vacuum impregnation liquids were in the order of $6 \pm 2 \times 10^8$ cfu/mL.

2.3. In Vitro Simulation of Gastrointestinal Digestion

The in vitro simulation of gastrointestinal digestion of the different food matrices was performed in sterile conditions following the protocol described by García-Hernández et al. [18], with some modifications. Each sample was digested three times in duplicate for the determination of both the microbial and the antioxidant properties.

For the oral stage simulation, apple samples were manually cut into pieces of about $2 \times 2 \times 5$ mm³ and mixed with human saliva in a ratio 1:1 (*w/v*) and 1:2 (*w/v*), for non-dehydrated (VI apples) and dehydrated samples (both AD and FD apples), respectively. After grinding with a T 25 digital ULTRA-TURRAX® (IKA®-Werke GmbH & Co. KG, Staufen, Germany) at 8500–9500 rpm for 1 min, the simulated oral bolus was mixed in a ratio 1:5 (*w/v*) with a 3 g/L solution of 3200–4500 U/mg pepsin porcine (Sigma-Aldrich, Madrid, Spain) in a sterile saline solution at 0.5% (*w/v*) adjusted to pH 2 with HCl (0.5 N). Following 2 h of constant agitation at 100 rpm and 37 °C on an orbital shaker (Optic Ivymen Systems TM), the resulting simulated chime was mixed in a ratio 1:1.8 (*w/w*) with a 1 g/L solution of 8 × USP porcine pancreatin (Sigma-Aldrich, Madrid, Spain) in a sterile saline solution at 0.5% (*w/v*) adjusted to pH 8 with NaOH (0.1 N). This mixture remained for another 4 h under constant agitation at 100 rpm and 37 °C on an orbital shaker (Optic Ivymen Systems™).

Sampling for the microbial counts was performed at the end of the oral phase at different times along gastric digestion (0, 10, 30, 60, and 120 min) and at different times along intestinal digestion (0, 30, 60, 120, 180, and 240 min). For each time point, 1 mL of the liquid phase was withdrawn from the reaction vessel. Sampling for the antioxidant properties was only performed at the end of both the gastric and the intestinal phases by separating 1 mL aliquots from the corresponding liquid phase.

2.4. Analytical Determinations

2.4.1. Moisture Content and Water Activity

The moisture content was obtained from the weight loss undergone by a certain amount of the sample when dried in a vacuum oven (Vaciotem-T, J.P. Selecta) at 60 °C and 200 mbar until a constant weight was reached. The water activity was measured in a CX-2 AquaLab dew point hygrometer (Decagon Devices, Inc., Pullman, WA, USA) at 25 °C.

2.4.2. Microbial Counts

Colony counts were measured by serial decimal dilution in sterile distilled water and plated on MRS Agar (Scharlau Chemie®, Barcelona, Spain). Since the microbial growth did not significantly increase under anaerobic conditions, the plates were incubated in aerobiosis at 37 °C for 24 h. The survival of the microbial strain during each stage of in vitro gastrointestinal digestion was then calculated as the ratio between the microbial concentration at the end of the stage and the microbial concentration at the end of the previous one, both referred to the same basis. Likewise, probiotic survival during the entire in vitro gastrointestinal digestion was calculated as the ratio between the microbial concentration in the liquid phase at the end of the intestinal stage and the microbial concentration of the undigested sample. All survival results were expressed as a percentage.

2.4.3. Antioxidant Properties

Measurements in the case of the undigested samples were carried out on the supernatants resulting from the mixing of 2 g of the VI apple (which was reduced to 0.35 g in the case of AD and FD apples) with 20 mL of an 80:20 (*v/v*) methanol in a water solution, dispersing for 2 min at 5000 rpm with a T 25 digital ULTRA-TURRAX® disperser (IKA®-Werke GmbH & Co. KG, Staufen, Germany), shaking in the dark for 1 h at 200 rpm on an orbital shaker (Rotabit, J.P. Selecta, Barcelona, Spain) and subsequent centrifugation at 10,000 rpm and 4 °C in a Medifriger-BL-S (J.P. Selecta, Barcelona, Spain) centrifuge. Measurements at the end of both the gastric and the intestinal stage were directly carried out on the resulting liquid phases, which were diluted with distilled water in a 1:5 (*v/v*) ratio for AD and FD apples.

The total phenol content, total flavonoid content, and the overall antioxidant activity of the target samples were spectrophotometrically determined following the procedures described in Burca-Busaga et al. [26].

The total phenol content, which was obtained from the blue color intensity that results after the reaction between the Folin–Ciocalteu reagent (Sigma-Aldrich, Madrid, Spain) and the phenolic compounds present in the sample, was measured at 760 nm and expressed as mg of the gallic acid equivalents per gram of the dried sample (mg GAE/g dw). Specifically, 125 μL of the sample, 125 μL of the Folin–Ciocalteu reagent, and 500 μL of double distilled water were mixed. Following 6 min of reaction in the dark, 1.25 mL of 7.5% (*w/v*) sodium bicarbonate and 1 mL of redistilled water were added and left to react for another 90 min.

The total flavonoid content, which was assessed following the aluminum chloride method, was measured at 368 nm and expressed as mg of the quercetin equivalents per gram of the dried sample (mg QE/g dw). For that purpose, 1.5 mL of sample reacted for 10 min with 1.5 mL of a 2% (*w/v*) solution of aluminum chloride in methanol.

Finally, the antioxidant activity was measured by the DPPH method, which consists of measuring the color change at 515 nm from deep-violet to pale-yellow undergone by radical 1,1-diphenyl-2-picrylhydrazyl when reduced by antioxidants or other radical species, and expressed as mg of the trolox equivalents per gram of the dried sample (mg TE/g dw). In this case, 100 µL of the sample, 900 µL of methanol, and 2000 µL of a 100 mM DPPH solution in methanol (*v/v*) reacted for 30 min in a spectrophotometry cuvette. A blank was used in which the sample was replaced by the same volume of redistilled water.

The bioaccessibility of each compound, defined as the portion that is released from the food matrix into the gastrointestinal tract and thus becomes available for intestinal absorption [17], was obtained as the ratio from its concentration in the liquid phase at the end of the intestinal stage and the concentration in the sample before digestion.

2.5. Statistical Analysis

The statistical analysis of the experimental data was performed by means of a one-way or multifactorial ANOVA (confidence level of 95%) carried out with the Statgraphics Centurion XVI tool.

3. Results and Discussion

3.1. Effect of the Food Matrix and the Vacuum Impregnation Solution on Probiotic Survival during In Vitro Digestion

The effect of the food matrix (VI, AD, or FD apples) and the vacuum impregnation liquid (0%_0 MPa, 0%_100 MPa, or 10%_0 MPa) on *L. salivarius* counts along the in vitro simulation of the gastrointestinal human digestion were recorded and are shown in Figure 1. The values obtained for the undigested samples (FOOD) were of the same order as those reported in a previous study by Burca-Busaga et al. [26].

Viable counts in both the VI and FD apples were hardly affected by the exposure to simulated mouth conditions. However, those in the AD apples decreased by 34 ± 4%, 47 ± 6%, and 68.1 ± 1.2% for liquids 0%_0 MPa, 0%_100 MPa, and 10%_0 MPa, respectively. Given that the microbial counts along the in vitro digestion were performed on the liquid phase, it is postulated that the microorganisms were completely released from the solid matrix to the salivary fluid in the VI and FD apples, but only partly in the AD samples. Leaching has previously been reported to be fast and excessive for FD apple cubes and a bit slower for AD apple cubes since freeze-drying causes a more extensive cell wall rupture than the convective process [29]. As for the significantly lower release of *L. salivarius* from the AD apples that were impregnated with the liquid that included 10% of trehalose in its composition, this is consistent with the ability of this disaccharide to replace the water of hydration at the membrane–fluid interface, thus preventing structural collapse as the tissue is dried [30].

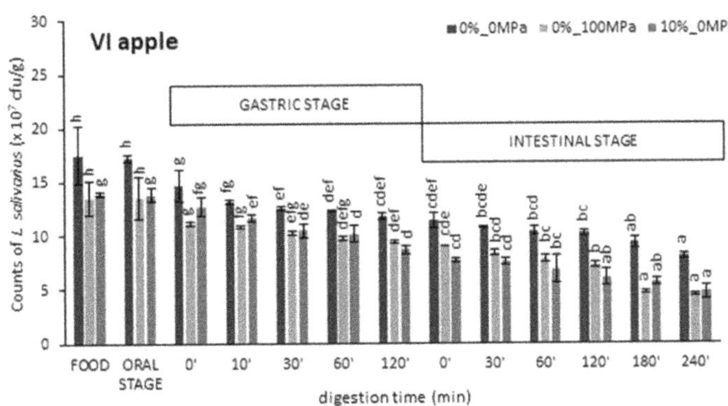

Figure 1. *Cont.*

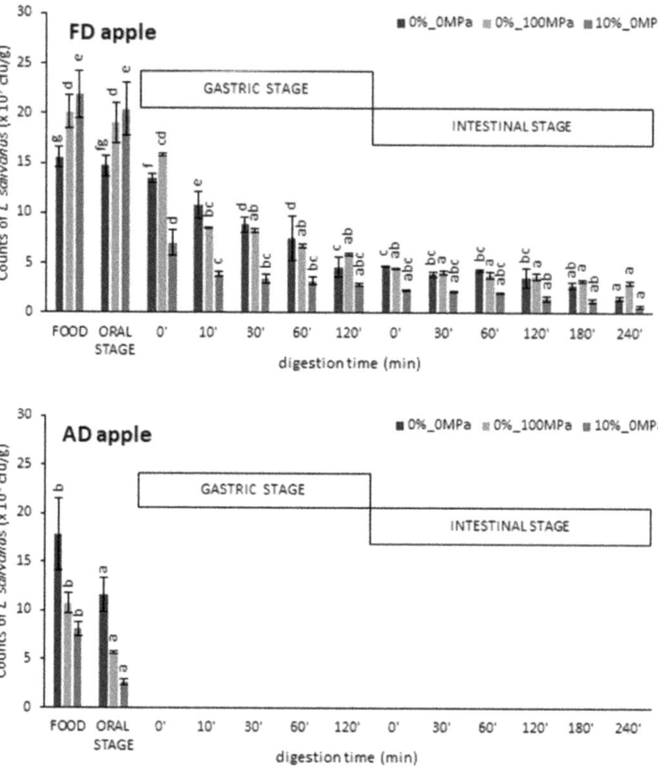

Figure 1. Microbial counts of *L. salivarius* along the gastrointestinal in vitro digestion of different food matrices (VI: vacuum impregnated apples; FD: freeze-dried apples; AD: air-dried apples) made with each of the impregnating solutions. Error bars represent the standard deviation of triplicates from two independent treatments. [a,b,c,d,e,f,g] different letters within the same series indicate statistically significant differences (*p*-value < 0.05).

A progressive decline in the microbial counts was observed in most cases along the second digestive step. These expected results are attributed to the ability of pepsin to destroy the peptide bond between amino acids and to degrade the microbial cell membrane [31]. However, as pointed out by García-Hernández et al. [18], microbial death due to the shock produced by the gastric juices is more likely to occur in those bacteria whose cell wall had been previously damaged by external factors. In accordance with this, a reduction in the viable number after the gastric stage was minimum in the VI apples (~38 ± 8% on average for samples impregnated with liquids 0%_0 MPa, 0%_100 MPa, and 10%_0 MPa) and increased with the application of a dehydration step (Table 1). Of the two dehydration techniques, freeze-drying was the one with the least impact on the microbial viability (a survival of 21 ± 8% on average for the samples impregnated with liquids 0%_0 MPa, 0%_100 MPa, and 10%_0 MPa). In the case of AD apples, the only contact with the simulated gastric juice reduced the number of viable cells to less than 10 cfu/g. As regards the vacuum impregnation liquid (Table 1), adding 10% of trehalose to its composition slightly reduced the survival of *L. salivarius* during in vitro gastric digestion possibly due to induced osmotic stress. On the contrary, homogenizing the microorganism at 100 MPa had no effect on its survival in VI apples after 120 min of exposure to the simulated gastric juice, but significantly increased that in FD apples. It seems that those cells that survived the freeze-drying step after their homogenization at 100 MPa were better prepared to withstand adverse stomach conditions. In a previous study [28], the survival of *L. salivarius* in gastric simulated conditions after the in vitro digestion of the vacuum impregnation liquids was reported to be $36 + 7\%$, $57 \pm 3\%$, and $51.34 \pm 0.13\%$ for liquids 0%_0 MPa, 0%_100 MPa, and 10%_0 MPa, respectively. Significantly higher values obtained in the present study after the in vitro digestion of VI apples ($63 \pm 7\%$ on average) confirms that including the lactobacillus into the porous structure of apple slices protects it as it passes through the upper gastrointestinal tract. In particular, the grip of microorganisms to apples mainly occurs in the intercellular spaces of the parenchymal tissue of the fruit [14]. This finding is particularly interesting in relation to the ability of *L. salivarius* to inhibit pro-inflammatory cytokine secretion from *Helicobacter pylori* (a well-known gastric pathogen) infected cells [32].

Table 1. Microbial concentration of the digested samples (final counts) and the survival percentage of *L. salivarius* to each stage (oral, gastric, and intestinal) and the entire simulated digestion (total) as affected by the food matrix and the growing media. Mean value $\pm$ standard deviation of triplicates from two independent treatments.

Treatment		Oral Stage $\frac{X^{lb}_{OS}}{X^{lb}_{UF}}\cdot 100$	Gastric Stage $\frac{X^{lb}_{GS}}{X^{lb}_{OS}}\cdot 100$	Intestinal Stage $\frac{X^{lb}_{IS}}{X^{lb}_{GS}}\cdot 100$	Total $\frac{X^{lb}_{IS}}{X^{lb}_{UF}}\cdot 100$	Final Counts (Log cfu/g)
MRS Broth		-	33 ± 2 [b]	48.50 ± 0.13 [bc]	15.8 ± 1.4 [b]	8.566 ± 0.004 [f]
VI apple	0%_0 MPa	97 ± 4 [de]	68 ± 3 [d]	68 ± 8 [d]	47 ± 5 [d]	7.904 ± 0.012 [e]
	0%_100 MPa	100 ± 2 [e]	65 ± 7 [d]	48 ± 2 [b]	31 ± 4 [c]	7.644 ± 0.015 [d]
	10%_0 MPa	93 ± 7 [d]	55 ± 9 [c]	54 ± 6 [bc]	28 ± 5 [c]	7.67 ± 0.06 [d]
FD apple	0%_0 MPa	99 ± 2 [de]	17 ± 5 [a]	53 ± 9 [bc]	9 ± 2 [a]	7.19 ± 0.05 [c]
	0%_100 MPa	94 ± 4 [de]	28 ± 5 [b]	58 ± 5 [c]	16 ± 3 [b]	7.438 ± 0.012 [b]
	10%_0 MPa	93 ± 2 [de]	14 ± 3 [a]	24 ± 3 [a]	3.1 ± 0.2 [a]	6.82 ± 0.07 [a]
AD apple	0%_0 MPa	66 ± 4 [c]	n d	n d	n d	n d
	0%_100 MPa	53 ± 6 [b]	n d	n d	n d	n d
	10%_0 MPa	31.9 ± 1.2 [a]	n d	n d	n d	n d

[abcde] different superscripts in the same column indicate statistically significant differences (p-value < 0.05). X^{lb}_{UF} are the colony counts of undigested food (cfu/g dw); X^{lb}_{OS} are the colony counts at the end of the oral stage (cfu/g dw); X^{lb}_{GS} are the colony counts at the end of the gastric stage (cfu/g dw); and X^{lb}_{IS} are the colony counts at the end of the intestinal stage (cfu/g dw).

Exposure to the small intestinal conditions (pancreatin solution at pH 8.0) resulted in further changes in the survival rate of *L. salivarius*. Again, the loss of viability was significantly higher (p-value < 0.05) in FD apples than in VI apples, and while homogenization at 100 MPa slightly increased *L. salivarius* survival in the simulated intestinal juice digestion in FD samples, it significantly reduced it in VI ones (p-value < 0.05). Adding 10% of trehalose to the vacuum impregnation liquid negatively affected the survival of lactobacillus during intestinal digestion in both VI and FD apples, but this was much greater in the case of the FD samples in which the disaccharide reached a higher concentration. Given that extracellular trehalose cannot provide sufficient protection for cells during dehydration and gastrointestinal digestion [33], the trehalose added to the growing media was neither imported nor did it induce the expression of trehalose-synthesizing genes to a sufficient extent.

The survival of *L. salivarius* during the entire gastrointestinal digestion was obtained, as explained in Section 2.4.2, from its survival during the simulated oral, gastric, and intestinal stages. As shown in Table 1, the survival of *L. salivarius* in VI apples after the simulated gastrointestinal digestion was similar to that previously reported by Barrera et al. [28] for vacuum impregnation liquids (between $26 \pm 5\%$ and $33 \pm 2\%$ for liquids 0%_0 MPa and 0%_100 MPa, respectively), but was significantly higher than that obtained for *L. salivarius* in FD apples or in MRS Broth. These differences, however, did not prevent the microorganism from reaching the end of the digestive process in a sufficient concentration ($>10^7$ cfu/g) to exert a beneficial effect on the consumer's health.

The survival of *L. salivarius* during the simulated gastrointestinal digestion of both VI and FD samples was significantly higher (~0.4 log reduction and ~1 log reduction, respectively) than that reported by Valerio et al. (2020) for *L. paracasei* IMPC2.1 in a pectin-coated dehydrated apple snack containing ≥9 log cfu/20 g portion (~2 log reduction). On the same microbial strain, air drying conducted at 60 °C for 1 h, 50 °C for 30 min, and 40 °C up to 24 h was less detrimental to its survival during the digestion process than one-stage drying at 40 °C [25]; on the contrary, the encapsulation of the lactobacillus by HPH at 70 MPa considerably reduced the survival of the strain during the digestion process in FD apple slices.

3.2. Effect of the Food Matrix and the Vacuum Impregnation Solution on the Bioaccesibility of Antioxidants during In Vitro Digestion

Total phenols (mg GAE/g dw), total flavonoids (mg QE/g dw), and overall antioxidant activity (mg TE/g dw) measured before and after each stage of the in vitro simulation of the gastrointestinal human digestion as affected by the food matrix (VI, AD, or FD apple) and the vacuum impregnation liquid (0%_0 MPa, 0%_100 MPa, or 10%_0 MPa) are shown in Figure 2. In accordance with the previous findings of the research group [26], the values obtained for the undigested vacuum impregnated apples were 5.6 ± 0.4 mg GAE/g dw, 1.4 ± 0.4 mg QE/g dw, and 6.5 ± 0.5 mg TE/G dw, regardless of the composition of the vacuum impregnation liquid used. These values remained almost constant after freeze-drying, but significantly increased (p-value < 0.05) after convective drying.

Following the gastric phase of the in vitro digestion, a significant increase (p-value < 0.05) in the content of both the total phenols and flavonoids was observed for VI samples. On average, the content of the total phenols and flavonoids increased by 52 ± 11% and 152 ± 37%, respectively, after the action of the simulated gastric juice. This is in agreement with previous findings [34,35] and indicates that the 3 g/L pepsin solution adjusted to pH 2 allowed more glycosidic bonds to be broken and enabled the release of more phenolic compounds (especially of the flavonoid type) than the chemical extraction with an 80% (*v/v*) solution of methanol in water. More specifically, apple samples impregnated with liquid 0%_0 MPa showed the greatest increase in total phenols (from 5.27 ± 0.07 to 8.7 ± 0.6 mg GAE/g dw) but the lowest increase in total flavonoids (from 1.78 ± 0.08 to 3.3 ± 0.5 mg QE/g dw). Unlike the phenolic and flavonoid contents, the DPPH values of the VI apples were 40 ± 4% lower after the gastric phase of digestion, regardless of the composition of the VI liquid. Such a decrease in the antiradical activity could be attributed to the presence of other antioxidant compounds different from polyphenols that are less soluble in the gastric juice and/or more sensitive to the acidic conditions of the stomach. This could be the case with carotenoids which, owing to the numerous double bonds of their chemical structure, are particularly susceptible to oxidation in acidic media [35]. As an example, the overall recovery of carotenoids after the gastric phase of mandarin pulp in vitro digestion was around 79% [36], but it decreased to 36–63% when a blended fruit juice containing orange, pineapple, and kiwi was digested [35].

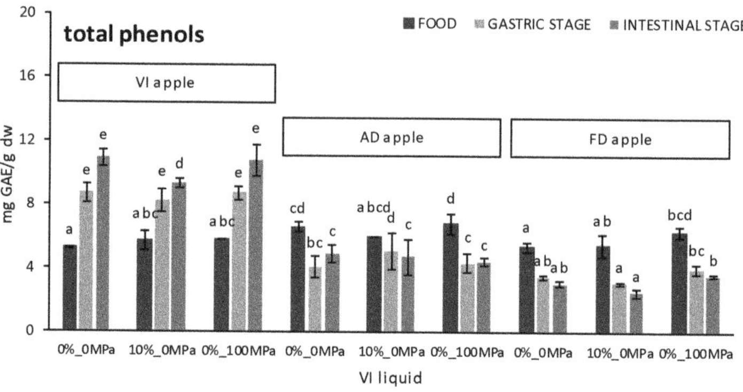

Figure 2. *Cont.*

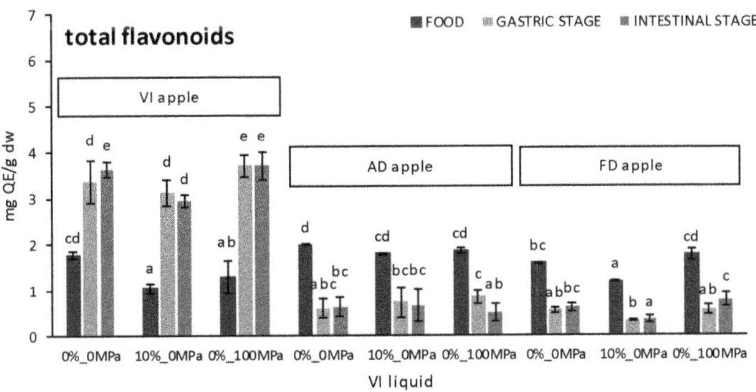

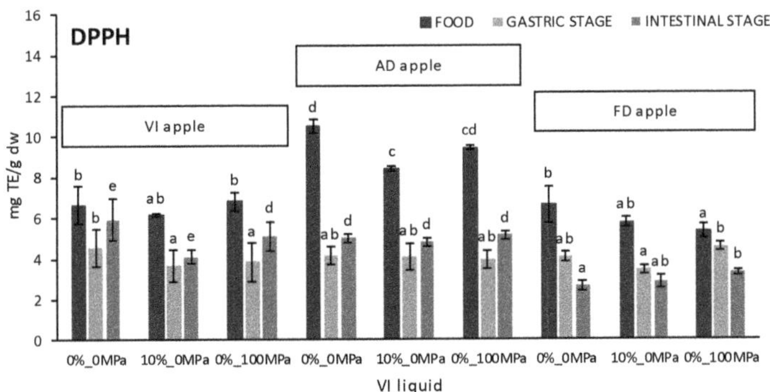

Figure 2. Antioxidant content at each stage of the gastrointestinal in vitro simulation (before digestion and after both the gastric and intestinal stages of digestion) as affected by the food matrix (VI, AD, and FD apples) and the vacuum impregnation liquid (0%_0 MPa, 10%_0 MPa, and 0%_100 MPa). Error bars represent the standard deviation of triplicates from two independent treatments. [a,b,c,d,e] different letters within the same series indicate statistically significant differences (p-value < 0.05) among samples analyzed at the same moment of the process.

As to the dried apples, all three antioxidant properties in the samples subjected to gastric conditions were significantly lower (p-value < 0.05) than in the undigested ones. A decrease in the total phenols in both AD and FD apples was of the same order (36 ± 4%) and, although in total flavonoids it was slightly higher in FD (68 ± 4%) than in AD apples (59 ± 8%), a reduction in the overall antioxidant activity was significantly higher (p-value < 0.05) in AD (57 ± 5%) than in the FD apples (38 ± 15%). Since neither the total content of phenols or flavonoids in the VI apples were negatively affected by the gastric conditions, the decrease observed in both the AD and FD apples could be attributed to the differences in the food matrix between the dry and wet samples. It follows that the release of polyphenols from the solid matrix to the gastric fluid was negatively affected by the structural changes derived from dehydration, regardless of the specific technique used. However, as aforementioned for *L. salivarius*, the leaching of other antioxidant compounds might be more effective from FD samples which are known to suffer a more extensive cell wall rupture. Of all the dehydrated samples, those impregnated with liquid 0%_100 MPa and subsequently FD showed the lowest decline in DPPH values after the gastric stage of the in vitro digestion. This would confirm the application of high-pressure processing on plant foods as a useful tool to improve the extractability and bioaccessibility of antioxidant compounds either through producing changes in the membrane permeability and the disruption of cell walls and cell organelles (as reported by Fernández-Jalao, et al. [17] for

"Golden Delicious" apples subjected to 400–600 MPa for 5 min) or through reducing the average particle size (as reported by Di Nunzio et al. [21] for mandarin juice homogenized at 20 MPa).

Following the intestinal phase of the in vitro digestion, the total phenol content of VI apples increased between 15% and 20%, depending on the composition of the vacuum impregnation liquid. However, that of AD and FD apple slices decreased by $11 \pm 3\%$, regardless of the drying technique or the vacuum impregnation liquid used. Based on these findings, it could be said that the total polyphenol release that took place during the gastric digestion of VI apples was uncompleted since vacuum impregnation allows these bioactive compounds to be better retained and protected in the intercellular spaces. This result would be in accordance with that reported by Liu et al. [37], who found that the amount of total extractable phenols released from freeze-dried apple pomace powder increased significantly (from 4.4 to 17.5 mg GAE/g) from the gastric to the jejunal phase due to alkaline hydrolysis causing the breaking of the ester bond linking phenolic acids to the cell wall. However, since Chen et al. [38] also observed that the phenolic content of red delicious apple extracts after the duodenal phase of in vitro digestion was 2.45 times higher than that obtained after the gastric phase, some other changes in polyphenolic compounds such as an interaction with other dietary components, as well as the modification of the chemical structure or solubility might happen. Interestingly, these same changes are considered responsible in other studies for the decrease in the total polyphenols observed in the mild alkaline intestinal conditions [39–41]. With regard to chlorogenic acid, the most abundant apple polyphenol, Bouayed et al. [40] reported that between 41% and 77% was degraded during the intestinal digestion of fresh apples, an amount that increased up to 100% in the case of flavanols and caffeic acid belonging to the group of hydroxycinnamic acids. In addition, isomers of chlorogenic acid, mainly cryptochlorogenic acid and neochlorogenic acid, were found to arise in the intestinal phase at concentrations almost comparable to that of residual chlorogenic acid. Nevertheless, the total polyphenolics in the intestinal medium was 40% lower than in the gastric medium. In a different study [41], the total polyphenol content of the air dried apple slices was reported to decrease from 121 ± 21 mg GAE/100 dw to 104 ± 25 mg GAE/100 dw (~14%) due to a pH change from acid (gastric digestion) to alkaline (intestinal digestion), and from 272 ± 8 mg GAE/100 dw to 188 ± 9 mg GAE/100 dw (~31%) when apple slices were enriched with grape juice by vacuum impregnation and ohmic heating at 50 °C before convective drying.

Regarding the total flavonoid content, it remained fairly stable after the intestinal phase simulation and only decreased by 7% on average, regardless of the food matrix and the vacuum impregnation liquid. Out of all the flavonoids in apples, epicatechin, procyanidin B2, and quercetin-3-O-galactoside were found to be the most degraded ones under the weak alkaline conditions of the intestinal digestion [37]. Of the three different matrices analyzed, FD apples were the ones that showed a significantly higher loss of flavonoids (p-value < 0.05), while that in VI apples was found to be the lowest. As regards the vacuum impregnation liquid, the 0%_100 MPa was the one that best prevented the degradation of the flavonoids. A similar positive effect on the bioaccessibility of flavonoids as a result of the ortanique low pulp juice homogenization at 20 MPa was previously observed by Di Nunzio et al. [21] and was explained in terms of the reduction in the particle size of the juice that facilitates the release of bioactives from the matrix.

Compared with the gastric phase of digestion, the overall antioxidant activity of both the VI and AD samples increased significantly (p-value < 0.05), but particularly in VI apples when liquid 0%_0 MPa was used as an impregnation solution, after the intestinal phase of digestion. On the contrary, the DPPH values of the FD samples decreased by $36 \pm 4\%$, $21 \pm 3\%$, and $28 \pm 3\%$ for vacuum impregnation liquids 0%_0 MPa, 0%_100 MPa, and 10%_0 MPa, respectively. An increase in the overall antioxidant activity of VI apples during the intestinal phase was consistent with the increase in the total phenol content. Likewise, a decline in the ability of FD apples to scavenge DPPH radicals matched with the decrease in both the total phenol and flavonoid content. However, the increase in the overall antioxidant activity of AD apples that took place during the intestinal phase despite the decrease in both the total phenol and flavonoid content suggests that novel compounds formed during the hot air-drying processing, such as Maillard-derived melanoidins responsible for browning during the drying process, might still be released to the soluble fraction under alkaline conditions.

All these changes affecting the antioxidant properties of the apple samples along the digestive process have a direct impact on their bioaccessibility, which is a prerequisite for their ability to be effectively absorbed from the intestinal tract into the blood circulation and delivered to the appropriate location within the body [42]. The bioaccessibility of the total phenols, the total flavonoids, and the total antioxidant activity as affected by the food matrix (VI, AD, or FD apples) and the vacuum impregnation liquid (0%_0 MPa, 0%_100 MPa, or 10%_0 MPa) are shown in Figure 3. As it can be observed, the bioaccessibility of the phenolic compounds, including flavonoids, and of all the other antioxidant compounds was significantly higher (p-value < 0.05) for VI apples. When liquids 0%_100 MPa or 10%_0 MPa were used as impregnating solutions, the total flavonoids released from VI to the liquid phase at the end of the intestinal stage increased 2.9-fold compared with their content in the undigested samples. This value was reduced to 2.0-fold for vacuum impregnation liquid 0%_0 MPa. A totally opposite effect on the bioaccessibility of both the total phenols and the antioxidants from the VI apples was observed, so that samples impregnated with liquid 0%_0 MPa showed the highest values. In any case, the bioaccesibility values obtained for the VI apples were considerably higher than expected; this could be related to the microbial activity of the lactobacillus strain. This is in line with the findings of Di Nunzio et al. [21], who reported a significant increase in narirutin and didymin bioaccessibility by adding 8 log cfu mL^{-1} of *L. salivarius* to ortanique juice homogenized at 20 MPa. As was also argued by these authors, this result can be attributed to the ability of the microorganism to enhance the release of phenolic compounds linked to fibers and other components of the food matrix. The total phenolic content, anthocyanin, and the DPPH radical scavenging capacity of Sohiong juice were also reported to increase in the presence of *L. plantarum* [43], possibly due to its β-glucosidase activity, as well as to its ability to biotransform the bioactive compounds into their metabolites or to chelate metal ions and to scavenge reactive oxygen species.

The bioaccessibility values obtained for both the AD and FD apples were of the same order: between 46.5% and 78.9% for total phenols, between 27.5% and 44.5% for total flavonoids, and between 39.7 and 61.2% for DPPH scavenging potential. In comparison, Bouayed et al. [40] found that the bioaccessibility of phenolic compounds in apples was 55%, which is in the range of the values obtained in the present study, and Gullon et al. [42] found that the bioaccessibility of polyphenolic compounds present in apple bagasse flour was 91.58%, which is much higher than the values obtained in the present study. Only significant differences (p-value < 0.05) were found between AD and FD apples for the bioaccessibility of the total phenols and the total antioxidant values and when liquid 10%_0 MPa was used as the impregnating solution, with AD apples showing higher values than FD ones. Regarding the vacuum impregnation liquid, it hardly affected the bioaccessibility of both the total phenols and flavonoids, but it did affect that of the total antioxidants. In general terms, both the addition of trehalose and the homogenization at 100 MPa increased the bioaccessibility of the total antioxidants in FD and AD apples.

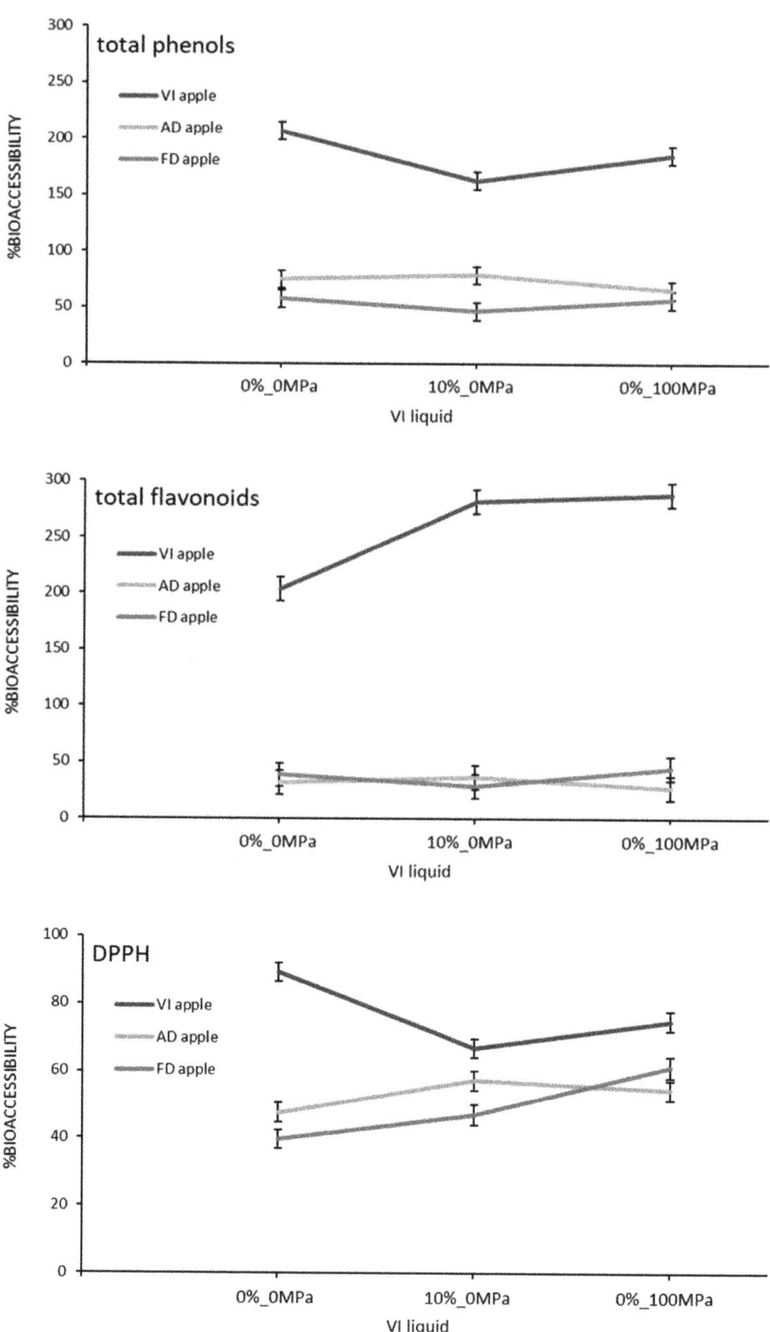

Figure 3. The bioaccessibility of antioxidants as affected by the food matrix (VI, AD, and FD apples) and the vacuum impregnation liquid (0%_0 MPa, 10%_0 MPa, and 0%_100 MPa). Mean values and LSD intervals with a 95% confident level.

4. Conclusions

The study shows that both the survival of *L. salivarius* and the bioaccessibility antioxidants during in vitro digestion were affected by the food matrix of which they are a part. The greatest viability was found when the microorganism was incorporated into the apple's porous structure by means of the vacuum impregnation technique, but it decreased significantly with the application of a dehydration step. The bioaccessibility of antioxidants also reached the highest values in vacuum impregnated samples, in which both the total phenol and total flavonoid release to the intestinal liquid phase doubled that which was present in the undigested food. The addition of trehalose and the homogenization at 100 MPa increased the bioaccessibility of the total antioxidants in FD and AD apples and the bioaccessibility of the total flavonoids in the VI samples. Homogenizing at 100 MPa also increased the survival of *L. salivarius* during in vitro digestion in the FD samples.

Author Contributions: Conceptualization, N.B., L.S. and C.B.; methodology and validation, N.B., L.S., J.G.-H., M.H. and C.B.; investigation, C.G.B.-B.; resources, N.B., L.S. and C.B.; data curation, C.G.B.-B.; writing-original draft preparation, C.G.B.-B., N.B. and C.B.; writing-review and editing, C.G.B.-B., N.B., L.S., J.G.-H., M.H. and C.B.; visualization, N.B., L.S. and C.B.; supervision, C.B.; project administration, C.B. All authors have read and agreed to the published version of the manuscript.

Funding: This research was funded by Generalitat Valenciana, project reference GV/2015/066 entitled "Mejora de la calidad funcional de un snack con efecto probiótico y antioxidante mediante la incorporación de trehalosa y la aplicación de altas presiones de homogeneización".

Data Availability Statement: The data presented in this study are available on request from the corresponding author.

Acknowledgments: The authors are grateful to the functional food lab team and colleagues in the Institute of Food Engineering for Development of the Universitat Politècnica de València (Spain). A special thanks goes to Guillermo Gaspar for his help in carrying out the experiments.

Conflicts of Interest: The authors declare no conflict of interest. The funders had no role in the design of the study; in the collection, analyses, or interpretation of data; in the writing of the manuscript, or in the decision to publish the results.

References

1. Janssen, M.; Chang, B.P.I.; Hristov, H.; Pravst, I.; Profeta, A.; Millard, J. Changes in Food Consumption During the COVID-19 Pandemic: Analysis of Consumer Survey Data from the First Lockdown Period in Denmark, Germany, and Slovenia. *Front. Nutr.* **2021**, *8*, 635859. [CrossRef] [PubMed]
2. Fortune Business Insights. Functional Food and Beverage Market Size, Share & COVID-19 Impact Analysis by Type (Functional Cereals & Grains, Functional Dairy Products, Functional Bakery Products, Functional Fats & Oils, an Other Functional/Fortidfied Foods), Distribution Channel (Supermarkets/Hypermarkets, Convenience Stores, Online Retail, an Others) and Regional Forecast 2021–2028. In *Market Research Report*; Fortune Business Insights: Pune, India, 2021. Available online: https://www. fortunebusinessinsights.com/functional-foods-market-102269 (accessed on 1 June 2021).
3. Aponte, M.; Murru, N.; Shoukat, M. Therapeutic, Prophylactic, and Functional Use of Probiotics: A Current Perspective. *Front. Microbiol.* **2020**, *11*, 2120–2136. [CrossRef]
4. Sharma, A. Importance of Probiotics in Cancer Prevention and Treatment. In *Recent Developments in Applied Microbiology and Biochemistry*; Buddolla, V., Ed.; Academic Press: Cambridge, MA, USA, 2019; Chapter 4; pp. 33–45. [CrossRef]
5. Ashaolu, T.J. Immune boosting functional foods and their mechanisms: A critical evaluation of probiotics and prebiotics. *Biomed. Pharmacother.* **2020**, *130*, 110625. [CrossRef]
6. Hajavi, J.; Esmaeili, S.-A.; Varasteh, A.; Vazini, H.; Atabati, H.; Mardani, F.; Momtazi-Borojeni, A.A.; Hashemi, M.; Sankian, M.; Sahebkar, A. The immunomodulatory role of probiotics in allergy therapy. *J. Cell. Physiol.* **2019**, *234*, 2386–2398. [CrossRef]
7. Terpou, A.; Papadaki, A.; Lappa, I.K.; Kachrimanidou, V.; Bosnea, L.A.; Kopsahelis, N. Probiotics in Food Systems: Significance and Emerging Strategies Towards Improved Viability and Delivery of Enhanced Beneficial Value. *Nutrients* **2019**, *11*, 1591. [CrossRef]
8. Boricha, A.A.; Shekh, S.L.; Pithva, S.P.; Ambalam, P.S.; Vyas, B.R.M. In vitro evaluation of probiotic properties of Lactobacillus species of food and human origin. *LWT* **2019**, *106*, 201–208. [CrossRef]
9. Terpou, A.; Papadaki, A.; Bosnea, L.; Kenallaki, M.; Kopsahelis, N. Novel frozen yogurt production fortified with sea buckthorn berries and probiotics. *LWT* **2019**, *105*, 242–249. [CrossRef]

10. Gomand, F.; Borges, F.; Burgain, J.; Guerin, J.; Revol-Junelles, A.-M.; Gaiani, C. Food Matrix Design for Effective Lactic Acid Bacteria Delivery. *Annu. Rev. Food Sci. Technol.* **2019**, *10*, 285–310. [CrossRef] [PubMed]
11. Betoret, N.; Puente, L.; Díaz, M.; Pagán, M.; García, M.; Gras, M.; Martínez-Monzó, J.; Fito, P. Development of probiotic-enriched dried fruits by vacuum impregnation. *J. Food Eng.* **2003**, *56*, 273–277. [CrossRef]
12. Akman, P.K.; Uysal, E.; Ozkaya, G.U.; Tornuk, F.; Durak, M.Z. Development of probiotic carrier dried apples for consumption as snack food with the impregnation of Lactobacillus paracasei. *LWT* **2019**, *103*, 60–68. [CrossRef]
13. Koutsos, A.; Lovegrove, J.A. An Apple a Day Keeps the Doctor Away—Inter-Relationship Between Apple Consumption, the Gut Microbiota and Cardiometabolic Disease Risk Reduction. In *Diet-Microbe Interactions in the Gut: Effects on Human Health and Disease*; Tuohy, K., Del Rio, D., Eds.; Academic Press: Cambridge, MA, USA, 2015; Chapter 12, pp. 173–194. [CrossRef]
14. Valerio, F.; Volpe, M.G.; Santagata, G.; Boscaino, F.; Barbarisi, C.; Di Biase, M.; Bavaro, A.R.; Lonigro, S.L.; Lavermicocca, P. The viability of probiotic Lactobacillus paracasei IMPC2.1 coating on apple slices during dehydration and simulated gastro-intestinal digestion. *Food Biosci.* **2020**, *34*, 100533. [CrossRef]
15. Serra, A.T.; Rocha, J.; Sepodes, B.; Matias, A.A.; Feliciano, R.P.; De Carvalho, A.; Bronze, M.; Duarte, C.; Figueira, M.E. Evaluation of cardiovascular protective effect of different apple varieties—Correlation of response with composition. *Food Chem.* **2012**, *135*, 2378–2386. [CrossRef]
16. Zhang, H.; Tsao, R. Dietary polyphenols, oxidative stress and antioxidant and anti-inflammatory effects. *Curr. Opin. Food Sci.* **2016**, *8*, 33–42. [CrossRef]
17. Fernández-Jalao, I.; Balderas, C.; Sánchez-Moreno, C.; De Ancos, B. Impact of an in vitro dynamic gastrointestinal digestion on phenolic compounds and antioxidant capacity of apple treated by high-pressure processing. *Innov. Food Sci. Emerg. Technol.* **2020**, *66*, 102486. [CrossRef]
18. García-Hernández, J.; Hernández-Pérez, M.; Peinado, I.; Andrés, A.; Heredia, A. Tomato-antioxidants enhance viability of L. reuteri under gastrointestinal conditions while the probiotic negatively affects bioaccessibility of lycopene and phenols. *J. Funct. Foods* **2018**, *43*, 1–7. [CrossRef]
19. Patrignani, F.; Vannini, L.; Kamdem, S.L.S.; Lanciotti, R.; Guerzoni, M.E. Effect of high pressure homogenization on Saccharomyces cerevisiae inactivation and physico-chemical features in apricot and carrot juices. *Int. J. Food Microbiol.* **2009**, *136*, 26–31. [CrossRef]
20. Burns, P.G.; Patrignani, F.; Tabanelli, G.; Vinderola, G.C.; Siroli, L.; Reinheimer, J.A.; Gardini, F.; Lanciotti, R. Potential of high pressure homogenisation on probiotic Caciotta cheese quality and functionality. *J. Funct. Foods* **2015**, *13*, 126–136. [CrossRef]
21. Di Nunzio, M.; Betoret, E.; Taccari, A.; Dalla Rosa, M.; Bordoni, A. Impact of processing on the nutritional and functional value of mandarin juice. *J. Sci. Food Agric.* **2020**, *100*, 4558–4564. [CrossRef]
22. Oikonomopoulou, V.P.; Krokida, M.K. Novel Aspects of Formation of Food Structure during Drying. *Dry. Technol.* **2013**, *31*, 990–1007. [CrossRef]
23. Betoret, E.; Betoret, N.; Arilla, A.; Bennár, M.; Barrera, C.; Codoñer, P.; Fito, P. No invasive methodology to produce a probiotic low humid apple snack with potential effect against Helicobacter pylori. *J. Food Eng.* **2012**, *110*, 289–293. [CrossRef]
24. Betoret, E.; Betoret, N.; Calabuig-Jiménez, L.; Patrignani, F.; Barrera, C.; Lanciotti, R.; Dalla Rosa, M. Probiotic survival and in vitro digestion of L. salivarius spp. salivarius encapsulated by high homogenization pressures and incorporated into a fruit matrix. *LWT* **2019**, *111*, 883–888. [CrossRef]
25. Betoret, E.; Betoret, N.; Calabuig-Jiménez, L.; Barrera, C.; Dalla Rosa, M. Effect of Drying Process, Encapsulation, and Storage on the Survival Rates and Gastrointestinal Resistance of L. salivarius spp. salivarius Included into a Fruit Matrix. *Microorganisms* **2020**, *8*, 654. [CrossRef]
26. Burca-Busaga, C.G.; Betoret, N.; Seguí, L.; Betoret, E.; Barrera, C. Survival of *Lactobacillus salivarius* CECT 4063 and Stability of Antioxidant Compounds in Dried Apple Snacks as Affected by the Water Activity, the Addition of Trehalose and High Pressure Homogenization. *Microorganisms* **2020**, *8*, 1095. [CrossRef]
27. Fito, P.; Chiralt, A.; Betoret, N.; Gras, M.; Cháfer, M.; Martínez-Monzó, J.; Andrés, A.; Vidal, D. Vacuum impregnation and osmotic dehydration in matrix engineering: Application in functional fresh food development. *J. Food Eng.* **2001**, *49*, 175–183. [CrossRef]
28. Barrera, C.; Burca, C.; Betoret, E.; García-Hernandez, J.; Hernández, M.; Betoret, N. Improving antioxidant properties and probiotic effect of clementine juice inoculated with *Lactobacillus salivarius* spp. salivarius (CECT 4063) by trehalose addition and/or sublethal homogenisation. *Int. J. Food Sci. Technol.* **2019**, *54*, 2109–2122. [CrossRef]
29. Lewicki, P.P.; Wiczkowska, J. Rehydration of Apple Dried by Different Methods. *Int. J. Food Prop.* **2007**, *9*, 217–226. [CrossRef]
30. Betoret, E.; Betoret, N.; Castagnini, J.M.; Rocculi, P.; Rosa, M.D.; Fito, P. Analysis by non-linear irreversible thermodynamics of compositional and structural changes occurred during air drying of vacuum impregnated apple (cv. Granny smith): Calcium and trehalose effects. *J. Food Eng.* **2015**, *147*, 95–101. [CrossRef]
31. Zhu, W.; Lyu, F.; Naumovski, N.; Ajlouni, S.; Ranadheera, C.S. Functional Efficacy of Probiotic Lactobacillus sanfranciscensis in Apple, Orange and Tomato Juices with Special Reference to Storage Stability and In Vitro Gastrointestinal Survival. *Beverages* **2020**, *6*, 13. [CrossRef]
32. Panpetch, W.; Spinler, J.K.; Versalovic, J.; Tumwasorn, S. Characterization of *Lactobacillus salivarius* strains B37 and B60 capable of inhibiting IL-8 production in Helicobacter pylori-stimulated gastric epithelial cells. *BMC Microbiol.* **2016**, *16*, 242. [CrossRef] [PubMed]
33. Gong, P.; Lin, K.; Zhang, J.; Han, X.; Lyu, L.; Yi, H.; Sun, J.; Zhang, L. Enhancing spray drying tolerance of Lactobacillus bulgaricus by intracellular trehalose delivery via electroporation. *Food Res. Int.* **2020**, *127*, 108725. [CrossRef]

34. Nayak, P.K.; Chandrasekar, C.M.; Sundarsingh, A.; Kesavan, R.K. Effect of *in-vitro* digestion on the bio active compounds and biological activities of fruit pomaces. *J. Food Sci. Technol.* **2020**, *57*, 4707–4715. [CrossRef]

35. Rodríguez-Roque, M.J.; Rojas-Graü, M.A.; Elez-Martinez, P.; Martín-Belloso, O. Changes in Vitamin C, Phenolic, and Carotenoid Profiles Throughout In Vitro Gastrointestinal Digestion of a Blended Fruit Juice. *J. Agric. Food Chem.* **2013**, *61*, 1859–1867. [CrossRef]

36. Petry, F.C.; Mercadante, A.Z. Impact of in vitro digestion phases on the stability and bioaccessibility of carotenoids and their esters in mandarin pulps. *Food Funct.* **2017**, *8*, 3951–3963. [CrossRef]

37. Liu, G.; Ying, D.; Guo, B.; Cheng, L.J.; May, B.; Bird, T.; Sanguansri, L.; Cao, Y.; Augustin, M. Extrusion of apple pomace increases antioxidant activity upon in vitro digestion. *Food Funct.* **2019**, *10*, 951–963. [CrossRef]

38. Chen, G.-L.; Chen, S.-G.; Zhao, Y.-Y.; Luo, C.-X.; Li, J.; Gao, Y.-Q. Total phenolic contents of 33 fruits and their antioxidant capacities before and after in vitro digestion. *Ind. Crop. Prod.* **2014**, *57*, 150–157. [CrossRef]

39. Bermúdez-Soto, M.J.; Tomás-Barberán, F.-A.; García-Conesa, M.T. Stability of polyphenols in chokeberry (Aronia melanocarpa) subjected to in vitro gastric and pancreatic digestion. *Food Chem.* **2007**, *102*, 865–874. [CrossRef]

40. Bouayed, J.; Deußer, H.; Hoffmann, L.; Bohn, T. Bioaccessible and dialysable polyphenols in selected apple varieties following in vitro digestion vs. their native patterns. *Food Chem.* **2012**, *131*, 1466–1472. [CrossRef]

41. Pavez-Guajardo, C.; Ferreira, S.R.S.; Mazzutti, S.; Guerra-Valle, M.E.; Sáez-Trautmann, G.; Moreno, J. Influence of In Vitro Digestion on Antioxidant Activity of Enriched Apple Snacks with Grape Juice. *Foods* **2020**, *9*, 1681. [CrossRef]

42. Gullon, B.; Pintado, M.E.; Barber, X.; Fernández-López, J.; Pérez-Alvarez, J.A.; Viuda-Martos, M. Bioaccessibility, changes in the antioxidant potential and colonic fermentation of date pits and apple bagasse flours obtained from co-products during simulated in vitro gastrointestinal digestion. *Food Res. Int.* **2015**, *78*, 169–176. [CrossRef]

43. Vivek, K.; Mishra, S.; Pradhan, R.C.; Jayabalan, R. Effect of probiotification with Lactobacillus plantarum MCC 2974 on quality of Sohiong juice. *LWT Food Sci. Technol.* **2019**, *108*, 55–60. [CrossRef]

Article

Dynamic In Vitro Gastric Digestion of Sheep Milk: Influence of Homogenization and Heat Treatment

Zheng Pan [1], Aiqian Ye [1,*], Siqi Li [1], Anant Dave [1], Karl Fraser [1,2] and Harjinder Singh [1]

1   Riddet Institute, Massey University, Private Bag 11 222, Palmerston North 4442, New Zealand; Z.Pan@massey.ac.nz (Z.P.); S.Li2@massey.ac.nz (S.L.); A.Dave@massey.ac.nz (A.D.); karl.fraser@agresearch.co.nz (K.F.); H.Singh@massey.ac.nz (H.S.)
2   AgResearch, Private Bag 11 008, Palmerston North 4442, New Zealand
*   Correspondence: A.M.Ye@massey.ac.nz

Abstract: Milk is commonly exposed to processing including homogenization and thermal treatment before consumption, and this processing could have an impact on its digestion behavior in the stomach. In this study, we investigated the in vitro gastric digestion behavior of differently processed sheep milks. The samples were raw, pasteurized (75 °C/15 s), homogenized (200/20 bar at 65 °C)–pasteurized, and homogenized–heated (95 °C/5 min) milks. The digestion was performed using a dynamic in vitro gastric digestion system, the human gastric simulator with simulated gastric fluid without gastric lipase. The pH, structure, and composition of the milks in the stomach and the emptied digesta, and the rate of protein hydrolysis were examined. Curds formed from homogenized and heated milk had much looser and more fragmented structures than those formed from unhomogenized milk; this accelerated the curd breakdown, protein digestion and promoted the release of protein, fat, and calcium from the curds into the digesta. Coalescence and flocculation of fat globules were observed during gastric digestion, and most of the fat globules were incorporated into the emptied protein/peptide particles in the homogenized milks. The study provides a better understanding of the gastric emptying and digestion of processed sheep milk under in vitro gastric conditions.

Keywords: sheep milk; protein; fat; pepsin; homogenization; heat treatment; protein coagulation; structure; gastric digestion

Citation: Pan, Z.; Ye, A.; Li, S.; Dave, A.; Fraser, K.; Singh, H. Dynamic In Vitro Gastric Digestion of Sheep Milk: Influence of Homogenization and Heat Treatment. *Foods* **2021**, *10*, 1938. https://doi.org/10.3390/foods10081938

Academic Editor: Tanja Cirkovic Velickovic

Received: 21 July 2021
Accepted: 17 August 2021
Published: 20 August 2021

Publisher's Note: MDPI stays neutral with regard to jurisdictional claims in published maps and institutional affiliations.

1. Introduction

Sheep milk is of high nutritional value and has potential for the development of nutritional and functional milk products, attracting a growing number of consumers worldwide [1]. Milk, as an important source of protein for humans, has been widely examined for its digestion behavior in both in vivo and in vitro studies [2–4]. The digestion of cow milk has been investigated extensively, whereas the digestion of noncow milk (i.e., sheep milk) is less studied.

Sheep milk and cow milk vary significantly in composition, physicochemical properties, and structures, which may potentially lead to different digestion behaviors within the gastrointestinal tract and the bioavailability of nutrients [1]. Jasińska [5], conducted a study to examine the hydrolysis of the casein micelles in the raw milks from 4 species (human, goat, mare, and two breeds of cow) and showed that the degrees of hydrolysis of the caseins by pepsin were 80%, 65%, 45%, 42%, and 23% for human, goat, mare, black and white cow, and red polish cow milk, respectively. Jasińska [5], attributed the differences in casein hydrolysis in the milks from different species to the different physicochemical properties and compositions of the caseins such as micellar structure and different levels of β-casein. Previous studies have also shown that the different compositions of milk proteins can result in different digestion behaviors [6–8]. For instance, goat milk has lower α_{s1}-casein content and higher β-casein content than cow milk, and infant formulas made from goat milk formed smaller flocs of aggregated proteins and fat globules during in vitro gastric digestion, resulting in faster protein digestion in the infant formula made with goat milk than in that made with cow milk [7,9]. Sheep milk has markedly higher levels of β- and α_{s2}-casein but lower levels of α_{s1}-casein than cow milk, which may potentially affect its coagulation behavior and protein hydrolysis in the stomach [10]. Previous research comparing the in vitro gastric digestions of cow, goat, and sheep milks found that the curds formed from sheep skim milk had higher total solids and lower moisture contents than those formed from cow and goat skim milks because of their different chemical compositions, resulting in a firmer curd from the sheep skim milk [6].

Milk is commonly exposed to different processing treatments (i.e., pasteurization and homogenization), which leads to structural changes in its components (i.e., protein and fat). For instance, the heat treatment of milk could result in a series of protein–protein and protein–lipid interactions and changes, depending on the heat intensity level [11–14]. The homogenization of milk increases the stability of the milk fat globules because of a decrease in fat globule size and the adsorption of caseins and whey proteins onto the surface of the newly formed milk fat globules [15]. Additionally, homogenization coupled with the heat treatment of milk increases the association of denatured whey proteins with the adsorbed caseins and milk fat globule membrane (MFGM) proteins via disulfide bonds, leading to alteration of the interfacial composition of the fat globules [16]. In turn, these changes in the milk components could have an impact on the digestion behavior of milk within the gastrointestinal tract.

Roy et al. [17] investigated the effect of pasteurization on the in vitro gastric digestion of milks from cow, goat, and sheep and found that all pasteurized milks formed less integrated curds than their raw milk counterparts, resulting in a greater extent of deformation and thus higher levels of fat release into the liquid phase. However, the effect of homogenization and intensive heat treatment on sheep milk has not been investigated. There has been extensive research on the in vitro digestion of cow milk treated with more intense heat treatment. The curds formed from intensively heated cow milk were more fragmented and crumbly compared with the more cohesive curds formed from unheated or pasteurized milk, which was attributed to the differences in the structural changes in the milk components that were induced by the different processing treatments [4,18]. The content and the structure of the curds formed from homogenized cow milk during gastric digestion also showed differences compared with those formed from raw cow milk. Ye et al. [19] reported that homogenized milk formed an integrated curd but with a more porous structure than that formed from untreated whole milk in the early stage of in vitro gastric digestion, and that the curd became less integrated and was separated into several small pieces at longer digestion times. Milks treated with a combination of homogenization and heat treatment were digested more effectively than those treated with either heat treatment or homogenization alone [4,19]. For example, Ye et al. [19] investigated the effects of homogenization and heat treatment on the formation of curds during the in vitro gastric digestion of whole cow milk and observed that homogenization of the milk followed by heat treatment resulted in the formation of curds with more fragmented and crumbly structures than those formed from raw and singly homogenized whole milk. These differences in the digestion behaviors of differently processed milks suggest that they are likely to have different physiological effects (i.e., level of satiety and secretion of cholecystokinin) within the gastrointestinal tract [20]. Therefore, the effect of homogenization and intensive heat treatment on the digestion behavior of sheep milk was investigated. Egger et al. [21,22], investigated the protein hydrolysis in cow skim milk powder using a simple in vitro static digestion model and in vivo pig digestion model and found that the protein patterns at the endpoints of gastric digestion were similar in both models. The comparison of the gastric digestion behaviors of pasteurized and UHT homogenized cow milks using a rat model and an in vitro dynamic human gastric simulator was assessed by Ye et al. [18]. The formation of curds in the stomach was found in both in vivo and in vitro and the protein digestion was following a similar trend in both models [18]. It suggests that the digestion behavior of milk observed in in vitro dynamic digestion model could generate a good approximation to the in vivo results.

In this study, both untreated and processed (homogenization at 200/20 bar and 65 °C, pasteurization at 75 °C for 15 s, and heat treatment at 90 °C for 5 min) sheep milks were digested using a dynamic in vitro gastric digestion model (a human gastric simulator, HGS) to investigate the effects of homogenization, heat treatment, and the combination of homogenization and heat treatment on their gastric digestion behaviors, including coagulation, curd structure, protein digestion, and the release of protein and fat from the stomach.

2. Materials and Methods

2.1. Milk Supply and Processing Treatments

Fresh sheep milk was collected from Spring Sheep Milk Co. (Auckland, New Zealand) and Maui Milk Co., Ltd., Waikato, New Zealand, during mid-lactation; the milks collected from the two companies were mixed at a ratio of 1:1. Pasteurization of the sheep milk was carried out at 75 °C for 15 s in a pilot-scale indirect UHT plant (Alfa-Laval, Huntingwood, NSW, Australia). The homogenized milk was obtained by homogenizing raw sheep milk at 200/50 bar and 65 °C in a 2-stage valve homogenizer in the Massey University pilot plant. In the experiments, the homogenized sheep milk was pasteurized at 75 °C for 15 s to make homogenized–pasteurized (homo–past) sheep milk; the homogenized and heated (homo–heat) sheep milk was obtained by heating to reach 95 °C in the UHT plant and then transferred to a water bath for holding for 5 min. The parameters for pasteurization selected for processing of sheep milk were based on the codes of practice documents of New Zealand Food Safety Authority [23]; the parameters for homogenization were commonly used in industrial processing of milk; the heat treatment at 95°C for 5 min is commonly used in the processing of yogurt [24]. After heat treatment, these milk samples were immediately cooled to 20 °C. The average fat globule sizes (d_{43}) of the milk samples, which were determined using a Mastersizer 2000 (Malvern Instruments Ltd., Malvern, UK), were 4.52 ± 0.14 μm and 0.62 ± 0.07 μm for the unhomogenized and homogenized sheep milks, respectively.

2.2. Chemicals for In Vitro Gastric Digestion

Pepsin from porcine gastric mucosa (EC 3.4.23.1; Catalogue No. P7000, Sigma Chemical Co., St. Louis, MO, USA) had an enzymatic activity of 550 units/mg solid, as tested in preliminary experiments. All other chemicals were of analytical grade and were purchased from BDH Chemicals (BDH Ltd., Poole, UK) and Sigma Chemical Co. (St. Louis, MO, USA) unless otherwise specified. All solutions were prepared using Milli-Q water purified by treatment with a Milli-Q apparatus (Millipore Corp., Bedford, MA, USA).

Simulated salivary fluid (SSF) and simulated gastric fluid (SGF) were prepared at 1.25× concentration according to the method described by Brodkorb et al. [25] with slight modifications. SSF was prepared based on the salt composition only, as described in Brodkorb et al. [25], because milk contains no starch. The SGF (pH 1.5) did not include gastric lipase because this study focused on the formation of protein curd and protein digestion induced by pepsin. $CaCl_2$ ($H_2O)_2$ was added into the SSF and the SGF immediately before the digestion experiment to achieve final concentrations of 1.5 mM and 0.15 mM, respectively. Furthermore, the SSF and SGF were supplemented with water to achieve a 1× concentration before addition into the stomach chamber.

2.3. In Vitro Gastric Digestion

A dynamic in vitro gastric model was used to mimic the dynamic gastric digestive process. The HGS developed by Kong and Singh [26] was employed for the in vitro gastric digestion of the sheep milks. The method described in Ye et al. [19] was used for the gastric digestion with slight modifications. A 200 g sheep milk sample pre-warmed at 37 °C was firstly mixed with an amount of SSF that equaled the total solids content of the milk samples (i.e., 18.5 g of solids in the sheep milk to 18.5 g of SSF), and then transferred and warmed in the HGS at 37 °C for 2 min; a 20 mL fasting solution containing SGF (16 mL) and pepsin solution (4 mL, 10,000 units/mL, prepared in water) was added into the mixture of milk sample and SSF [27,28]; the SGF and the pepsin solution (10,000 units/mL) were then added using two separate pumps at flow rates of 2.4 mL/min and 0.6 mL/min, respectively, to achieve a 1× concentration of SGF and a pepsin activity of 2000 units/mL; the gastric emptying rate was 3.6 mL/min; the emptied digesta were removed from the bottom of the stomach chamber at 20-min intervals for accurate control of the gastric emptying. To mimic the contraction and temperature of the stomach, the contraction frequency of the HGS was set at 3 times/min and the temperature was maintained at 37 °C using a heater and a thermostat. The gastric digestion time was up to 240 min, but most of the experiments were stopped at different times to collect the coagulated milk curds for further analysis. The digesta removed from the HGS at each time interval were filtered through a mesh with a pore size diameter of 1 mm for further analysis, and the solid mass with size greater than 1 mm was put back into the HGS for further digestion. The processing and the digestion experiments were triplicated with 3 different batches of raw sheep milk.

2.4. pH Measurement

The pH of the mixture of milk and SSF was defined as the initial pH in the HGS. The pHs of the emptied digesta at different digestion time points refer to the pH in the HGS as the simulated gastric contraction prevented easy access into the gastric chamber for direct determination using the pH meter.

2.5. Weight of Curds

The content within the HGS at different digestion time points was collected and filtered through a sieve mesh with a 1-mm pore size to separate the curd and the aqueous phase. Curd larger than 1 mm was then immediately rinsed with pepsin-free SGF and weighed to obtain the weight of the curds. Subsequently, these curds were heated at 90 °C for 5 min to inactivate the pepsin and then dried at 105 °C for 24 h in a vacuum oven, so that the weight of total solids of the curd could be determined. The measurements of the curd weight were duplicated with 2 different batches of sheep milk.

2.6. Calcium, Protein, and Fat Content Analysis

The total calcium content of the dried curds was determined by inductively coupled plasma optical emission spectroscopy after the curd powder had been dissolved and digested by nitric and hydrochloric acids. The protein and fat contents of the curds and digesta obtained at different digestion times from different milk samples were determined using the Dumas method (AOAC 968.06) and the Mojonnier method (AACC 30-10) [29]. The protein content was calculated using a conversion factor of 6.25 for multiplication of the nitrogen content of the curds and digesta. The analyses of calcium, protein, and fat content of curds were duplicated with 2 different batches of sheep milk.

2.7. Protein Hydrolysis

The protein compositions of the curds and the emptied digesta were determined using tricine sodium dodecyl sulfate-polyacrylamide gel electrophoresis (tricine SDS-PAGE). The sample buffer (containing 40% glycerol, 2% SDS, 0.04% Coomassie Brilliant Blue G-250, 5% β-mercaptoethanol, and 20% 0.2 M Tris-HCl, pH 6.8) was mixed with liquid digesta to achieve a protein concentration of 1 mg/mL. For the solid curd samples, lyophilized and ground curd powder was dissolved in 1 mL of sample buffer to achieve a protein concentration of 1 mg/mL. These mixtures were then heated at 90 °C for 5 min, and 7 μL of these heated mixtures was loaded in the wells of an electrophoresis gel that was pre-made using a Mini-PROTEAN electrophoresis system. The electrophoresis gel was composed of resolving gel (16% acrylamide, glycerol, 3 M Tris-HCl buffer, pH 8.45) and 4% stacking gel (4% acrylamide, 3 M Tris-HCl buffer, pH 6.8). Two running buffers were employed to separate low-molecular-mass proteins/peptides with high resolution: an anode buffer (0.2 M Tris-HCl, pH 8.9) and a cathode buffer (0.1 M Tris base, 0.1 M tricine, and 0.1% SDS, pH 8.25). After running at a constant voltage of 120 V for around 2 h, the gel was fixed using 5% glutaraldehyde with constant gentle shaking for 25 min. The fixed gel was stained for 20 min using 0.03% Coomassie Brilliant Blue G-250 solution (0.3 g of Coomassie Brilliant Blue G-250 in 1 L of 10% glacial acetic acid solution) and then destained with destaining solution (10% glacial acetic acid solution) for 2 h to remove excess dye. The protein patterns of the gels were imaged and analyzed using a Molecular Dynamics Model PD-SI computing densitometer (Molecular Dynamics Inc., Sunnyvale, CA, USA).

2.8. Microstructure of Curds and Digesta

A confocal laser scanning microscope (Leica Microsystems Pty Ltd., Heidelberg, Germany) was used to study the microstructure of the digesta obtained from the digestion of the milks. The curd and digesta samples were stained and observed immediately without inactivating the pepsin after collection, according to the method described by Wang et al. [30]. The fluorescent dyes Nile Red and Fast Green were used to stain the oil (argon laser with an excitation line at 488 nm) and the protein (He–Ne laser with excitation line at 633 nm), respectively. A 200 μL aliquot of digesta was transferred into an Eppendorf tube and then mixed with 5 μL of 1.0% (w/v) Fast Green and 10 μL of 0.1% (w/v) Nile Red. The samples were stained for at least 5 min. For curd samples, an aliquot was taken using a blade and stained with 1.0% Fast Green and 0.1% Nile Red for at least 10 min. The stained samples were placed on concave confocal microscope slides (Sail; Sailing Medical-Lab Industries Co. Ltd., Suzhou, China), covered with a coverslip, and examined with ×40 and ×100 oil immersion lenses. The microstructures of digesta and curds were measured once at each time point with multiple fields.

2.9. Statistical Analysis

Each experiment was performed triplicates unless specified in methods using freshly prepared milk samples. Data plotting and statistical analysis were performed using GraphPad Prism 8.4.0 (GraphPad Software, San Diego, CA, USA). Statistical analysis was performed using 1-way analysis of variance and Tukey's multiple comparison test at a significance level of $p < 0.05$. Relationships between protein release and fat release from the curds were determined using nonlinear regression analysis. The results are presented as the mean ± standard deviation.

3. Results and Discussion

3.1. Gastric pH Profile

The pH profiles of the milk samples in the HGS are shown in Figure 1. The pHs of all milk samples decreased progressively, reaching a pH of between 2 and 3 at 240 min of digestion. However, the pHs at 80–200 min of digestion showed statistically significant differences ($p < 0.05$) among the milk samples, except for the pHs of the homo–past and homo–heat milks ($p > 0.05$). The homogenized milk samples (homo–past and homo–heat) had a markedly slower decrease in pH with the digestion time, reaching about pH 2.6 at 240 min of digestion, compared with the unhomogenized milk samples (pH 2.0 and pH 2.3 for the raw and pasteurized milks, respectively). There was also a difference in the pH-decrease patterns between the unheated and heated milk samples; the effect of stronger heating conditions on the pH profiles was more marked, resulting in a slower decrease in pH. These differences were probably related to the formation of curds with different structures in the differently treated milk samples, which might lead to different diffusion rates of molecules and ions from the SGF into and out of the curds.

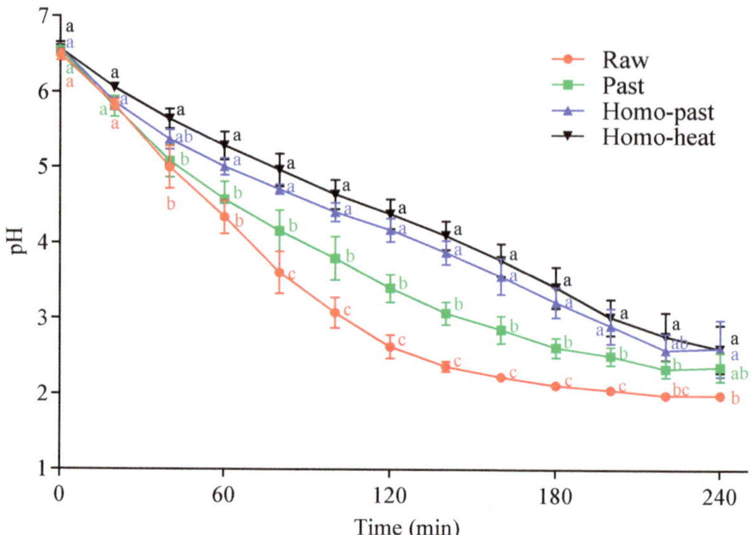

Figure 1. pH changes during the in vitro gastric digestion of differently processed sheep milks: ●, raw milk; ■, pasteurized (Past) milk; ▲, homogenized and pasteurized (Homo–past) milk; ▼, homogenized and heated (Homo–heat) milk. Different lowercase letters indicate significant difference ($p < 0.05$). Error bars represent standard deviations.

3.2. Gastric Coagulation of Sheep Milk

The appearance of the curds formed at different digestion times from differently processed sheep milks is shown in Figure 2. In all milk samples, protein coagulation was visible immediately after the addition of the milk into the SGF (20 mL of SGF with a pepsin activity of 2000 units/mL). After 60 min of digestion, a firm curd with a smooth surface was observed in both the raw milk and the pasteurized milk; at this stage, the serum phase became clear, indicating that most of the casein micelles and fat globules were incorporated into the curd. With further digestion to 180 min, the firm curds formed from the raw milk samples became smaller but remained integrated, whereas the curds formed from the pasteurized milk samples became less integrated and were broken into several small pieces with various sizes (Figure 2). However, the curds from the homo–past and homo–heat milks appeared to be more fragmented and looser than those from the raw and pasteurized milks throughout the gastric digestion. In the homo–heat sheep milk, the curd crumbles were much looser, smaller, and more evenly fragmented than those from the homo–past milk. These results indicated that homogenization and different levels of heat treatment of sheep milk could lead to the formation of differently structured curds.

Previous studies have suggested that the formation of curds in the stomach is initially driven by the enzymatic action of pepsin on κ-casein [3,31,32]. The present study showed that the coagulation of sheep milk occurred immediately after the mixing with SGF, in which the pH in the system was greater than 6. The result was in agreement with previous findings for cow milk [3,19].

Figure 2. Appearance of curds collected at 60 and 180 min during the in vitro gastric digestion of raw, pasteurized (Past), homogenized and pasteurized (Homo–past), and homogenized and heated (Homo–heat) sheep milks.

3.3. Microstructure of Curds

The structures of the milk curds formed from the raw, pasteurized, homo–past, and homo–heat sheep milks at different digestion times were determined using confocal microscopy (Figure 3). For all milk samples before digestion, the fat globules were evenly distributed in the protein aqueous phase, and the fat globule size in the homogenized milks appeared to be much smaller than that in the unhomogenized milks. At 60 min of digestion, a closely-knit network of protein matrix was observed in all milk samples, and many fat globules were also found within the protein matrix. Additionally, the size of the fat globules within the curd became larger during the digestion progress, compared with that in undigested milk samples, suggesting that aggregation and/or coalescence of the fat globules occurred because of the effect of hydrolysis by pepsin on the membrane proteins surrounding the fat globules [33,34].

The small fat globules of the homo–past and homo–heat sheep milks appeared to be well embedded within the protein matrix, whereas the fat globules of the raw and pasteurized milks seemed to be entrapped within the curd without much contact with the protein network (Figure 3). The difference between the homogenized and unhomogenized sheep milk samples was probably related to the changes in the structure of the casein micelles and the interfacial protein composition of the fat globules that were caused by homogenization and heat treatment. Homogenization of milk leads to increases in the total surface area of the fat globules and the adsorption of caseins and whey proteins [15,35]. These proteins that coat the smaller globules may interact with the protein network within the curd, leading to the structural changes in the curds during the gastric digestion [4]. Further, more intense heat treatment of the homogenized sheep milk could result in a higher level of association of denatured whey proteins with the casein micelles and with MFGM proteins, which may reduce the casein–casein interactions and casein–fat interactions and thus hinder the aggregation and coagulation of milk proteins [18]. It appears that these structural changes of the fat globules and proteins that were induced by homogenization and more severe heat treatment were responsible for the formation of the differently structured curds, as observed in Figure 2.

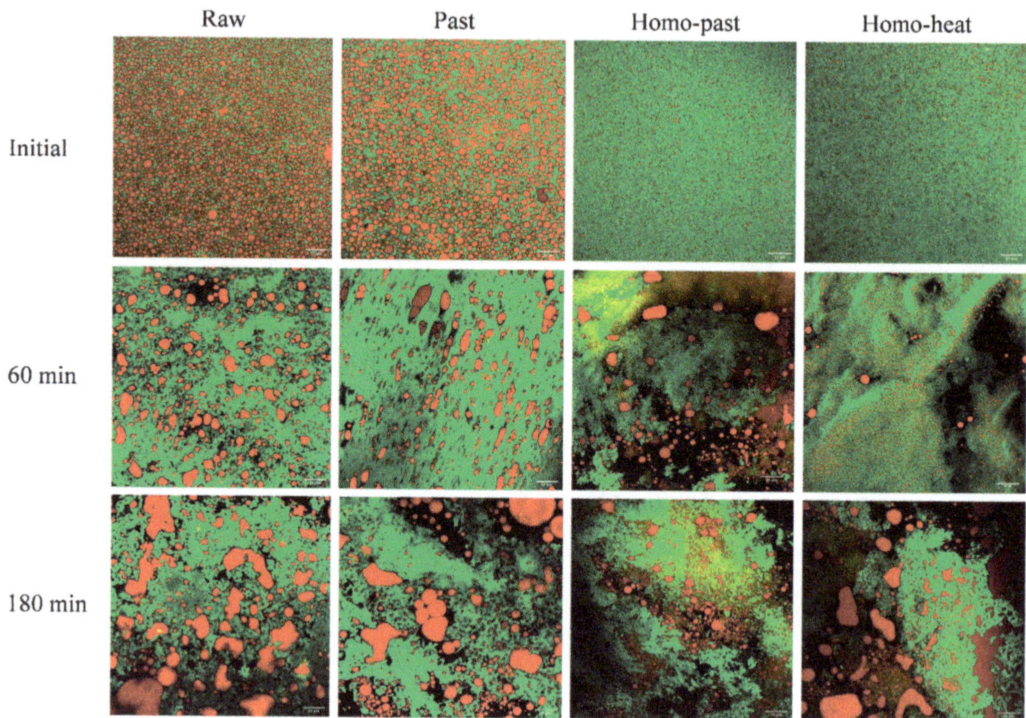

Figure 3. Confocal micrographs of curds obtained at 60 and 180 min during in vitro gastric digestion of raw, pasteurized (Past), homogenized and pasteurized (Homo–past), and homogenized and heated (Homo–heat) sheep milks. Red shows the fat and green shows the protein. The scale bar in all images is 25 μm.

3.4. Disintegration of Curds

3.4.1. Weight of Curds

The weights of the curd and the total solids of the curds in all milk samples decreased gradually throughout the gastric digestion (Figure 4A,B). The initial (20 min) weights of the curds and the total solids of the curds followed the order raw < pasteurized < homo–past < homo–heat, indicating that homogenization and more intense heating resulted in the incorporation of more milk components into the curds. This was attributed to the whey proteins that associated with the casein micelles during the heat treatment [18]. Statistical analysis showed that there were significant differences in the weights of the curds between the homogenized (homo–past and homo–heat) milks and the unhomogenized (raw and pasteurized) milks ($p < 0.01$) and between the homo–past and homo–heat milks at 20 min ($p < 0.05$), whereas there were no significant differences in the weights of the total solids of the curds ($p > 0.05$), suggesting that more moisture might be retained in the curds formed from the homogenized and heated sheep milks. This result was in agreement with previous results, which showed that curds with a looser structure contained more liquid [18,36]. Thus, a curd with higher moisture content might contain more SGF and pepsin. With further digestion, the weights of the curds and the total solids of the curds decreased much faster in the homo–past and homo–heat sheep milks than in the raw and pasteurized sheep milks. Significant differences between each time point in the weights of the curds and the total solids were observed in the homo–past and homo–heat sheep milks ($p < 0.05$) but not in the raw and pasteurized sheep milks. At 240 min of digestion, both the weight of the curds and the weight of the total solids showed a reverse order compared with their initial weights: raw > pasteurized > homo–past > homo–heat. These results suggested that pasteurization alone of sheep milk did not have much impact on the breakdown of the curd compared with the raw sheep milk; pasteurization combined with homogenization significantly accelerated the disintegration of the sheep milk curds; more intense heat treatment of the homogenized sheep milk did not further impact the breakdown of the curds compared with the homo–past sheep milk but resulted in a significantly higher weight of the curds at 20 min.

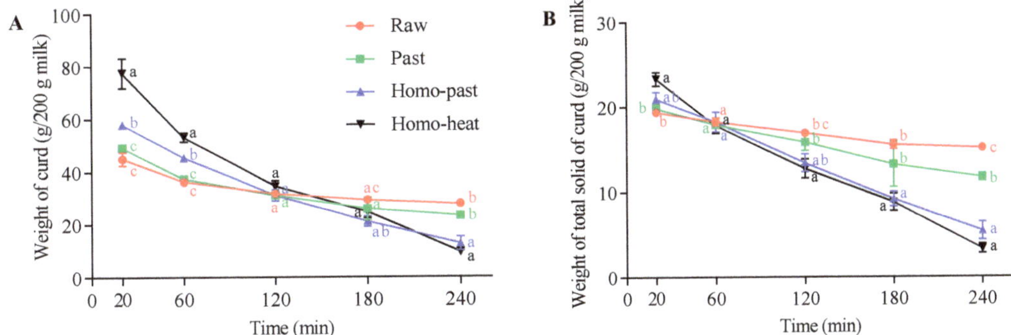

Figure 4. Changes in (**A**) the weight of the curds and (**B**) the weight of the total solids of the curds at different time points (20–240 min) during the in vitro gastric digestion of differently processed sheep milks: •, raw milk; ■, pasteurized (Past) milk; ▲, homogenized and pasteurized (Homo–past) milk; ▼, homogenized and heated (Homo–heat) milk. Different lowercase letters indicate significant difference ($p < 0.05$). Error bars represent standard deviations.

3.4.2. Changes in Protein, Fat, and Calcium Contents in the Curd

Figure 5A shows the protein content of the curds as the function of digestion time in the 4 different sheep milks. The protein contents of the raw and pasteurized milks remained nearly constant at around 43% during the digestion period, whereas those of the homo–past and homo–heat sheep milks showed a decreasing trend as the digestion progressed, especially after 60 min. For the homo–past milk, the protein content of the curds decreased in the first 120 min, after which it remained roughly unchanged at around 36% until the end of digestion. The decreasing trend for the protein content of the homo–heat milk continued to 180 min (from ~40–~30%). The fat content of the curds is shown in Figure 5B. Similar to the protein content, there were no obvious changes in the fat content of the curds of the unhomogenized milks throughout the gastric digestion, with a fat content of ~45% for the raw milk and of ~46% for the pasteurized milk. However, the fat contents of the curds of the homogenized milks showed an opposite trend compared with their protein counterparts. The fat contents of the curds increased for 120 min and 180 min of digestion for the homo–past milk (from 47.8–55.6%) and the homo–heat milk (from 43–63.8%), respectively, after which the level remained almost unchanged.

The relationships between the amounts of protein and fat in the curds are shown in Figure 5C. The slope of the regression line for the raw and pasteurized sheep milks was close to 1, indicating a strong correlation between fat and protein in the curds at different time points. Similar results have been reported for the gastric digestion of unheated and heated cow milks [36]; the slopes of the regression lines for the fat–protein profiles of the curds formed from these milk samples were close to 1. This indicates that the fat was evenly distributed in the curds and was released from the curds in equal proportion to the protein. However, a nonlinear correlation between the fat and protein in the curds was found for the homo–past and homo–heat sheep milks, suggesting that fat and protein had different rates of release from the curd. The amounts of fat and protein in the homo–past and homo–heat sheep milks were significantly lower than those in the raw and pasteurized sheep milks at 240 min of digestion (the leftmost points in Figure 5C) ($p < 0.05$), indicating that the release of fat and protein occurred more rapidly in homogenized sheep milk than in unhomogenized sheep milk.

The homo–past and homo–heat sheep milks had higher initial weights of the fat in the curds than the raw and pasteurized sheep milks (the rightmost point in Figure 5C). Previous studies have suggested that newly formed fat globules in milk after homogenization are covered by caseins and whey proteins and are able to act as pseudo-protein particles that can interact with the protein phase during coagulation, becoming an integral part of the protein matrix and thus resulting in higher fat contents of the curds [37,38]. The microstructures of the curds (Figure 3) also confirmed these findings. Further, the inclusion of smaller fat globules in the protein matrix can soften the casein network because the homogeneously distributed small fat globules in the protein network result in a larger intermicellar distance and thus hinder the fusion and syneresis of the casein matrix [35,39]. It should be noted that the gastric lipase was not included in SGF in this study because we mainly focused on the investigation of the formation of protein curd and protein digestion under gastric conditions. Previous studies have shown that the gastric lipolysis represents 10% to 30% of total lipid digestion throughout the gastrointestinal tract [2], but the reciprocal effects of gastric lipase and pepsin on the milk digestion behavior are unclear. The inclusion of gastric lipase in the gastric digestion of milk may have an impact on the breakdown of curds and the release of fat from the curds, further investigations need to be taken in the future.

The calcium contents of the dried curds obtained from the 4 different types of sheep milk are presented in Figure 5D. The calcium content in all curds decreased with increasing digestion time, but the rates of decrease were different between the unhomogenized and homogenized milks. The calcium contents of the dried curds in the raw and pasteurized milks decreased steadily over the digestion time, whereas those in the homo–past and homo–heat milks decreased rapidly in the first 60 min and then slowly in the following 120 min. These results suggest that calcium was gradually released from the curd to the liquid phase as the digestion time increased. It has been suggested that bound calcium solubilizes when the pH of the curds decreases to below 5.6, resulting in the release of the solubilized calcium into the liquid phase with the progress of the digestion [40]. The fast release of calcium from the curds of the homo–past and homo–heat sheep milks could be attributed to the faster diffusion of the acid (SGF) into the curds because of their fractured and looser structure (Figure 2). In comparison, the lower rate of calcium release from the curds of the raw and pasteurized sheep milks was due to the formation of an integrated curd surface barrier (Figure 2) that probably impeded the flowing SGF.

These results suggest that the release of the curd contents (protein, fat, and calcium) was markedly dependent on the structure of the curds. The curd formed from the raw and pasteurized sheep milks was more integrated and firmer, which may have impeded the diffusion of the SGF and pepsin into the curds, resulting in a slower breakdown of the curd structure and a slower release of the curd contents. In contrast, the looser and crumbled structures of the curds of the homo–past and homo–heat sheep milks were easily deformed by the physical forces of gastric contraction, leading to a faster release from the curds. These results are in line with previous results on cow milk reported by Ye et al. [18], who showed that the homogenization of cow milk followed by heat treatment made the structure of the curds more open, leading to rapid loss of total solids and rapid release of fat from the curds during gastric digestion. The faster release of calcium and fat from curds of homogenized sheep milk may lead to faster fat digestion and thus enhance uptake of calcium and fat in the small intestine.

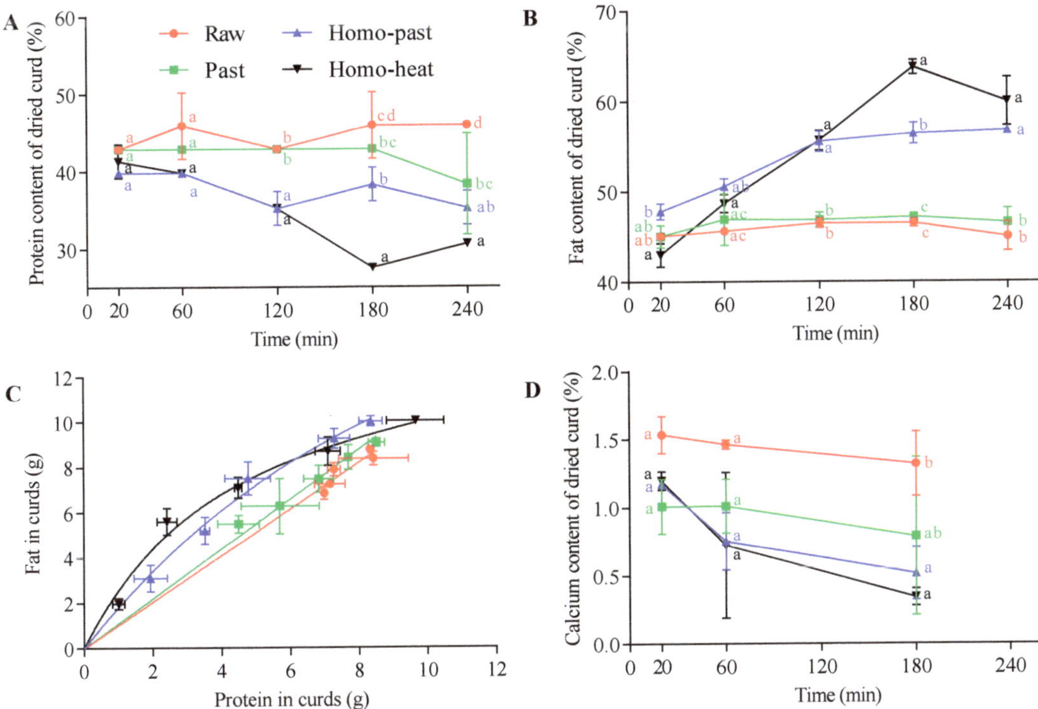

Figure 5. Changes in (**A**) the protein content of the dried curds and (**B**) the fat content of the dried curds, (**C**) the relationship between the amounts of fat and protein in the curds, and (**D**) the calcium content of the dried curds at different time points (20–240 min) during the in vitro gastric digestion of differently processed sheep milks: •, raw milk; ■, pasteurized (Past) milk; ▲, homogenized and pasteurized (Homo–past) milk; ▼, homogenized and heated (Homo–heat) milk. Different lowercase letters indicate significant difference ($p < 0.05$). Error bars represent standard deviations.

3.5. SDS-PAGE Protein Patterns of Curds

The protein hydrolysis by pepsin in the curds was determined using tricine SDS-PAGE (Figure 6A). The amount of each individual protein in the curds is shown in Figure 6B. The κ-casein (κ-CN) band disappeared and 2 new bands at ~23 kDa (macropeptides) and 14 kDa (para-κ-CN) appeared in all samples. Interestingly, unlike the α-lactalbumin (α-La) band, which disappeared in all curd samples, a band corresponding to β-lactoglobulin (β-Lg) was observed throughout the digestion in all types of sheep milk. At 20 min, the amounts of β-Lg in the curds were raw (~0.26 g) < pasteurized (~0.31 g) < homo–past (~0.4 g) < homo–heat (~1.03 g) (Figure 6B), indicating that greater heat treatment incorporated more β-Lg into the curds. The small amount of whey proteins observed in the curds of the raw, pasteurized, and homo–past milk samples could have been caused by the entrapment of whey proteins during the formation of the curds; they could be expelled from the curds and hydrolyzed gradually as the digestion progressed [17,36]. The markedly higher β-Lg content observed in the curd formed from the homo–heat milk was due to the higher level of denatured whey proteins associated with the casein micelles and the MFGM proteins [12,15].

The changes in the intensities of all intact protein bands, including αs-CN, β-CN, and β-Lg, showed marked differences among the sheep milk samples. The intensities of these intact protein bands decreased slowly in the unhomogenized milks with increasing digestion time, but faded away much faster in the homogenized milks (Figure 6A). Bands corresponding to peptides were detected in all milk samples after 60 min of digestion, with the homogenized milks having more intense peptide bands than the unhomogenized milks. This indicates that the homogenization of sheep milk followed by heat processing led to a greater digestibility of proteins during gastric digestion. Faster degradation of proteins has been shown to be caused by the looser and crumbly structure of curds, which allows pepsin to diffuse into the curds rapidly and to hydrolyze the proteins [19]. These findings further support the faster breakdown of the curds and the release of fat and calcium from the curds in the homogenized and heated sheep milk because of the faster hydrolysis of proteins by pepsin, as observed earlier (Figures 4 and 5).

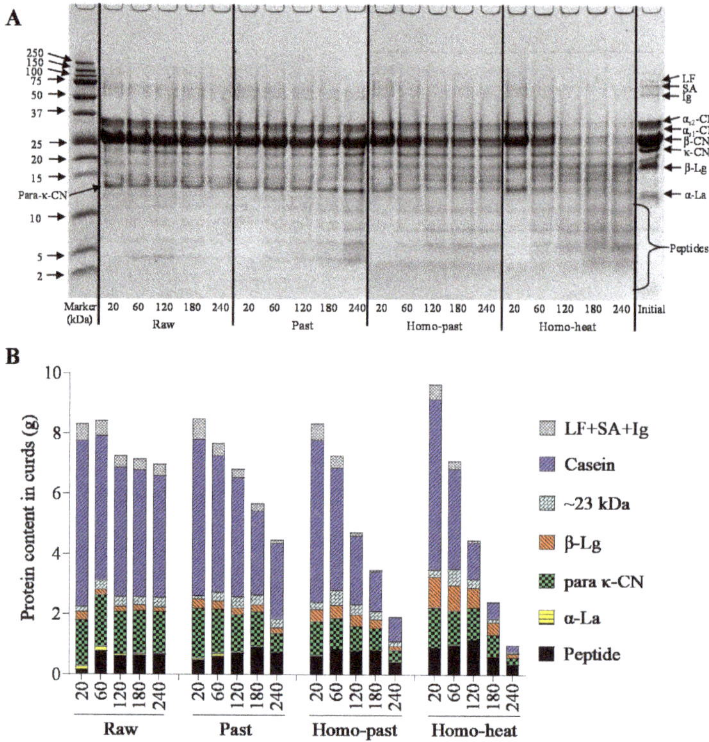

Figure 6. (**A**) Reducing tricine SDS-PAGE patterns and (**B**) protein contents of curds obtained from raw, pasteurized (Past), homogenized and pasteurized (Homo–past), and homogenized and heated (Homo–heat) sheep milks. The numbers (20, 60, 120, 180, 240) refer to different gastric emptying times (min). Initial refers to before digestion. The protein concentration in each sample for the tricine SDS-PAGE was 1 mg/mL.

3.6. Protein and Fat Contents of Emptied Digesta

The protein contents of the digesta emptied from the stomach at different digestion times from the different milk samples are shown in Figure 7A. The protein contents of the unhomogenized milks decreased gradually during 120 min of digestion, after which they remained roughly constant at ~0.41% for raw milk and ~0.69% for pasteurized milk towards the end of the digestion. However, the homo–past and homo–heat sheep milks showed the opposite trend, with the protein contents of the digesta initially increasing in the first 120 min of digestion, with little change on prolonged digestion. At digestion times of longer than 120 min, the protein contents of the digesta of the homo–past and homo–heat sheep milks were significantly higher than those of the raw and pasteurized milks ($p < 0.05$). The fat contents of the digesta are shown in Figure 7B. Those in the raw and pasteurized sheep milks decreased slightly during the first 120 min of digestion and then stayed almost unchanged on prolonged digestion, whereas the homo–past and homo–heat sheep milks showed an increasing trend over the whole digestion.

The changes in the protein contents of the digesta were in reasonable agreement with the release of protein from the curds (Figure 5). The rapid release of protein from the curds of the homo–past and homo–heat sheep milks (Figure 5) resulted in higher protein contents in the gastric chyme, thus leading to increased protein contents in the emptied digesta. In contrast, the slowly digested protein in the curds of the raw and pasteurized sheep milks could have resulted in lower levels of protein in the emptied digesta because of the dilution caused by the continuous addition of SGF. For the fat content, the trends were similar for the curds (Figure 5B) and the digesta (Figure 7B), which showed that the fat content increased throughout the digestion progress in the homo–past and homo–heat sheep milks, whereas it changed little in the raw and pasteurized sheep milks. The continuous increases in the fat contents of the emptied digesta for the homo–past and homo–heat sheep milks may have been caused by the phase separation between the water and oil phases in the stomach. Mulet-Cabero et al. [4] found that the fat content at the end of digestion was generally higher in homogenized milk than in unhomogenized milk, as a cream layer formed at longer digestion times. Thus, the cream layer could remain in the stomach until the end of digestion, resulting in increases in the fat contents of both the curds and the digesta of the homo–past and homo–heat sheep milks at longer digestion times.

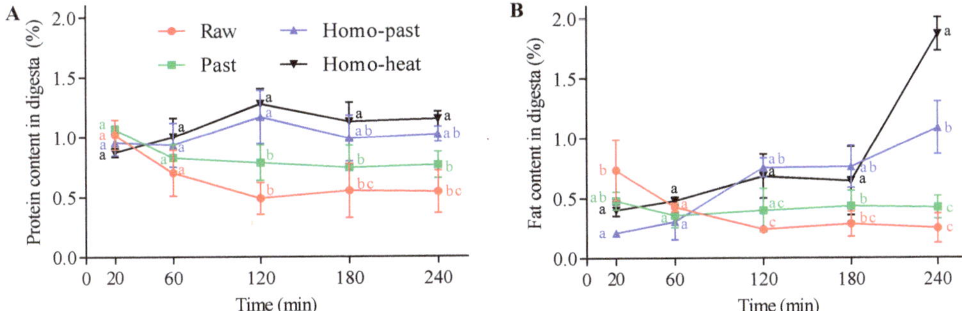

Figure 7. (**A**) Protein and (**B**) fat contents in the digesta emptied at different time points (0–240 min) during the in vitro gastric digestion of differently processed sheep milks: ●, raw milk; ■, pasteurized (Past) milk; ▲, homogenized and pasteurized (Homo–past) milk; ▼, homogenized and heated (Homo–heat) milk. Different lowercase letters indicate significant difference ($p < 0.05$). Error bars represent standard deviations.

3.7. SDS-PAGE Protein Patterns of Digesta

Figure 8A shows the protein patterns of the digesta obtained from the differently processed sheep milks. Faint intact casein bands were present at 20 min of digestion but disappeared after 60 min of digestion in the raw, pasteurized, and homo–past sheep milk samples, and bands corresponding to peptides (between 2 and 10 kDa) appeared concomitantly. This finding is consistent with previous studies, which reported that the casein bands in the digesta could be due to the delivery of very fine casein particles to the digesta when the curd lost its mass, and that increasing amounts of peptides were evacuated from the stomach because of hydrolysis of the caseins [17,19,41]. In contrast, the homo–heat milk samples showed more conspicuous intact casein bands in the digesta, the intensities of which were significantly higher than for the raw, pasteurized, and homo–past milk samples ($p < 0.01$) (Figure 8B). The difference can be attributed to the looser and crumblier structure of the curds formed from the homo–heat milk, leading to the delivery of crumbly particles (containing mostly caseins) smaller than 1 mm to the small intestine.

Differences were also found for the bands at ~14, 18, and 23 kDa, which correspond to α-La, β-Lg, and macropeptides, respectively. Within the first 120 min of digestion, there were low intensities of these bands (especially β-Lg) in the homo–heat sheep milk, whereas markedly higher intensities of these bands were observed in the other three types of sheep milk (Figure 8A). About 0.32, 0.34, 0.28, and 0.08 g of β-Lg were detected in the digesta obtained at 20 min for the raw, pasteurized, homo–past, and homo–heat milks, respectively (Figure 8B). The lower β-Lg content in the digesta of the homo–heat milk was attributed to the higher amounts of β-Lg that were incorporated in the curds (Figure 6) because of their association with the casein micelles after more intense heat treatment [42]. The β-Lg and α-La bands in all types of milk faded away gradually during the digestion and disappeared in the digesta emptied at 240 and 120 min, respectively, which is in agreement with the previous findings of Roy et al. [17]. The decreased intensities of the β-Lg and α-La bands could have been due to the dilution by the continuous addition of SGF and the hydrolysis by pepsin when the pH was less than 4 [43].

These results suggested that, in the raw, pasteurized, and homo–past sheep milks, the protein emptied from the stomach was composed mainly of whey proteins in the early stages of digestion and consisted mainly of peptides at longer digestion times because of the digestion of protein by pepsin. In comparison, the protein in the homo–heat sheep milk was digested faster, leading to a predominant content of peptides in the digesta from the beginning of digestion.

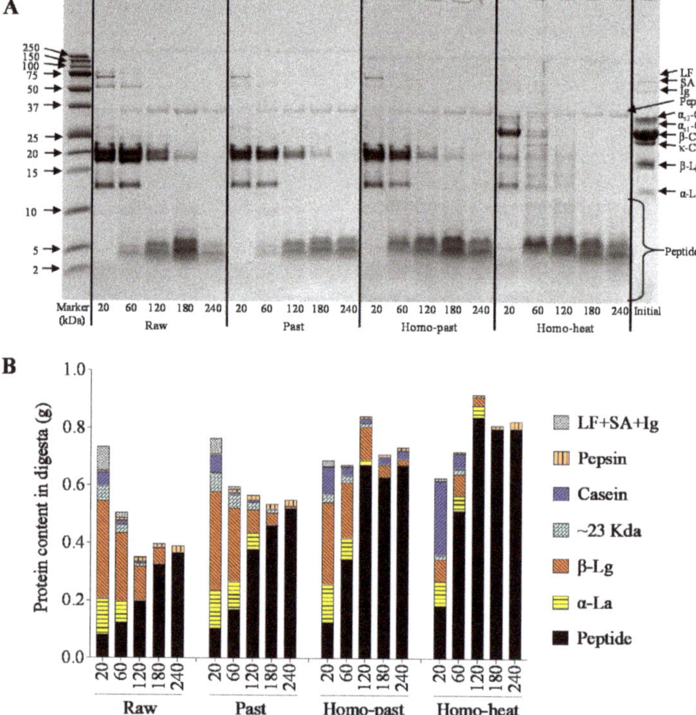

Figure 8. (**A**) Reducing tricine SDS-PAGE patterns and (**B**) protein contents of the digesta obtained from raw, pasteurized (Past), homogenized and pasteurized (Homo–past), and homogenized and heated (Homo–heat) sheep milks. The numbers (20, 60, 120, 180, 240) refer to different gastric emptying times (min). Initial refers to before digestion. The protein concentration in each sample for tricine SDS-PAGE was 1 mg/mL.

3.8. Microstructure of Digesta

The microstructures of the fat and protein in the digesta were observed using confocal microscopy (Figure 9). There were large amounts of fat globules in the digesta of the raw and pasteurized sheep milks throughout the gastric digestion; the fat globule size appeared to remain unchanged during the first 120 min of gastric digestion but increased slightly at 240 min of gastric digestion. This is consistent with the previous report of Roy et al. [17], who also found that there were a few larger fat globules in the digesta of raw and pasteurized sheep milks at 90–240 min, which was caused by their coalescence after the MFGM proteins had been hydrolyzed by pepsin. Small-sized fat globules were found in the digesta of the homo–past and homo–heat sheep milks at 60 min of digestion, accompanied by some protein/peptide particles. There was an increasing number of protein/peptide particles with various sizes in the digesta of the homo–past and homo–heat sheep milks when the gastric digestion time was extended beyond 120 min. Interestingly, the small fat globules were still well embedded in the protein/peptide particles in the digesta of the homo–past and homo–heat sheep milks at digestion times of longer than 120 min. In agreement with the present results, a previous study demonstrated that the peptides resulting from the hydrolysis of protein by pepsin formed a new surface layer on the fat globules, which was unable to create strong interfacial layers with sufficient electrostatic repulsion and steric barriers, thus resulting in flocculation of fat globules via protein–peptide or peptide–peptide interactions [34]. Moreover, no individual fat globules were observed in the digesta of the homo–past and homo–heat sheep milks at digestion times longer than 120 min, suggesting that all fat globules had been incorporated into the protein/peptide particles. However, the incorporation of fat globules into protein/peptide particles may change the density of the particles, leading to the creaming of the less dense particles. This may support the occurrence of phase separation in the stomach, as mentioned earlier.

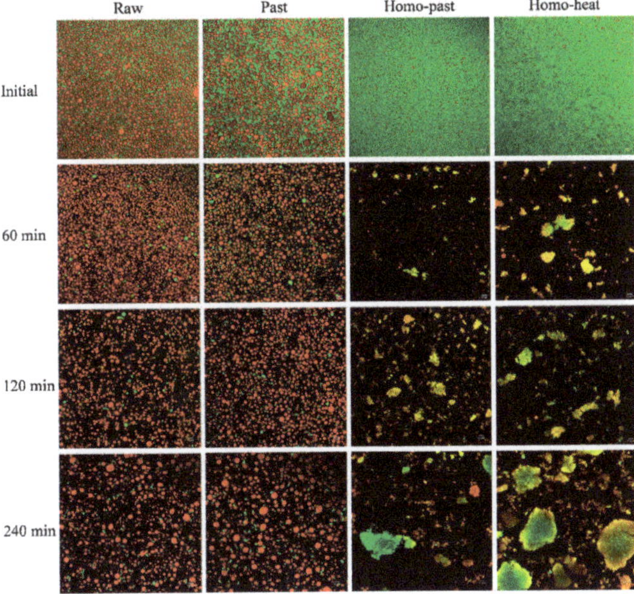

Figure 9. Confocal micrographs of digesta obtained at different time points (0–240 min) during in vitro gastric digestion of raw, pasteurized (Past), homogenized and pasteurized (Homo–past), and homogenized and heated (Homo–heat) sheep milks. Red shows the fat and green shows the protein. The scale bar in all images is 10 μm.

4. Conclusions

This study demonstrated the effect of heat treatment and homogenization on the dynamic in vitro gastric digestion of sheep milk. All milk samples formed structured curds in the stomach. The curds formed from the homogenized milks had a much looser and fractured structure than those formed from the unhomogenized milks because of the inclusion of smaller fat globules into the curd; the homogenization of sheep milk followed by heating at 90 °C for 5 min resulted in crumblier curds compared with homogenization coupled with pasteurization because of the incorporation of more whey proteins into the curd. The differently structured curds led to different rates of disintegration of the curds and of protein hydrolysis by pepsin, and thus impacted the gastric emptying rate. The relative rates of fat release from the curds to the emptied digesta were dependent on whether or not the sheep milk was homogenized. The release of protein, fat, and calcium from the curds occurred at much faster rates in the homogenized sheep milks than in the unhomogenized sheep milks, leading to higher fat and protein contents in the emptied digesta of the homogenized sheep milks. Flocculation of the fat globules was observed in the digesta of the homogenized sheep milks, and most of the fat globules were incorporated into the protein/peptide particles. These findings in the present study suggest that raw and pasteurized sheep milk could give persistent satiety because of its longer gastric emptying time, which could be suitable for the overweight population. In contrast, the homogenized and intensely heated sheep milk would be beneficial for elderly individuals and athletes by enhancing muscle building because of its fast protein digestion.

Author Contributions: Conceptualization, Z.P., A.Y., S.L., A.D. and H.S.; methodology, Z.P.; A.Y., S.L., A.D., software, Z.P.; formal analysis, Z.P.; investigation, Z.P.; resources, A.Y., H.S.; data curation, Z.P.; writing—original draft preparation, Z.P.; writing—review and editing, A.Y., S.L., A.D., K.F. and H.S.; supervision, A.Y., A.D., K.F. and H.S.; project administration, A.Y.; funding acquisition, A.Y. All authors have read and agreed to the published version of the manuscript.

Funding: Ministry of Business, Innovation and Employment-New Zealand Milks Mean More (NZ3M) and Massey University-Doctoral Scholarship, New Zealand and the Riddet Institute, a New Zealand Centre of Research Excellence (CoRE) funded by the Tertiary Education Commission.

Data Availability Statement: The datasets used and/or analyzed during the current study are available from the corresponding author on request.

Acknowledgments: The authors would like to acknowledge the support of Spring Sheep Milk Co. and Maui Milk Co., Ltd. (Waikato, New Zealand) for supplying the sheep milks. They would also like to thank Jian Cui, Sihan Ma, and X. Zhu for their technical assistance with the experiments, and Claire Woodhall for proofreading the manuscript.

Conflicts of Interest: The authors declare no conflict of interest.

References

1. Wendorff, W.; Haenlein, G.F. Sheep milk–composition and nutrition. In *Handbook of Milk of Non-Bovine Mammals*, 2nd ed.; Park, Y.W., Haenlein, G.F., Wendorff, W.L., Eds.; John Wiley & Sons: Chichester, UK, 2017; pp. 210–221.
2. Gallier, S.; Zhu, X.Q.; Rutherfurd, S.M.; Ye, A.; Moughan, P.J.; Singh, H. In vivo digestion of bovine milk fat globules: Effect of processing and interfacial structural changes. II. Upper digestive tract digestion. *Food Chem.* **2013**, *141*, 3215–3223. [CrossRef]
3. Ye, A.; Cui, J.; Dalgleish, D.; Singh, H. Formation of a structured clot during the gastric digestion of milk: Impact on the rate of protein hydrolysis. *Food Hydrocoll.* **2016**, *52*, 478–486. [CrossRef]
4. Mulet-Cabero, A.-I.; Mackie, A.R.; Wilde, P.J.; Fenelon, M.A.; Brodkorb, A. Structural mechanism and kinetics of in vitro gastric digestion are affected by process-induced changes in bovine milk. *Food Hydrocoll.* **2019**, *86*, 172–183. [CrossRef]
5. Jasińska, B. The comparison of pepsin and trypsin action on goat, cow, mare and human caseins. *Rocz. Akad. Med. Bialymst.* **1995**, *40*, 486–493.
6. Roy, D.; Ye, A.; Moughan, P.J.; Singh, H. Structural changes in cow, goat, and sheep skim milk during dynamic in vitro gastric digestion. *J. Dairy Sci.* **2021**, *104*, 1394–1411. [CrossRef]
7. Ye, A.; Cui, J.; Carpenter, E.; Prosser, C.; Singh, H. Dynamic in vitro gastric digestion of infant formulae made with goat milk and cow milk: Influence of protein composition. *Int. Dairy J.* **2019**, *97*, 76–85. [CrossRef]
8. Shen, L.; Dael, P.V.; Luten, L.; Deelstra, H. Estimation of selenium bioavailability from human, cow's, goat and sheep milk by an in vitro method. *Int. J. Food Sci. Nutr.* **1996**, *47*, 75–81. [CrossRef] [PubMed]
9. Claeys, W.L.; Verraes, C.; Cardoen, S.; De Block, J.; Huyghebaert, A.; Raes, K.; Dewettinck, K.; Herman, L. Consumption of raw or heated milk from different species: An evaluation of the nutritional and potential health benefits. *Food Control* **2014**, *42*, 188–201. [CrossRef]

10. Balthazar, C.F.; Pimentel, T.C.; Ferrão, L.L.; Almada, C.N.; Santillo, A.; Albenzio, M.; Mollakhalili, N.; Mortazavian, A.M.; Nascimento, J.S.; Silva, M.C.; et al. Sheep milk: Physicochemical characteristics and relevance for functional food development. *Compr. Rev. Food Sci. Food Saf.* **2017**, *16*, 247–262. [CrossRef]

11. Remeuf, F.; Lenoir, J.; Duby, C.; Letilly, M.-T.; Normand, A. Etude des relations entre les caractéristiques physico-chimiques des laits de chèvre et leur aptitude à la coagulation par la présure. *Lait* **1989**, *69*, 499–518. [CrossRef]

12. Anema, S.G.; Li, Y. Association of denatured whey proteins with casein micelles in heated reconstituted skim milk and its effect on casein micelle size. *J. Dairy Res.* **2003**, *70*, 73–83. [CrossRef]

13. Anema, S.G. Role of κ-casein in the association of denatured whey proteins with casein micelles in heated reconstituted skim milk. *J. Agric. Food Chem.* **2007**, *55*, 3635–3642. [CrossRef] [PubMed]

14. Ye, A.; Singh, H.; Taylor, M.W.; Anema, S. Interactions of whey proteins with milk fat globule membrane proteins during heat treatment of whole milk. *Lait* **2004**, *84*, 269–283. [CrossRef]

15. Ye, A.; Anema, S.G.; Singh, H. Changes in the surface protein of the fat globules during homogenization and heat treatment of concentrated milk. *J. Dairy Res.* **2008**, *75*, 347–353. [CrossRef] [PubMed]

16. Michalski, M.-C.; Januel, C. Does homogenization affect the human health properties of cow's milk? *Trends Food Sci. Technol.* **2006**, *17*, 423–437. [CrossRef]

17. Roy, D.; Ye, A.; Moughan, P.J.; Singh, H. Impact of gastric coagulation on the kinetics of release of fat globules from milk of different species. *Food Funct.* **2021**, *12*, 1783–1802. [CrossRef]

18. Ye, A.; Liu, W.; Cui, J.; Kong, X.; Roy, D.; Kong, Y.; Han, J.; Singh, H. Coagulation behaviour of milk under gastric digestion: Effect of pasteurization and ultra-high temperature treatment. *Food Chem.* **2019**, *286*, 216–225. [CrossRef]

19. Ye, A.; Cui, J.; Dalgleish, D.; Singh, H. Effect of homogenization and heat treatment on the behavior of protein and fat globules during gastric digestion of milk. *J. Dairy Sci.* **2017**, *100*, 36–47. [CrossRef]

20. Mulet-Cabero, A.-I.; Mackie, A.R.; Brodkorb, A.; Wilde, P.J. Dairy structures and physiological responses: A matter of gastric digestion. *Crit. Rev. Food Sci. Nutr.* **2020**, *60*, 3737–3752. [CrossRef]

21. Egger, L.; Ménard, O.; Baumann, C.; Duerr, D.; Schlegel, P.; Stoll, P.; Vergères, G.; Dupont, D.; Portmann, R. Digestion of milk proteins: Comparing static and dynamic in vitro digestion systems with in vivo data. *Food Res. Int.* **2019**, *118*, 32–39. [CrossRef]

22. Egger, L.; Schlegel, P.; Baumann, C.; Stoffers, H.; Guggisberg, D.; Brügger, C.; Dürr, D.; Stoll, P.; Vergères, G.; Portmann, R. Physiological comparability of the harmonized INFOGEST in vitro digestion method to in vivo pig digestion. *Food Res. Int.* **2017**, *102*, 567–574. [CrossRef]

23. Authority, N.Z.F.S. Heat Treatment Code of Practice. 2009. Available online: https://www.mpi.govt.nz/dmsdocument/46228-Dairy-Heat-treatment-Code-of-practice (accessed on 1 August 2019).

24. Jørgensen, C.E.; Abrahamsen, R.K.; Rukke, E.-O.; Johansen, A.-G.; Schüller, R.B.; Skeie, S.B. Improving the structure and rheology of high protein, low fat yoghurt with undenatured whey proteins. *Int. Dairy J.* **2015**, *47*, 6–18. [CrossRef]

25. Brodkorb, A.; Egger, L.; Alminger, M.; Alvito, P.; Assuncao, R.; Ballance, S.; Bohn, T.; Bourlieu-Lacanal, C.; Boutrou, R.; Carriere, F.; et al. INFOGEST static in vitro simulation of gastrointestinal food digestion. *Nat. Protoc.* **2019**, *14*, 1–24. [CrossRef] [PubMed]

26. Kong, F.; Singh, R.P. A human gastric simulator (HGS) to study food digestion in human stomach. *J. Food Sci.* **2010**, *75*, E627–E635. [CrossRef]

27. Lydon, A.; Murray, C.; McGinley, J.; Plant, R.; Duggan, F.; Shorten, G. Cisapride does not alter gastric volume or pH in patients undergoing ambulatory surgery. *Can. J. Anesth.* **1999**, *46*, 1181–1184. [CrossRef] [PubMed]

28. Vertzoni, M.; Dressman, J.; Butler, J.; Hempenstall, J.; Reppas, C. Simulation of fasting gastric conditions and its importance for the in vivo dissolution of lipophilic compounds. *Eur. J. Pharm. Biopharm.* **2005**, *60*, 413–417. [CrossRef] [PubMed]

29. International Dairy Federation. *IDF Standard?* IDF: Brussels, Belgium, 1987.

30. Wang, X.; Lin, Q.; Ye, A.; Han, J.; Singh, H. Flocculation of oil-in-water emulsions stabilised by milk protein ingredients under gastric conditions: Impact on in vitro intestinal lipid digestion. *Food Hydrocoll.* **2019**, *88*, 272–282. [CrossRef]

31. Delfour, A.; Alais, C.; Jollès, P. Structure of cows kappa-caseino-glycopeptide-n-terminal octadecapeptide. *Chimia* **1966**, *20*, 148–150.

32. Mulvihill, D.M.; Fox, P.F. Proteolytic Specificity of Chymosins and Pepsins on Beta-Caseins. *Irish J. Food Sci. Technol.* **1979**, *2*, 135–139. Available online: http://www.jstor.org/stable/25557963 (accessed on 18 February 2021).

33. Gallier, S.; Ye, A.; Singh, H. Structural changes of bovine milk fat globules during in vitro digestion. *J. Dairy Sci.* **2012**, *95*, 3579–3592. [CrossRef]

34. Ye, A.; Cui, J.; Singh, H. Proteolysis of milk fat globule membrane proteins during in vitro gastric digestion of milk. *J. Dairy Sci.* **2011**, *94*, 2762–2770. [CrossRef] [PubMed]

35. Kelly, A.L.; Huppertz, T.; Sheehan, J.J. Pre-treatment of cheese milk: Principles and developments. *Dairy Sci. Technol.* **2008**, *88*, 549–572. [CrossRef]

36. Ye, A.; Cui, J.; Dalgleish, D.; Singh, H. The formation and breakdown of structured clots from whole milk during gastric digestion. *Food Funct.* **2016**, *7*, 4259–4266. [CrossRef]

37. Lopez, C. Focus on the supramolecular structure of milk fat in dairy products. *Reprod. Nutr. Dev.* **2005**, *45*, 497–511. [CrossRef]

38. Ong, L.; Dagastine, R.R.; Kentish, S.E.; Gras, S.L. Microstructure of milk gel and cheese curd observed using cryo scanning electron microscopy and confocal microscopy. *LWT* **2011**, *44*, 1291–1302. [CrossRef]

39. Guinee, T.; O'Brien, B. The quality of milk for cheese manufacture. In *Technology of Cheesemaking*, 2nd ed.; Law, B.A., Tamime, A.Y., Eds.; John Wiley & Sons: Chichester, UK, 2010; Volume 67, pp. 1–67.
40. Everett, D.W.; Auty, M.A.E. Cheese structure and current methods of analysis. *Int. Dairy J.* **2008**, *18*, 759–773. [CrossRef]
41. Awad, S.; Lüthi-Peng, Q.-Q.; Puhan, Z. Proteolytic activities of chymosin and porcine pepsin on buffalo, cow, and goat whole and β-casein fractions. *J. Agric. Food Chem.* **1998**, *46*, 4997–5007. [CrossRef]
42. Anema, S.G. The whey proteins in milk: Thermal denaturation, physical interactions, and effects on the functional properties of milk. In *Milk Proteins*, 3rd ed.; Boland, M., Singh, H., Eds.; Academic Press: London, UK, 2020; pp. 325–384.
43. Wang, X.; Ye, A.; Lin, Q.; Han, J.; Singh, H. Gastric digestion of milk protein ingredients: Study using an in vitro dynamic model. *J. Dairy Sci.* **2018**, *101*, 6842–6852. [CrossRef] [PubMed]

Article

Impact of Ultra-High-Pressure Homogenization of Buttermilk for the Production of Yogurt

Louise Krebs [1,2], Amélie Bérubé [1], Jean Iung [1], Alice Marciniak [1], Sylvie L. Turgeon [1] and Guillaume Brisson [1,*]

1 STELA Dairy Research Centre, Department of Food Sciences, Institute of Nutrition and Functional Foods (INAF), Université Laval, Québec, QC G1V 0A6, Canada; louise.krebs.1@ulaval.ca (L.K.); amelie.berube.3@ulaval.ca (A.B.); jean.iung.1@ulaval.ca (J.I.); alice.marciniak.1@ulaval.ca (A.M.); sylvie.turgeon@fsaa.ulaval.ca (S.L.T.)
2 Faculty of Health, Medicine and Life Sciences, Maastricht University, 6211 LK Maastricht, The Netherlands
* Correspondence: guillaume.brisson@fsaa.ulaval.ca

Abstract: Despite its nutritional properties, buttermilk (BM) is still poorly valorized due to its high phospholipid (PL) concentration, impairing its techno-functional performance in dairy products. Therefore, the objective of this study was to investigate the impact of ultra-high-pressure homogenization (UHPH) on the techno-functional properties of BM in set and stirred yogurts. BM and skimmed milk (SM) were pretreated by conventional homogenization (15 MPa), high-pressure homogenization (HPH) (150 MPa), and UHPH (300 MPa) prior to yogurt production. Polyacrylamide gel electrophoresis (PAGE) analysis showed that UHPH promoted the formation of large covalently linked aggregates in BM. A more particulate gel microstructure was observed for set SM, while BM gels were finer and more homogeneous. These differences affected the water holding capacity (WHC), which was higher for BM, while a decrease in WHC was observed for SM yogurts with an increase in homogenization pressure. In stirred yogurts, the apparent viscosity was significantly higher for SM, and the pretreatment of BM with UHPH further reduced its viscosity. Overall, our results showed that UHPH could be used for modulating BM and SM yogurt texture properties. The use of UHPH on BM has great potential for lower-viscosity dairy applications (e.g., ready-to-drink yogurts) to deliver its health-promoting properties.

Keywords: buttermilk; yogurt; MFGM; ultra-high-pressure homogenization

Citation: Krebs, L.; Bérubé, A.; Iung, J.; Marciniak, A.; Turgeon, S.L.; Brisson, G. Impact of Ultra-High-Pressure Homogenization of Buttermilk for the Production of Yogurt. *Foods* **2021**, *10*, 1757. https://doi.org/10.3390/foods10081757

Academic Editors: Harjinder Singh and Alejandra Acevedo-Fani

Received: 29 June 2021
Accepted: 27 July 2021
Published: 29 July 2021

Publisher's Note: MDPI stays neutral with regard to jurisdictional claims in published maps and institutional affiliations.

1. Introduction

Buttermilk (BM), which is the serum phase generated during the production of butter, is produced in approximately equal parts as butter [1]. The Canadian dairy industry produced 118,235 metric tons of butter in 2020 [2], leading to an estimated equal volume of BM, while global BM production has been estimated at about 3.2 million tons per annum [3]. Despite the large quantities produced every year, BM is still undervalued. Indeed, BM is mainly used in animal feeds, for its emulsifying capacity or for providing a milk flavor in various food applications [4,5]. As with skim milk (SM), BM is composed of the main milk solids-not-fat, namely caseins (CNs), whey proteins (WPs), lactose, and minerals [6]. The main difference between SM and BM is the phospholipid (PL) content, which, in BM, is up to ten times higher than in SM [7] and four times higher than in raw milk [8]. PLs are part of a very specific structure found in dairy products, the milk fat globule membrane (MFGM) [9], which is released into BM during the churning of cream into butter. The unique composition of PLs in MFGM (phosphatidylcholine-PC, phosphatidylethanolamine-PE, phosphatidylserine-PS, phosphatidylinositol-PI, and sphingomyelin-SM), along with the MFGM proteins, is responsible for its various health benefits [10–12]. For example, daily dietary supplementation with BM led to a decrease in concentration of total and LDL cholesterol in moderately hypercholesterolemic men and women [10]. Furthermore, sphingomyelin-fortified milk was found to have a positive influence on infants' gut microbiota and neurocognitive development [13].

Despite its nutritional and biological properties, compared to SM, BM possesses limited technological properties in dairy applications due to its high PL content [4,14,15]. Thus, the incorporation of BM into dairy matrices, such as yogurt, presents different technological challenges, such as a decrease in firmness [16] and lower apparent viscosity [17], which impacts the quality of the final product. Nevertheless, the incorporation of very small amounts of BM or BM powder (BMP) into yogurt matrices produced interesting results, with a reported decrease in whey separation and syneresis [18], and an increase in water-holding capacity (WHC) [16]. Contradictory impacts on firmness have been reported, which could be due to the various sources and quantities of BM used [16,19]. Thus, the use of mechanical treatment to improve BM's functional properties and facilitate its incorporation into various dairy matrices is of high interest in order to provide the benefits of its health-promoting activities. In yogurt production, conventional homogenization using pressures between 20 and 60 MPa [20] has already been shown to affect yogurt properties. As a matter of fact, it stabilizes emulsions and decreases the creaming phenomenon of milk [21] by reducing the size of fat globules and simultaneously increasing their surface area [22]. This leads to lipids being denser and more homogeneously dispersed throughout the liquid, resulting in an improved viscosity and WHC of yogurt [23]. The fat globules from heated homogenized milk are known to interact with the proteins in the acid gel network as active filler particles [24]. However, Ji et al. (2011) reported that the extent of interaction of the fat globules with the protein network depends on their size and, therefore, the homogenization pressure [25]. They showed that milk recombined at low homogenization pressures resulted in larger fat globules with less active interaction with the network, while milk treated at higher pressures resulted in smaller globules more tightly bound to the protein network. Similarly, high-pressure homogenization (HPH) (between 150 and 200 MPa) produced even smaller fat globules with increased surface area [20] and, thus, improved properties within the yogurt matrix. A few studies have also reported that the size of homogenized fat globules affects their incorporation in the protein network, especially at higher homogenization pressures and when using microfluidization [25–27]. Recently, with the development of high-pressure intensifiers and valve materials (stainless steel, ceramic, and seals) resistant to extremely high homogenization, pressures up to 400 MPa can be reached [20,28,29]. This process is known as ultra-high-pressure homogenization (UHPH), which ranges from 200 to 400 MPa. Applied to whole milk, it causes fat globule reduction to submicron sizes. However, above that pressure, fat globules form clusters through the aggregation of CN and WP at their surface [30], which induces higher WHC and changes in the gel firmness [23,31]. In addition, some authors have reported decreases in CN micelle sizes from 5% to 33% in the UHPH range [21,32,33].

Given the interesting results from the application of UHPH for yogurt production from SM, as well as the high nutritional quality of BM, there is much interest in the use of UHPH for the production of BM yogurt. Hence, the aim of this study was to examine the impact of UHPH as a pre-treatment for the production of yogurts from BM compared to those produced from SM. In this paper, set and stirred yogurts were produced and characterized using different homogenization treatments (15, 150, and 300 MPa). This new approach of using UHPH for improving BM use in yogurt is of great interest for BM valorization through the production of a potentially highly functional product enriched in MFGM health-promoting components.

2. Materials and Methods

2.1. Materials

Whole raw milk and raw cream were provided by a local supplier (Quebec City, QC, Canada) and skim milk powder (SMP) was obtained from Agropur (Quebec City, QC, Canada). The thermophilic yogurt culture YC-X11 Yo-Flex® (Chr. Hansen A/S, Hørsholm, Denmark) was composed of *Streptococcus thermophilus* and *Lactobacillus delbrueckii* subsp.

Bulgaricus. Analytical-grade sodium hydroxide for the preparation of 0.1 M of NaOH was obtained from Fisher Chemical (Ottawa, ON, Canada). Mini-PROTEAN TGX Stain-Free Gels (12%, 15-well comb, 15 µL), 2× Laemmli sample buffer, native sample buffer, Precision Plus ProteinTM All Blue Standards, 10× Tris/glycine/sodium dodecyl sulfate (SDS) buffer, and 10× Tris/glycine buffer were all obtained from BioRad (Hercules, CA, USA). 2-Mercaptoethanol was provided by Sigma-Aldrich (St. Louis, MO, USA). Methanol was obtained from Fisher Chemical (Ottawa, ON, Canada) and glacial acetic acid from Anachemia (Radnor, PA, USA). Fast Green FCF and Nile Red were obtained from Sigma-Aldrich (Oakville, ON, Canada).

2.2. Production of BM and SM Yogurts

The production steps of BM yogurts and SM yogurts (control) are presented in Figure 1. Briefly, raw cream and whole raw milk were pasteurized (Chalinox/Hydro-Québec CFI-25, QC, Canada) at 85 °C for 30 s and 72 °C for 15 s, respectively. Both pasteurized cream and pasteurized milk were matured overnight at 10 °C. Then, to obtain BM, the matured pasteurized cream was churned at 75 rpm at approximately 12 °C using a pilot plant-scale butter churn with a capacity of 8–15 L (Qualtech Equipment, QC, Canada). The pasteurized BM and whole milk were heated for 10 min to 40 °C and cream was separated using a cream separator (Westfalia LWA-205-DeLaval, Lund, Sweden). Skimmed BM and SM (approximately 10% total solids) were enriched with 30 g of SMP per liter, to reach a solid content of approximately 12% (±0.25%) and achieve a similar yogurt firmness as commercial yogurts. The final composition of BM and SM (Table 1) was determined with a LactoScope FTIR milk analyzer (Delta Instruments, Drachten, The Netherlands). Pasteurized BM and SM were aliquoted into 3 batches of 2 L. Each batch was used for one homogenization parameter of 15 MPa, 150 MPa, and 300 MPa using a UHPH system (Nano Debee Model 45-4, Bee International, South Easton, MA, USA).

Table 1. Composition of the standardized buttermilk (BM) and skimmed milk (SM) mix used for yogurt production.

	Buttermilk	Skimmed Milk
Total solids (%)	11.88 ± 0.03 *	11.89 ± 0.11
Lipids (%)	0.59 ± 0.07	0.13 ± 0.05
Proteins (%)	4.28 ± 0.06	4.51± 0.08
Lactose (%)	6.26 ± 0.05	6.49 ± 0.11

* Mean values (n = 4) ± standard deviation.

After UHPH treatment, BM and SM were heated at 85 °C on a stove for 15 min (temperature rise time of approximately 10 min), with continuous stirring, and then cooled on ice to 42 °C. Starter cultures (YC-X11 Yo-Flex®) were added according to the manufacturer's directions [34]. Briefly, 0.05 g of the frozen culture was added to 2 L of treated BM or SM, stirred for several minutes until complete dissolution, and subsampled for further analysis. All samples were incubated at 42 °C to reach a pH of 4.6 (approximately 6 h for BM yogurts and 8 h for SM yogurts) and were then stored at 4 °C overnight. For set yogurt, incubation took place in 100 mL plastic containers and 50 mL tubes, and yogurts were stored at 4 °C until further analysis. Stirred yogurts were produced by manual stirring with a metal spoon (30 times clockwise and 30 times anticlockwise) to break down the gel. The six stirred yogurts (3 pressure levels-15, 150, and 300 MPa, and 2 sources-BM and SM) were stored at 4 °C until further analysis.

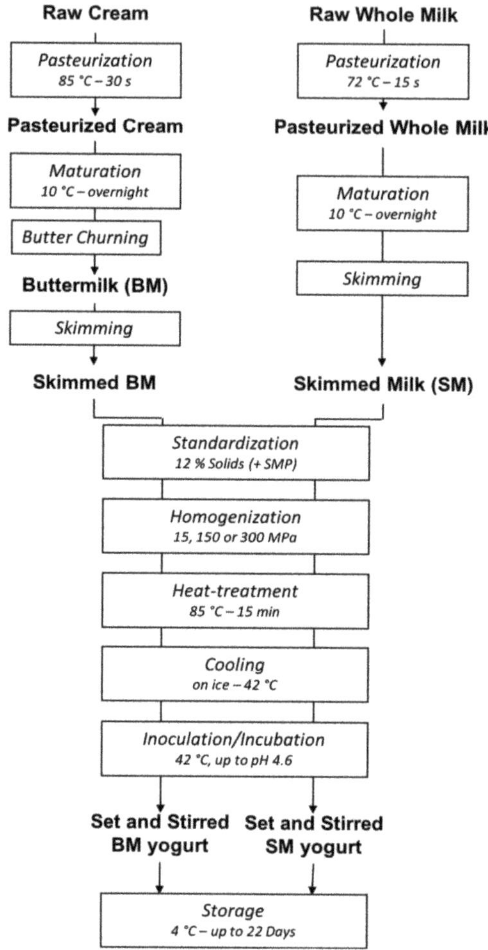

Figure 1. Experimental design of the production of set and stirred yogurts from buttermilk (BM) and skimmed milk (SM) treated by ultra-high-pressure homogenization (UHPH).

2.3. Protein Profiles of Homogenized BM and SM

The protein profiles of all fluid BM and SM samples after homogenization treatment were determined by native polyacrylamide gel electrophoresis (PAGE) and SDS-PAGE under nonreducing conditions. In addition, SM and BM controls were prepared with samples treated at 15 MPa, under reducing conditions (50 μL of 2-mercaptoethanol and 950 μL of Laemmli buffer). All samples were diluted in distilled water (1:9), and 25 μL of each dilution was mixed with 25 μL of their respective sample buffer. Solutions were then loaded onto precast 12% acrylamide gels in a Mini PROTEAN® Tetra Cell (BioRad, Hercules, CA, USA). Precision Plus Protein™ All Blue Standards with molecular weights ranging from 10 to 250 kDa were used as the molecular weight marker. The electrophoresis was conducted at a constant voltage of 120 V for 1 h. The gels were then stained with Coomassie blue solution (BioRad, Hercules, CA, USA) for 1 h, followed by de-staining with a mixture of methanol, acetic acid, and distilled water (1:1:8) overnight. The gels were scanned the next day using a ChemiDoc™ MP imaging system (Bio-Rad, Hercules, CA, USA).

2.4. Physico-Chemical Characteristics of Set and Stirred Yogurts

Physico-chemical characterization of set and stirred yogurts was performed using the following analysis and was repeated after storage at 4 °C on days 1, 8, 15, and 22 after yogurt production.

2.4.1. pH

The pH of set and stirred yogurts was measured using a pH meter (Orion Star T910, Thermo Fisher Scientific, Waltham, MA, USA) calibrated with standardized buffer solutions (pH 4.0 and 7.0).

2.4.2. Firmness

The firmness of set yogurts was assessed according to Le et al. (2011) with slight modifications [16]. A penetration test was carried out on set yogurts at 4 °C using a texturometer (TA.XT2, Texture Technologies, New York, NY, USA). A 25 mm-diameter cylindrical probe was used at a constant rate of 1 mm/s for a distance of 40 mm. The maximum force recorded in real-time represents the firmness (N).

2.4.3. Water-Holding Capacity

The WHC of set yogurts was measured on days 1, 8, 15, and 22 after yogurt production according to Le et al. (2011) [16]. Samples were centrifuged (IEC Centra CL2 centrifuge, Thermo Fisher Scientific Inc., Milford, MA, USA) at $1200 \times g$ for 15 min at 4 °C. The top layer (whey) was removed and weighed, and the WHC was calculated according to the following equation.

$$\text{WHC } (\%) = \left[\frac{(\text{sample weight (g)} - \text{expelled whey(g))}}{\text{sample weight (g)}} \right] \times 100 \qquad (1)$$

2.4.4. Confocal Laser Scanning Microscopy

The microstructure of set and stirred yogurts was investigated using confocal laser scanning microscopy (CLSM) based on the methods of Lucey et al. (1998) and Zhao et al. (2016) with some adaptations [24,35]. Following milk inoculation, milks were incubated at 42 °C for 1 h. A volume of 960 μL of BM or SM was then mixed with 20 μL of 1% Fast Green (proteins) and 20 μL of 2% Nile Red (PLs). After a 15 min waiting period with intermittent stirring, 3 mL of BM or SM was added and thoroughly mixed. Mixtures were poured into a 35 mm culture dish with a 15 mm glass bottom and placed in the incubator for fermentation (approximately 9 h). Stirred yogurts were produced by manual stirring with a metal spoon. Imaging was performed using a STELLARIS5 confocal microscope (LEICA, Mannheim, Germany) equipped with an HC PL APO CS $40 \times / 0.853$ NA air objective. The sample was mounted in a MatTek coverslip bottom 35 mm culture dish with a 14 mm glass diameter (MatTek life science, Ashland, MA, USA) and excited with lasers at 552 nm and 638 nm, separately, for Nile red and Fast Green FCF, respectively. The emission was acquired with a variable dichroic selecting wavelength over 540–600 nm for Nile red and over 640–800 nm for Fast Green FCF.

2.4.5. Titratable Acidity

The titratable acidity of stirred yogurt was determined at 20 °C according to the AOAC method 947.05 [36]. Samples were prepared by mixing 10 g of yogurt in 40 mL of distilled water. Samples were continuously stirred and titrated using 0.1 M of NaOH to a pH of 8.3 using an automatic titrator (Orion Start T910, Thermo Fisher Scientific, Waltham, MA, USA). The amount of titrant needed to reach pH 8.3 was noted, and the titratable acidity (% lactic acid) was calculated according to the following equation.

$$\textit{Titrable acidity (\% lactic acid)} = \frac{\text{titrant (mL)}}{\text{sample weight (g)}} \times 0.9 \qquad (2)$$

2.4.6. Apparent Viscosity

A temperature-controlled rheometer (TA instruments, model ARES-G2, New Castle, DE, USA) was used for the measurements of apparent viscosity of stirred yogurts at 10 °C, according to Yu, Wang, and McCarthy (2016) with some adaptations [37]. The measuring system consisted of a 40 mm, 0.04 radius cone and plate geometry. Shear rate sweep (1 to 120 s^{-1}) and shear stress response tests were performed. Apparent viscosity was measured at a shear rate of 50 s^{-1}.

2.4.7. Drained Syneresis

Drained syneresis of the stirred yogurts was measured at 4 °C, according to Hassan et al. (1996) with some modifications [38]. A mesh screen (Cell strainer, pluriSelect USA, El Cajon, CA, USA) with a mesh size of 200 μm and a radius of 20 mm was used, and mesh tension was released by running water through the mesh. The yogurt sample (4 g) was poured on a mesh screen placed above a previously weighed empty centrifuge tube. Samples were left for 2 h at 4 °C in order to collect the whey resulting from the syneresis. The weight of the centrifuge tubes was measured after sample draining, and the drained syneresis was calculated according to the following equation.

$$Drained\ synerisis = \frac{drained\ whey\ (g)}{sample\ weight\ (g)} \times 100 \tag{3}$$

2.5. Statistical Analysis

Four independent productions of yogurts (replicates) were performed with different batches of raw cream and whole raw milk. Analyses were conducted for each replicate. Data were processed using a multi-factor ANOVA and SAS software (SAS University Edition) to compare BM and SM yogurt properties. Significant differences between pressure treatments and storage time were evaluated with the Tukey test. Evaluations were based on a significance level of $p < 0.05$.

3. Results and Discussion

3.1. Impact of UHPH Treatment on Protein Profiles of BM and SM

To study the impact of UHPH on BM and SM proteins, their profiles were analyzed using native PAGE and nonreducing SDS-PAGE (Figure 2a,b). Different profiles were observed for BM and SM samples for native PAGE (Figure 2a). UHPH treatment did not have as large an impact on SM as it did on BM. Similar band patterns were observed for SM between all the homogenization pressures tested, but more drastic changes were observed for the BM samples. Indeed, an overall decrease was observed in the intensity of the BM protein bands migrating within the gel as the homogenization pressure increased from 150 MPa to 300 MPa compared to the BM treated with a conventional homogenization pressure (15 MPa). This decrease indicates that the main milk proteins underwent denaturation and aggregation upon UHPH, the severity of which was dependent on the pressure used. Concomitantly, a proportional increase in the intensities of the protein signals in the loading wells was observed for the BM samples homogenized at 150 MPa and 300 MPa, confirming that large protein aggregates formed during higher homogenization treatments. The BM and SM samples were then analyzed by SDS-PAGE (Figure 2b) under nonreducing (lanes 2–4 for SM and 6–8 for BM) and reducing conditions (lane 1 for SM and lane 5 for BM) to determine the nature of these interactions. UHPH treatment of SM (lanes 2–4) did not show any effects among the pressures tested (150 MPa and 300 MPa), as can be seen from their similar profiles. The SM protein band intensities were similar for all pressure treatments and were comparable to the samples under reducing conditions (lane 1), indicating that a very low polymerization and aggregation of milk proteins occurred in SM during UHPH. However, gel electrophoresis demonstrated an important impact of the UHPH treatment on the protein aggregation in BM. First, bands corresponding to WP (α-lactalbumin (α-LA) and β-lactoglobulin (β-LG)) decreased in intensity as the homogenization pressure increased. This result is in agreement with Lopez-Fandiño, Carrascosa, and Olano (1996), who observed increasing denaturation and aggregation of WP, more precisely β-LG, at pressures from 100 to 400 MPa [39]. A possible explanation for the lower impact of UHPH on SM protein aggregation could be that SM underwent lower pasteurization temperatures (72 °C/15 s) than those of the cream used for the production of BM (85 °C/30 s). It is known that the pasteurization of cream induces important WP denaturation and the formation of aggregates, especially for β-LG, through intermolecular disulfide interactions with the CN and the MFGM proteins, the extent of which depends on the severity of the thermal treatment [40]. Finally, bands corresponding to MFGM proteins (visible on lane 5) are not detected in UHPH-treated BM samples (lanes 6–8). This suggests strong covalent interactions between WPs, especially β-LG, CN micelles, and MFGM proteins, which are also attributable to the increase in temperature during pressure treatment [41]. These results indicate that more protein denaturation takes place in BM than in SM due to BM's higher MFGM content, and, as mentioned above, due to the higher temperature applied during the pasteurization of cream in the production of BM. Thus, the more severe pasteurization treatment for cream leads to more potential interactions between MFGM fragments, notably through the MFGM proteins and WPs under thermal treatment of cream, as observed by Morin et al. (2007) [42].

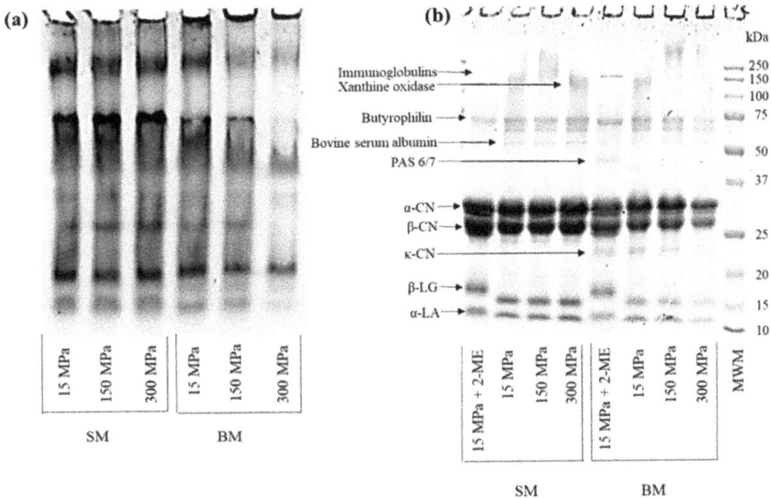

Figure 2. Acrylamide gels (12%) of skimmed milk (SM) and buttermilk (BM) following pressure treatments ((**a**) native polyacrylamide gel electrophoresis (PAGE) pattern, and (**b**) sodium dodecyl sulfate (SDS)-PAGE pattern under reducing (lanes 1 and 5) and nonreducing conditions). PAS 6/7 = periodic acid Schiff 6/7, MWM = molecular weight markers, 2-ME = 2-mercaptoethanol.

3.2. Impact of UHPH Treatment on Physico-Chemical Properties of Set BM and SM Yogurt

Figure 3 presents the physico-chemical and textural properties ((a) pH, (b) WHC, and (c) firmness) of set yogurt as a function of dairy source (SM: plain line, and BM: stippled line), pressure treatment (15, 150, and 300 MPa), and storage time (days 1, 8, 15, and 22). The pH of set yogurt was not influenced by the dairy source or homogenization pressure applied (Figure 3a). However, a significant decrease in pH was noticed during the storage time. After storage at 4 °C, and regardless of the dairy source and pressure treatment, a significant drop in pH was observed from 4.58 at day 1 to 4.42 at day 8, finally reaching 4.33 at day 22. This expected effect is attributed to fermentation of residual lactose by the starter cultures [43].

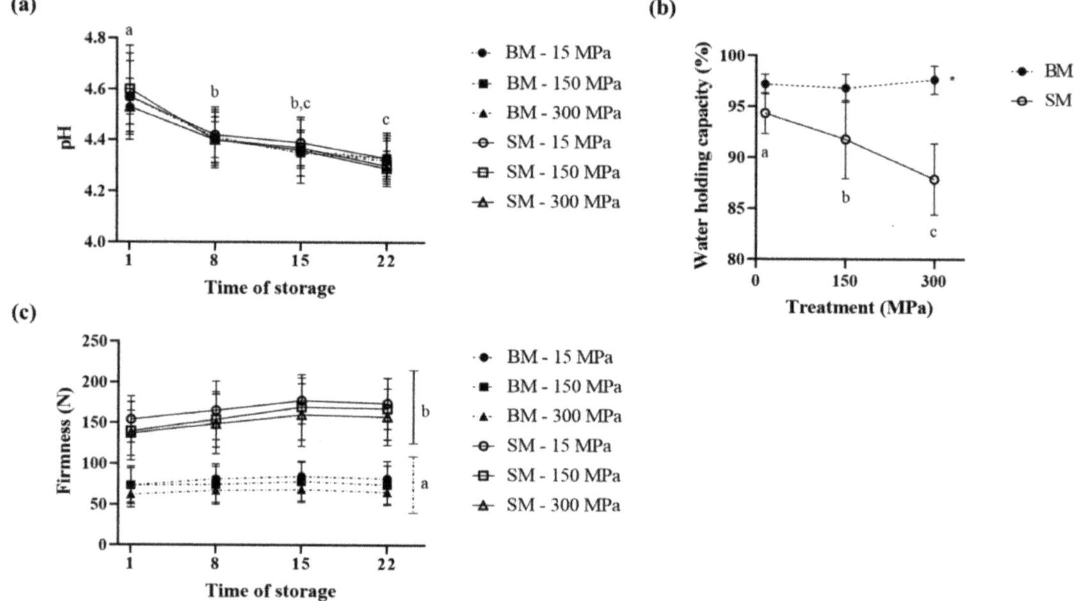

Figure 3. Evolution of physico-chemical properties ((**a**) pH, (**b**) water-holding capacity-WHC, and (**c**) firmness) for set buttermilk (BM-stippled line) and skimmed milk (SM-plain line) yogurts as a function of time of storage (1, 8, 15, and 22 days) and pressure levels (15, 150, and 300 MPa).

The WHC of set yogurt indicates the ability of the yogurt gel structure to retain water [44]. A low value for WHC is associated with an unstable yogurt gel network [45]. As observed in Figure 3b, the dairy source and pressure treatment applied highly influence the WHC of set yogurt in an interactive way. BM yogurts had a higher WHC than SM yogurts did regardless of the storage time and treatment, with an average of 97.26% for BM in contrast to 92.14% for SM. These results agreed with the study of Le et al. (2011). These authors observed increases in WHC of 7, 15, 21, and 31%, when they replaced 1, 2, 3, and 4% of the original total SM solids in their yogurt mix with equivalent amounts of solids from a MFGM isolate, whereas substituting SM with BMP at the same ratios did not impact the WHC [16]. However, other authors reported an increase in WHC when low-fat yogurts (12% total solids) were enriched with 1% and 2% of BMP [19]. In our study, 8% of the total 12% solids originated from BM solids. We also calculated that MFGM represented approximately 0.37% of the 12% total solids of our yogurt mix (based on an average MFGM extraction yield of 3.5 g of MFGM/L of BM using a common method [46], results not shown). Overall, the higher WHC of BM yogurt can be explained by its composition, more precisely, the PL content, which is higher in BM than in SM. Indeed, PLs show amphiphilic characteristics, allowing increased retention of water [16,47], while simultaneously, milk proteins have excellent WHCs [48]. Interactions between PLs and WPs or β-CN via electrostatic and hydrophobic connections (Gallier et al., 2012), as well as interactions between MFGM proteins and CNs or WPs via covalent disulfide bonds occurring during pasteurization of the cream [42], might have contributed to a more compact gel with reinforced interactions, increasing water retention within the yogurt gel for the BM-based yogurt [19]. In addition, SM treated at 300 MPa exhibited the lowest WHC value, which contrasts with previous studies that associated higher UHPH pressures to enhanced water retention due to increased interactions between WPs, CNs, and lipids [31,44,49]. However, these authors used whole milk rather than SM, and the homogenized fat globules are known to participate in increasing the strength of the gel network by participating as active filler particles, increasing the WHC [50,51].

In order to complete the characterization of set yogurts, firmness, which represents the gel network strength, was monitored through storage time. Statistical analysis showed that even though pressure treatment and time of storage did not impact firmness, it was significantly ($p < 0.0001$) influenced by the dairy source (BM or SM) used for yogurt production (Figure 3c). BM yogurts had a lower firmness than SM yogurts did, with average values of 70.34 N and 156.62 N, respectively (regardless of the time of storage and pressure level). These results agree with those of Le et al. (2011) who found that SM yogurts had a higher firmness than those fortified with 1, 2, 3, or 4% BMP [16]. However, very recently, Zhao, Feng, and Mao (2020) observed a higher firmness in yogurts fortified with 1–2% BMP than in a control SMP-yogurt [19], while 4% BMP yogurts had a lower firmness. In fact, exceeding a certain BM concentration might have adverse effects on the development of a stable yogurt gel, and it is assumed that PLs take more space and interact with the proteins, thereby disrupting the gel network [52]. Despite the fact that the pressure level did not significantly impact the firmness ($p = 0.0901$), we observed a decreasing tendency for both BM and SM yogurts. This tendency is in contrast with the results of yogurts obtained from milk treated between 100 and 300 MPa, which showed increased firmness with increasing pressure [31,44,53,54]. Globally, these studies show that while UHPH enhanced the interactions between WPs and CNs, in BM, the presence of smaller fat globules embedded in the protein network led to a higher firmness.

The final physico-chemical property evaluated was microstructure. The microstructure of set yogurts (SM and BM, each treated with 15, 150, and 300 MPa) was analyzed by CLSM and is shown in Figure 4, where the green color refers to proteins, and red to PLs. Overall, SM and BM yogurts exhibited different microstructures, which were impacted by UHPH treatments. The set yogurts made from SM exhibited protein clusters of larger size for all pressure treatments, whereas those made from BM had smaller protein particles, homogeneously distributed. The main difference in yogurts at 15 MPa was the PL content, which was, as expected, higher in BM than in SM. The few PLs present in SM seem to be of larger size than in BM, which might be due to the destructive effect of butter churning on the MFGM, resulting in smaller MFGM fragments in BM. In addition, PLs in BM yogurt were widely distributed throughout the gel matrix at 15 MPa. Increasing the pressure from 15 MPa to 150 MPa and 300 MPa largely changed the SM yogurt microstructure. As a matter of fact, we observed larger protein clusters forming large serum pores within the SM yogurt gels. However, for BM yogurts at 150 MPa and 300 MPa, the gel structures had very fine and continuous protein networks, which is supported by the decrease in particle size distribution of UHPH BM (unpublished data). In addition, for those pressures, PLs seemed to be distributed more heterogeneously and were bound to the protein network. The results for BM's gel microstructure are in line with a previous study, which reported a dense structure with irregularly clustered protein aggregates for yogurt fortified with BMP due to a high content of proteins interacting with MFGM components [47]. Especially at 300 MPa, visible CN micelles can interact with MFGM fragments, trapping them within the yogurt protein network upon coagulation and preventing them from forming a more stranded gel [55]. These interactions between casein and MFGM would explain the lower firmness and viscosity observed for BM yogurts compared to the particulate gel observed for SM yogurts. The results of Le et al. (2011) support our observations of aggregated MFGM fragments within a homogeneous finely particulate casein network in BM yogurts [16]. The lower firmness observed for BM yogurts could be associated with the more homogeneous and finer gel observed and the occurrence of MFGM fragments within the gel, as indicated by Le et al. (2011) [16]. SM yogurts had stronger stranded networks, as can be seen from the higher contrast between the serum (black) and CN (green) phases. These microstructural changes in the protein gel network support the different physico-chemical and texture properties observed for both BM and SM set yogurts. For example, these differences could explain the enhanced WHC of BM yogurt as the homogeneously distributed PLs within the protein network increases water retention while reducing the firmness of the gel network [16].

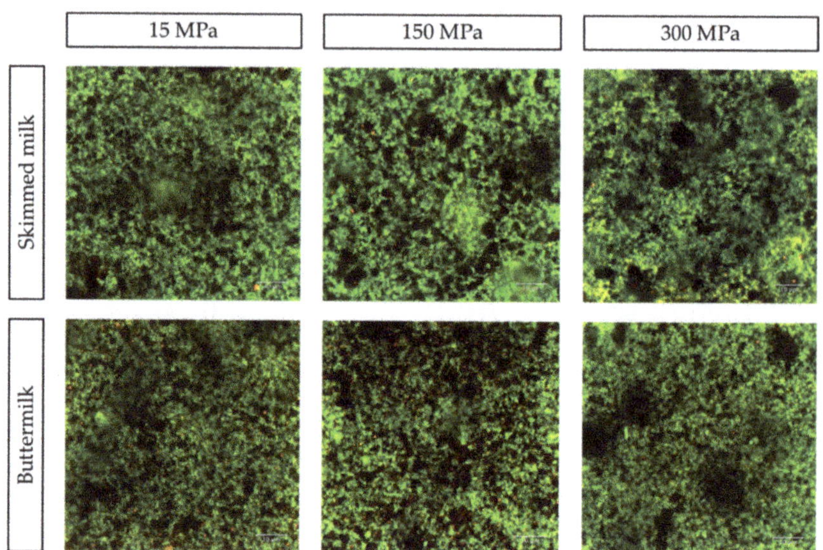

Figure 4. Confocal laser scanning microscopy (CLSM) images of set skimmed milk (SM) and buttermilk (BM) yogurts with different pressure applications (15, 150, and 300 MPa). Red color represents the phospholipids (PLs) labeled with Nile Red; green color represents the milk proteins labeled with Fast Green FCF. Scale bar (10 μm).

3.3. Impact of UHPH Treatment on Physico-Chemical Properties of Stirred BM and SM Yogurt

Figure 5 represents the physico-chemical properties ((a) pH, (b) titratable acidity, (c) apparent viscosity, and (d) drained syneresis) of stirred yogurt as a function of the dairy source (BM and SM), homogenization pressure (15, 150, and 300 MPa), and storage time (days 1, 8, 15, and 22). No significant differences were observed in the pH of stirred yogurts (Figure 5a), between BM and SM dairy sources or between the different homogenization pressures. These results agree with those of Serra et al. (2008) who treated whole milk with pressures of 200 and 300 MPa prior to yogurt production [49]. In addition, an expected decrease in pH was observed upon storage at 4 °C, regardless of the source and pressure treatment. Indeed, the pH value decreased significantly ($p < 0.0001$) from 4.48 on day 1 to an average of 4.31 for days 8, 15, and 22. The drop in pH within the first week of yogurt storage agrees with results from a previous study by Moschopoulou et al. (2018), who noticed a drop in pH from day 1 (around pH 4.45) to day 7 (around pH 4.2) in semi-skimmed cow milk yogurt [43]. This decrease in pH was attributed to the residual lactose fermentation. However, Yildiz and Bakirci (2019) did not observe differences in pH in their BM- and WP-enriched yogurts with increasing storage time [17]. This can be explained by the presence of components such as WP, which are known to have great buffering capacity [56].

Interestingly, titratable acidity (Figure 5b), which is defined as the total acid concentration in a sample, was not impacted by storage time or pressure treatment. However, this property was influenced by the dairy source used for yogurt production. The titratable acidity of BM yogurt was significantly lower than that of SM yogurt ($p = 0.0002$), with averages of 1.01% and 1.06%, respectively. In contrast, another study, in which buffalo SM was replaced with 25, 50, 75, and 100% BM (fortified with 3% SMP), reported lower values for titratable acidity for the control SM yogurt (0.92%) compared to yogurt fortified with BM [18]. The authors demonstrated that a replacement with 100 and 75% BM (titratable acidity of 0.97%) led to even higher values for titratable acidity than 50% (0.95%) or 25% (0.93%). These differences between our study and the literature could be due to a slightly (however nonsignificant) higher protein content in SM yogurts (4.51%) compared to BM (4.28%), which might have induced the higher buffering capacity of SM, as observed by Trachoo and Mistry (1998) [57]. Indeed, the addition of ultrafiltered BM increased the protein content and, therefore, the buffering capacity followed by the titratable acidity of low-fat yogurt (titratable acidity: 1.39%) compared to the use of BMP (1.29%) [57].

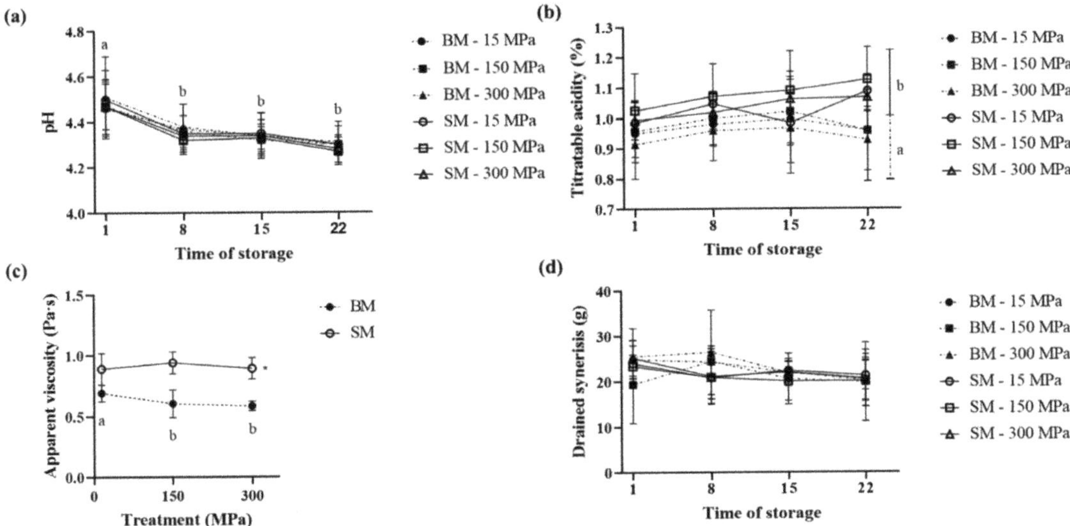

Figure 5. Evolution of physico-chemical properties ((**a**) pH, (**b**) titratable acidity, (**c**) apparent viscosity, and (**d**) drained syneresis) for stirred buttermilk (BM stippled line) and skimmed milk (SM. plain line) yogurts as a function of storage time (1, 8, 15, and 22 days) and pressure levels (15, 150, and 300 MPa).

The pressure, storage time, and dairy source parameters studied had different impacts on the physical properties of stirred yogurt. The apparent viscosity (Figure 5c) of stirred yogurts was affected by an interaction between pressure treatment and dairy source; however, no further differences were detected throughout the time of storage. This nonsignificance of the storage time contrasts with prior studies. Yildiz and Bakirci (2019), for example, observed irregular changes in apparent viscosity with storage time [17]. The interaction between homogenization pressure and dairy source was found to be significant ($p = 0.015$) with higher values for SM (0.94 Pa·s) than BM (0.63 Pa·s), whereas for SM yogurt, the apparent viscosity did not change with pressure; a slight but constant decrease was observed for BM yogurt from 15 MPa to 300 MPa. The higher values for SM compared to BM are in line with Yildiz and Bakirci (2019), who measured a lower viscosity for yogurts fortified with 2% BMP (+1% SMP) compared to those fortified with 3% SMP [17]. Recently, in contrast to our observations, Zhao, Feng, and Mao (2020) found that the viscosity of low-fat SMP-yogurt depends on the level of BMP fortification (0.5–4.0%) [19]. They found that viscosity increased significantly (up to ~60%) as the level of BMP incorporation increased to 2.0%, whereas the addition of 4% BMP resulted in a loss in viscosity (~13%). This suggests that the higher BM component content contributed to lowering yogurt viscosity, as observed in our study where BM was mainly used for yogurt manufacture. While the slightly higher protein content of the SM yogurt mix (Table 1) could have influenced the rheological properties of yogurts [58,59], the impact of BM addition seems to be related to increases in the amount of MFGM constituents. Studies have shown that while the fortification of small amounts of BMP (1–2%) increases viscosity, probably due to the emulsifying [48] and amphiphilic [47] properties of PLs and proteins, using higher amounts of BMP (4%) decreases viscosity [19]. The reinforced interactions between MFGM proteins and CNs or WPs via noncovalent or disulfide bonds [60], as well as interactions between PLs and WPs or β-CN, mainly via hydrophobic and electrostatic links (Gallier, 2012), probably contribute to the beneficial effect of MFGM fortification on physico-chemical yogurt properties, at least until a critical BMP concentration is reached. Consistent with our study, Le et al. (2011) concluded that BM supplementation contributes to the lower firmness of low-fat yogurts compared to the controls (12% SMP), which they also explained was due to the higher concentration of PLs from BMP [16]. Treatment of whole milk with UHPH has also been reported to induce interactions between denatured WP, lipids, and water, as well as interactions between CNs or between CNs and lipids, which enhances viscosity [61,62].

Finally, drained syneresis of the stirred yogurt was measured throughout the time of storage (Figure 5d). Drained syneresis measures the serum released due to shrinkage of the yogurt gel network [17], which is related to a textural defect of yogurts [63]. In our study, no difference was observed on the drained syneresis, regardless of the dairy source, pressure level, and storage time, with an average of 22.09 g of whey released. Our results are in contrast with those of Yildiz and Bakirci (2019) who found lower drained syneresis for control (3% SMP) yogurts compared to BMP-enriched yogurt [17].

Just as for set yogurts, the microstructure of stirred yogurts (SM and BM, each treated with 15, 150, and 300 MPa) was also studied using CLSM (Figure 6). Again, different microstructures were observed for stirred SM and BM yogurts subjected to different pressure treatments. For SM yogurts, changes within the microstructure with increasing pressure seem to be less distinct than for BM yogurts. At all three pressures, PLs were evenly distributed throughout the gel. However, at 150 MPa, more serum pores of larger size were observed than at 15 and 300 MPa. In addition, SM yogurts pretreated with 300 MPa had a more homogeneous gel with smaller protein particles than those pretreated with 15 MPa. In BM yogurts, however, more PLs were present in the gel, which exhibited finer and more homogeneously distributed protein aggregates, the difference being particularly visible at 300 MPa. At 15 MPa, BM gels contained large serum phases, which lessened with increasing pressure. Indeed, at 300 MPa, serum phases were virtually absent, whereas SM yogurt still exhibited larger serum pores at 300 MPa. A very fine gel with small protein particles was observed for BM 300 MPa. Further, PL particles were bound to protein particles and were of smaller size compared to those in gels treated with 15 and 150 MPa. This might be due to the effect of pressure-induced particle size reduction on proteins and PLs, as previously explained. The higher apparent viscosity of SM over BM stirred yogurts can probably be traced back to the stronger gel network of SM compared to BM set yogurts. For BM yogurts, the decrease in apparent viscosity with increasing pressure could be attributed to the wider distribution of smaller PLs at 150 and 300 MPa, possibly impacting protein gel strength in a way that decreases apparent viscosity. As observed for set yogurts, the interaction between PLs and the protein network at higher pressures (150 and 300 MPa), combined with the more denatured and aggregated protein in BM, might have impaired the formation of a stable gel [55].

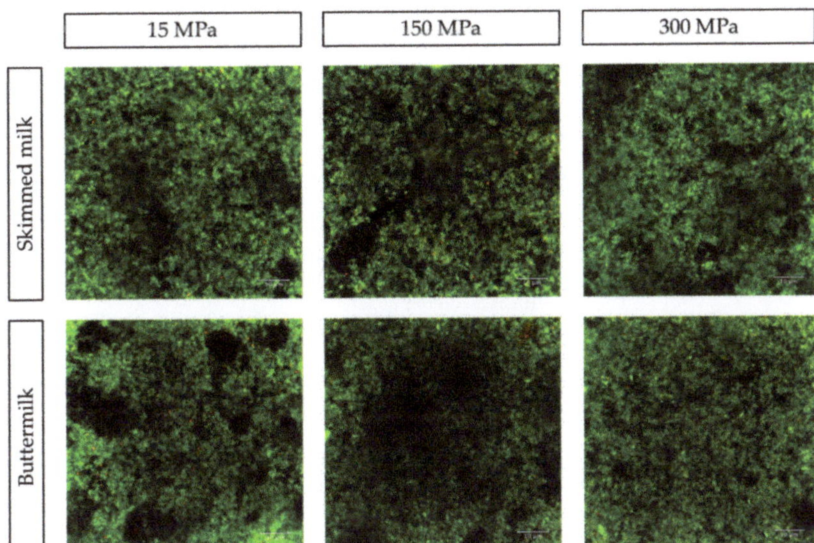

Figure 6. Confocal laser scanning microscopy (CLSM) images of stirred skimmed milk (SM) and buttermilk (BM) yogurts with different pressure applications (15, 150, and 300 MPa). The red color represents the phospholipids (PLs) labeled with Nile Red; the green color represents the milk proteins labeled with Fast Green FCF. Scale bar (10 μm).

4. Conclusions

This work studied the impact of UHPH on BM to improve its techno-functionality for incorporation into yogurt applications. The results showed that UHPH treatment more drastically impacted BM proteins, which underwent more denaturation and aggregation than SM proteins did, thus impacting the physico-chemical and textural properties of the set and stirred yogurts produced from BM and SM. In addition, the gel microstructure was influenced by the UHPH-treatment and depended on the dairy source. Indeed, at the highest pressure (300 MPa), set SM yogurts presented large protein clusters with large serum pores, while set BM yogurts produced finer and more homogeneously distributed protein particles that interacted with PL and correlated with lower firmness. This work represents the first step in understanding the impact of UHPH on BM for the production of yogurt. It can support the development of new technology for the valorization of BM in order to take advantage of the beneficial effects of MFGM components on human health. Future research could focus on enriching BM with cream prior to UHPH treatment to produce a full-fat yogurt and investigating the impact of lipids on UHPH-treated yogurt properties. It could also focus on the interaction effect between lipids and proteins by studying their impact on the gel network and yogurt microstructure.

Author Contributions: Conceptualization, G.B. and L.K.; methodology, L.K.; software, A.M.; validation, G.B., A.M. and L.K.; formal analysis, A.M. and L.K.; investigation, L.K., A.B. and J.I.; resources, G.B.; data curation, L.K. and A.M.; writing—original draft preparation, L.K.; writing—review and editing, A.M., S.L.T. and G.B.; visualization, L.K. and A.M.; supervision, G.B.; project administration, G.B.; funding acquisition, G.B. All authors have read and agreed to the published version of the manuscript. The authors are thankful to Barb Conway for editing this manuscript.

Funding: This research was funded by the Natural Sciences and Engineering Research Council of Canada (NSERC) [grant number CRDPJ/537396-2018], the Quebec Consortium for Industrial Bioprocess Research and Innovation (CRIBIQ), and Novalait Inc.

Institutional Review Board Statement: Not applicable.

Informed Consent Statement: Not applicable.

Data Availability Statement: Not applicable.

Acknowledgments: The authors thank Diane Gagnon and Pascal Lavoie (Food Science Department, Université Laval) for technical support. The authors also thank Alexandre Bastien (Department of Animal Sciences, Université Laval) for technical support and assistance with the confocal laser scanning microscope. We also acknowledge the Natural Sciences and Engineering Research Council of Canada (NSERC), the Quebec Consortium for Industrial Bioprocess Research and Innovation (CRIBIQ), and Novalait for the financial support of this project.

Conflicts of Interest: The authors declare no conflict of interest. The funders had no role in the design of the study; in the collection, analyses, or interpretation of data; in the writing of the manuscript, or in the decision to publish the results.

References

1. Morin, P.; Britten, M.; Jiménez-Flores, R.; Pouliot, Y. Microfiltration of Buttermilk and Washed Cream Buttermilk for Concentration of Milk Fat Globule Membrane Components. *J. Dairy Sci.* **2007**, *90*, 2132–2140. [CrossRef]
2. Statistics Canada. Table 32-10-0109-01 Supply and Disposition of Milk Products in Canada. Available online: https://doi.org/10.25318/3210010901-eng (accessed on 13 April 2021).
3. Kumar, R.; Kaur, M.; Garsa, A.K.; Shrivastava, B.; Reddy, V.; Tyagi, A.T. Natural and cultured buttermilk. In *Fermented Milk and Dairy Products*, 1st ed.; CRC Press, Taylor and Francis: Boca Raton, FL, USA, 2015; pp. 203–225.
4. Vanderghen, C.; Bodson, P.; Danthine, S.; Paquot, M.; Deroanne, C.; Blecker, C. Milk Fat Globule Membrane and Buttermilks: From Composition to Valorization. *Biotechnol. Agron. Soc. Environ.* **2010**, *14*, 485–500.
5. Barukcic Irena, I. Whey and Buttermilk Neglected Sources of Valuable Beverages. *Nat. Beverages* **2019**, *13*, 209.
6. Corredig, M.; Dalgleish, D.G. Isolates from Industrial Buttermilk: Emulsifying Properties of Materials Derived from the Milk Fat Globule Membrane. *J. Agric. Food Chem.* **1997**, *45*, 4595–4600. [CrossRef]
7. Britten, M.; Lamothe, S.; Robitaille, G. Effect of Cream Treatment on Phospholipids and Protein Recovery in Butter-Making Process. *Int. J. Food Sci. Technol.* **2008**, *43*, 651–657. [CrossRef]
8. Rombaut, R.; Camp, J.V.; Dewettinck, K. Analysis of Phospho- and Sphingolipids in Dairy Products by a New HPLC Method. *J. Dairy Sci.* **2005**, *88*, 482–488. [CrossRef]
9. Walstra, P.; Wouters, J.T.M.; Geurts, T.J. *Dairy Science and Technology*, 2nd ed.; CRC Press: Boca Raton, FL, USA, 2006.
10. Conway, V.; Couture, P.; Richard, C.; Gauthier, S.F.; Pouliot, Y.; Lamarche, B. Impact of buttermilk consumption on plasma lipids and surrogate markers of cholesterol homeostasis in men and women. *Nutr. Metab. Cardiovasc. Dis.* **2013**, *23*, 1255–1262. [CrossRef]
11. Conway, V.; Couture, P.; Gauthier, S.; Pouliot, Y.; Lamarche, B. Effect of buttermilk consumption on blood pressure in moderately hypercholesterolemic men and women. *Nutrition* **2014**, *30*, 116–119. [CrossRef]
12. Fuller, K.L.; Kuhlenschmidt, T.B.; Kuhlenschmidt, M.S.; Jimenez-Flores, R.; Donovan, S.M. Milk fat globule membrane isolated from buttermilk or whey cream and their lipid components inhibit infectivity of rotavirus in vitro. *J. Dairy Sci.* **2013**, *96*, 3488–3497. [CrossRef]
13. Tanaka, K.; Hosozawa, M.; Kudo, N.; Yoshikawa, N.; Hisata, K.; Shoji, H.; Shinohara, K.; Shimizu, T. The pilot study: Sphingomyelin-fortified milk has a positive association with the neurobehavioural development of very low birth weight infants during infancy, randomized control trial. *Brain Dev.* **2013**, *35*, 45–52. [CrossRef]
14. Evers, J.M. The milkfat globule membrane—Compositional and structural changes post secretion by the mammary secretory cell. *Int. Dairy J.* **2004**, *14*, 661–674. [CrossRef]
15. Singh, H. The milk fat globule membrane—A biophysical system for food applications. *Curr. Opin. Colloid Interface Sci.* **2006**, *11*, 154–163. [CrossRef]
16. Le, T.T.; van Camp, J.; Pascual, P.A.L.; Meesen, G.; Thienpont, N.; Messens, K.; Dewettinck, K. Physical properties and microstructure of yoghurt enriched with milk fat globule membrane material. *Int. Dairy J.* **2011**, *21*, 798–805. [CrossRef]
17. Yildiz, N.; Bakirci, I. Investigation of the use of whey powder and buttermilk powder instead of skim milk powder in yogurt production. *J. Food Sci. Technol.* **2019**, *56*, 4429–4436. [CrossRef] [PubMed]
18. El-Nour, A.A.M.; El-Kholy, A.M.; El-Safty, M.S.; Mokbel, S.M. Using Buttermilk in Making Fat-Free Yogurt. *J. Dairy Sci. Technol.* **2014**, *1*, 1–9.
19. Zhao, L.L.; Feng, R.; Mao, X.Y. Addition of buttermilk powder improved the rheological and storage properties of low-fat yogurt. *Food Sci. Nutr.* **2020**, *8*, 3061–3069. [CrossRef]
20. Dumay, E.; Chevalier-Lucia, D.; Picart-Palmade, L.; Benzaria, A.; Gràcia-Julià, A.; Blayo, C. Technological aspects and potential applications of (ultra) high-pressure homogenisation. *Trends Food Sci. Technol.* **2013**, *31*, 13–26. [CrossRef]
21. Sandra, S.; Dalgleish, D.G. Effects of ultra-high-pressure homogenization and heating on structural properties of casein micelles in reconstituted skim milk powder. *Int. Dairy J.* **2005**, *15*, 1095–1104. [CrossRef]

22. McCrae, C.H.; Hirst, D.; Law, A.J.R.; Muir, D.D. Heat stability of homogenized milk: Role of interfacial protein. *J. Dairy Res.* **1994**, *61*, 507–516. [CrossRef]
23. Tamime, A.Y.; Deeth, H.C. Yogurt: Technology and Biochemistry. *J. Food Prot.* **1980**, *43*, 939–977. [CrossRef]
24. Lucey, J.A.; Teo, C.T.; Munro, P.A.; Singh, H. Microstructure, permeability and appearance of acid gels made from heated skim milk. *Food Hydrocoll.* **1998**, *12*, 159–165. [CrossRef]
25. Ji, Y.-R.; Lee, S.K.; Anema, S.G. Effect of heat treatments and homogenisation pressure on the acid gelation properties of recombined whole milk. *Food Chem.* **2011**, *129*, 463–471. [CrossRef] [PubMed]
26. Ciron, C.I.E.; Gee, V.L.; Kelly, A.L.; Auty, M.A.E. Comparison of the effects of high-pressure microfluidization and conventional homogenization of milk on particle size, water retention and texture of non-fat and low-fat yoghurts. *Int. Dairy J.* **2010**, *20*, 314–320. [CrossRef]
27. Ciron, C.I.E.; Gee, V.L.; Kelly, A.L.; Auty, M.A.E. Modifying the microstructure of low-fat yoghurt by microfluidisation of milk at different pressures to enhance rheological and sensory properties. *Food Chem.* **2012**, *130*, 510–519. [CrossRef]
28. Zamora, A.; Ferragut, V.; Jaramillo, P.D.; Guamis, B.; Trujillo, A.J. Effects of ultra-high pressure homogenization on the cheese-making properties of milk. *J. Dairy Sci.* **2007**, *90*, 13–23. [CrossRef]
29. Patrignani, F.; Lanciotti, R. Applications of High and Ultra High Pressure Homogenization for Food Safety. *Front. Microbiol.* **2016**, *7*, 1132. [CrossRef]
30. Thiebaud, M.; Dumay, E.; Picart, L.; Guiraud, J.P.; Cheftel, J.C. High-pressure homogenisation of raw bovine milk. Effects on fat globule size distribution and microbial inactivation. *Int. Dairy J.* **2003**, *13*, 427–439. [CrossRef]
31. Serra, M.; Trujillo, A.J.; Quevedo, J.M.; Guamis, B.; Ferragut, V. Acid coagulation properties and suitability for yogurt production of cows' milk treated by high-pressure homogenisation. *Int. Dairy J.* **2007**, *17*, 782–790. [CrossRef]
32. Hayes, M.G.; Kelly, A.L. High pressure homogenisation of raw whole bovine milk (a) effects on fat globule size and other properties. *J. Dairy Res.* **2003**, *70*, 297–305. [CrossRef]
33. Chevalier-Lucia, D.; Blayo, C.; Gràcia-Julià, A.; Picart-Palmade, L.; Dumay, E. Processing of phosphocasein dispersions by dynamic high pressure: Effects on the dispersion physicochemical characteristics and the binding of α-tocopherol acetate to casein micelles. *Innov. Food Sci. Emerg. Technol.* **2011**, *12*, 416–425. [CrossRef]
34. CHR Hansen. YC-X11 Information Produit Version: 4 PI EU FR 03-03-2018. Available online: https://www.fromagex.com/pub/media/itm/magb1/Productfiles/I200CCL712/1618403567_I200CCL712_YCX11_Fiche_Technique.pdf (accessed on 13 April 2021).
35. Zhao, L.L.; Wang, X.L.; Tian, Q.; Mao, X.Y. Effect of casein to whey protein ratios on the protein interactions and coagulation properties of low-fat yogurt. *J. Dairy Sci.* **2016**, *99*, 7768–7775. [CrossRef]
36. AOAC. *Official Methods of Analysis of AOAC International*, 15th ed.; Association of Official Analytical Chemists: Arlington, VA, USA, 1990.
37. Yu, H.-Y.; Wang, L.; McCarthy, K.L. Characterization of yogurts made with milk solids nonfat by rheological behavior and nuclear magnetic resonance spectroscopy. *J. Food Drug Anal.* **2016**, *24*, 804–812. [CrossRef] [PubMed]
38. Hassan, A.N.; Frank, J.F.; Schmidt, K.A.; Shalabi, S.I. Textural Properties of Yogurt Made with Encapsulated Nonropy Lactic Cultures. *J. Dairy Sci.* **1996**, *79*, 2098–2103. [CrossRef]
39. Lopez-Fandiño, R.; Carrascosa, A.V.; Olano, A. The Effects of High Pressure on Whey Protein Denaturation and Cheese-Making Properties of Raw Milk. *J. Dairy Sci.* **1996**, *79*, 929–936. [CrossRef]
40. Corredig; Corredig, M.; Dalgleish, D.G. Effect of Heating of Cream on the Properties of Milk Fat Globule Membrane Isolates. *J. Agric. Food Chem.* **1998**, *46*, 2533–2540. [CrossRef]
41. Ye, A.; Anema, G.; Singh, H. High-Pressure-Induced Interactions Between Milk Fat Globule Membrane Proteins and Skim Milk Proteins in Whole Milk. *J. Dairy Sci.* **2004**, *87*, 4013–4022. [CrossRef]
42. Morin, P.; Jiménez-Flores, R.; Pouliot, Y. Effect of processing on the composition and microstructure of buttermilk and its milk fat globule membranes. *Int. Dairy J.* **2007**, *17*, 1179–1187. [CrossRef]
43. Moschopoulou, E.; Sakkas, L.; Zoidou, E.; Theodorou, G.; Sgouridou, E.; Kalathaki, C.; Liarakou, A.; Chatzigeorgiou, A.; Politis, I.; Moatsou, G. Effect of milk kind and storage on the biochemical, textural and biofunctional characteristics of set-type yoghurt. *Int. Dairy J.* **2018**, *77*, 47–55. [CrossRef]
44. Serra, M.; Trujillo, A.J.; Guamis, B.; Ferragut, V. Evaluation of physical properties during storage of set and stirred yogurts made from ultra-high pressure homogenization-treated milk. *Food Hydrocoll.* **2009**, *23*, 82–91. [CrossRef]
45. Lucey, J.A. The relationship between rheological parameters and whey separation in milk gels. *Food Hydrocoll.* **2001**, *15*, 603–608. [CrossRef]
46. Corredig, M.; Roesch, R.R.; Dalgleish, D.G. Production of a novel ingredient from buttermilk. *J. Dairy Sci.* **2003**, *86*, 2744–2750. [CrossRef]
47. Romeih, E.A.; Abdel-Hamid, M.; Awad, A.A. The addition of buttermilk powder and transglutaminase improves textural and organoleptic properties of fat-free buffalo yogurt. *Dairy Sci. Technol.* **2014**, *94*, 297–309. [CrossRef]
48. Romeih, E.A.; Moe, K.M.; Skeie, S. The influence of fat globule membrane material on the microstructure of low-fat Cheddar cheese. *Int. Dairy J.* **2012**, *26*, 66–72. [CrossRef]
49. Serra, M.; Trujillo, A.J.; Jaramillo, P.D.; Guamis, B.; Ferragut, V. Ultra-High Pressure Homogenization-Induced Changes in Skim Milk: Impact on Acid Coagulation Properties. *J. Dairy Res.* **2008**, *75*, 69–75. [CrossRef]

50. Lucey, J.A.; Munro, P.A.; Singh, H. Rheological properties and microstructure of acid milk gels as affected by fat content and heat treatment. *J. Food Sci.* **1998**, *63*, 660–664. [CrossRef]
51. Keogh, M.K.; O'Kennedy, B.T. Rheology of stirred yogurt as affected by added milk fat, protein and hydrocolloids. *J. Food Sci.* **1998**, *63*, 108–112. [CrossRef]
52. Saffon, M.; Britten, M.; Pouliot, Y. Thermal aggregation of whey proteins in the presence of buttermilk concentrate. *J. Food Eng.* **2011**, *103*, 244–250. [CrossRef]
53. Patrignani, F.; Iucci, L.; Lanciotti, R.; Vallicelli, M.; Maina Mathara, J.; Holzapfel, W.H.; Guerzoni, M.E. Effect of High-Pressure Homogenization, Nonfat Milk Solids, and Milkfat on the Technological Performance of a Functional Strain for the Production of Probiotic Fermented Milks. *J. Dairy Sci.* **2007**, *90*, 4513–4523. [CrossRef]
54. Amador-Espejo, G.G.; Suarez-Berencia, A.; Juan, B.; Barcenas, M.E.; Trujillo, A.J. Effect of moderate inlet temperatures in ultra-high-pressure homogenization treatments on physicochemical and sensory characteristics of milk. *J. Dairy Sci.* **2014**, *97*, 659–671. [CrossRef]
55. Morin, P.; Pouliot, Y.; Britten, M. Effect of Buttermilk Made from Creams with Different Heat Treatment Histories on Properties of Rennet Gels and Model Cheeses. *J. Dairy Sci.* **2008**, *91*, 871–882. [CrossRef]
56. González-Martínez, C.; Becerra, M.; Cháfer, M.; Albors, A.; Carot, J.M.; Chiralt, A. Influence of substituting milk powder for whey powder on yoghurt quality. *Trends Food Sci. Technol.* **2002**, *13*, 334–340. [CrossRef]
57. Trachoo, N.; Mistry, V.V. Application of ultrafiltered sweet buttermilk and sweet buttermilk powder in the manufacture of nonfat and low fat yogurts. *J. Dairy Sci.* **1998**, *81*, 3163–3171. [CrossRef]
58. Jumah, R.Y.; Shaker, R.R.; Abu-Jdayil, B. Effect of milk source on the rheological properties of yogurt during the gelation process. *Int. J. Dairy Technol.* **2001**, *54*, 89–93. [CrossRef]
59. Lee, W.J.; Lucey, J.A. Formation and Physical Properties of Yogurt. *Asian Australas. J. Anim. Sci.* **2010**, *23*, 1127–1136. [CrossRef]
60. Lopez, C.; Camier, B.; Gassi, J.-Y. Development of the milk fat microstructure during the manufacture and ripening of Emmental cheese observed by confocal laser scanning microscopy. *Int. Dairy J.* **2007**, *17*, 235–247. [CrossRef]
61. Lanciotti, R.; Vannini, L.; Pittia, P.; Guerzoni, M.E. Suitability of high-dynamic-pressure-treated milk for the production of yoghurt. *Food Microbiol.* **2004**, *21*, 753–760. [CrossRef]
62. Shah, N.P. Functional cultures and health benefits. *Int. Dairy J.* **2007**, *17*, 1262–1277. [CrossRef]
63. Amatayakul, T.; Sherkat, F.; Shah, N.P. Syneresis in Set Yogurt as Affected by EPS Starter Cultures and Levels of Solids. *Int. J. Dairy Technol.* **2006**, *59*, 216–221. [CrossRef]

Article

The Application of Pureed Butter Beans and a Combination of Inulin and Rebaudioside A for the Replacement of Fat and Sucrose in Sponge Cake: Sensory and Physicochemical Analysis

Aislinn M. Richardson [1,2], Andrey A. Tyuftin [1], Kieran N. Kilcawley [3], Eimear Gallagher [4], Maurice G. O'Sullivan [2,*] and Joseph P. Kerry [1,*]

[1] Food Packaging Group, School of Food and Nutritional Sciences, University College Cork, College Road, Cork, Ireland; 108003904@umail.ucc.ie (A.M.R.); a.tiuftin@ucc.ie (A.A.T.)

[2] Sensory Group, School of Food and Nutritional Sciences, University College Cork, College Road, Cork, Ireland

[3] Food Quality and Sensory Science Department, Teagasc Food Research Centre, Moorepark, Fermoy, Cork, Ireland; kieran.kilcawley@teagasc.ie

[4] Ashtown Food Research Centre, Teagasc, Ashtown, D15, Dublin, Ireland; eimear.gallagher@teagasc.ie

* Correspondence: maurice.osullivan@ucc.ie (M.G.O.); Joe.Kerry@ucc.ie (J.P.K.); Tel.: +353-02-14903-798 (J.P.K.)

Citation: Richardson, A.M.; Tyuftin, A.A.; Kilcawley, K.N.; Gallagher, E.; O'Sullivan, M.G.; Kerry, J.P. The Application of Pureed Butter Beans and a Combination of Inulin and Rebaudioside A for the Replacement of Fat and Sucrose in Sponge Cake: Sensory and Physicochemical Analysis. *Foods* **2021**, *10*, 254. https://doi.org/10.3390/foods10020254

Academic Editor: Alessandra Marti
Received: 15 December 2020
Accepted: 21 January 2021
Published: 26 January 2021

Publisher's Note: MDPI stays neutral with regard to jurisdictional claims in published maps and institutional affiliations.

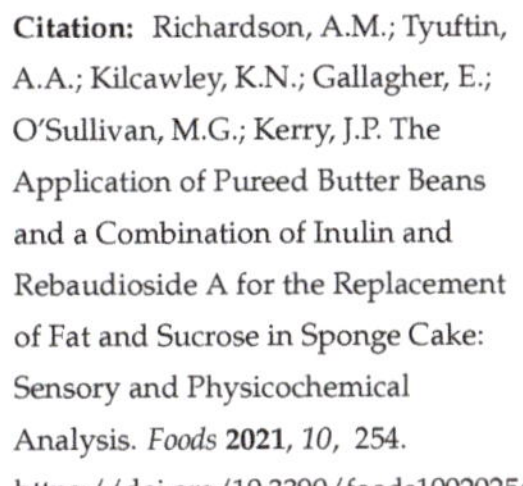

Abstract: Determining minimum levels of fat and sucrose needed for the sensory acceptance of sponge cake while increasing the nutritional quality was the main objective of this study. Sponge cakes with 0, 25, 50 and 75% sucrose replacement (SR) using a combination of inulin and Rebaudioside A (Reb A) were prepared. Sensory acceptance testing (SAT) was carried out on samples. Following experimental results, four more samples were prepared where fat was replaced sequentially (0, 25, 50 and 75%) in sucrose-replaced sponge cakes using pureed butter beans (Pbb) as a replacer. Fat-replaced samples were investigated using sensory (hedonic and intensity) and physicochemical analysis. Texture liking and overall acceptability (OA) were the only hedonic sensory parameters significantly affected after a 50% SR in sponge cake ($p < 0.05$). A 25% SR had no significant impact on any hedonic sensory properties and samples were just as accepted as the control sucrose sample. A 30% SR was chosen for further experiments. After a 50% fat replacement (FR), no significant differences were found between 30% sucrose-replaced sponge cake samples in relation to all sensory (hedonic and intensity) parameters investigated. Flavour and aroma intensity attributes such as buttery and sweet and, subsequently, liking and OA of samples were negatively affected after a 75% FR ($p < 0.05$). Instrumental texture properties (hardness and chewiness (N)) did not discriminate between samples with increasing levels of FR using Pbb. Moisture content increased significantly with FR ($p < 0.05$). A simultaneous reduction in fat (42%) and sucrose was achieved (28%) in sponge cake samples without negatively affecting OA. Optimised samples contained significantly more dietary fibre ($p < 0.05$).

Keywords: inulin; sugar reduction; sucrose replacement; fat reduction; Rebaudioside A; sponge cake; butter beans; sensory; physicochemical properties; bakery

1. Introduction

Sponge cake, which starts as a fluid batter, is classified by its solid porous structure after baking. It is prepared using four main ingredients, namely, eggs, sugar, wheat flour and fat, which are responsible for multiple functions in the batter and during baking. The presence of fat and sugar in food products yields positive hedonic responses which may override metabolic responses such as satiety [1].

According to WHO (World Health Organization) guidelines (2011), adults and children should consume only 10% of calories taken from sugars in their daily diet, which is about

10–14 teaspoons of sugar. In Ireland, this number achieved 14.6%, which is 31.5% higher than the recommended dosage [2]. This contributes to dietary imbalances associated with overweight and obesity and consequently the prevalence of non-communicable diseases [3]. As cakes contribute greatly to the dietary intake of sugar and fat in developed countries [4,5], it is necessary to find ways to reduce/replace sugar and fat in these products.

Before we consider sucrose and fat replacement/reduction strategies, we must first understand the functions performed by these ingredients in cakes. Sucrose is the most common sugar used in baking [6]. Sucrose is responsible for the tenderisation of cakes by competing with gluten for available water during batter mixing, thus impeding on the formation of a strong gluten network [7]. During baking, sucrose increases the temperature of starch gelatinisation and protein denaturation by binding water and therefore limiting water availability to starch granules [8]. In doing so, sucrose increases the volume and bulk of the final product. The most commonly known function of sucrose is that it provides sweetness to cakes and also acts a flavour enhancer [6]. In relation to the function of fats, fat entraps air in the batter during the creaming step which provides a structure for leavening gases [7]. Thus, fat also affects the volume and bulk of the final product. Fat tenderises and adds lubricity to the texture of cakes by coating the protein and starch particles, thereby interfering with the protein matrix [9]. Fat also emulsifies liquid in cake production which adds further moisture and softness to the product [10]. Fat sources such as butter contribute to flavour and can act as a flavour enhancer [10].

Thus, as sugar and fat perform many key functions in cakes, it is difficult to reduce and replace these ingredients without affecting the physical quality and important sensory attributes associated with liking and overall acceptability.

Several authors have studied the impact of sucrose replacement (SR) using a variety of different replacers such as polyols, fibres and artificial intense sweeteners on the overall quality of cakes [11]. Fibres such as inulin are of particular interest for use as sugar replacers as they possess prebiotic properties [12]. Inulin (Figure 1) is a plant-derived storage polysaccharide which is predominantly produced by chicory roots on a commercial basis.

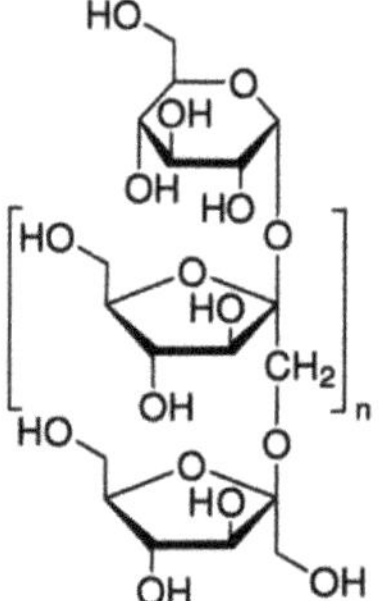

Figure 1. Chemical structure of inulin.

Roberfroid [13] reported that fructans such as inulin may have positive effects regarding colon cancer prevention. Furthermore, the incorporation of dietary fibre into foods is now considered as a principal prevention strategy against the risk of non-communicable disease [14]. Fibres such as inulin provide the bulk properties (texture and volume) by binding water [15] and therefore competing with gluten proteins for available water and delaying starch gelatinisation. Inulin contains 25–35% of the energy of digestible carbohydrates and its sweetness level is 10% that of sucrose [12]. Cakes containing 30% sugar replacement (SR) with inulin were shown to be similar to control sugar samples with regard to bubble size distribution and physical and sensory properties in a study conducted by [16]. As mentioned, inulin has a low sweetness value relative to that of sucrose, and therefore it is necessary to combine inulin with an intense sweetener. In a study conducted by Struck et al. (2016) [11,17], a 30% SR in muffins with inulin and Rebaudioside A (Reb A) produced cakes with a similar descriptive sensory profile to reference muffins. Reb A is a steviol glycoside (Figure 2) whose sweetness is 300 times greater in comparison with a sucrose [18,19]. Reb A represents white odorless crystals. Steviol glycosides such as Reb A were approved by the European Union for use in foods and are considered natural intense sweeteners [20]. These sweeteners are a special class of intense sweeteners as they are considered natural unlike most other intense sweeteners [20] because they are produced from Stevia leaves [21].

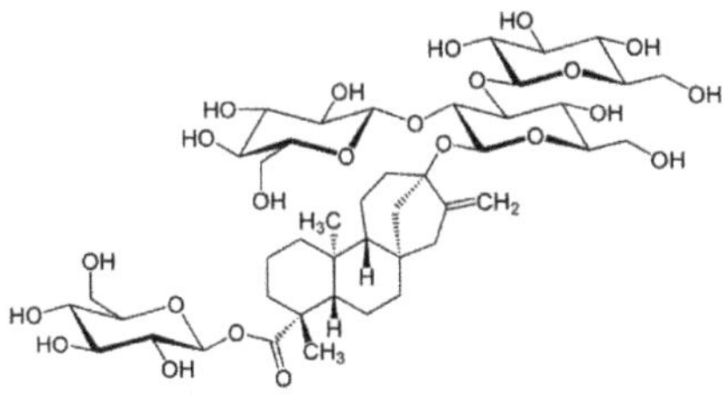

Figure 2. Structure of Rebaudioside A.

Therefore, the present study investigated the impact of a sequential SR (0, 25, 50 and 75) using a combination of inulin and Reb A on the sensory acceptance of sponge cake to determine the minimum levels of sucrose needed in the formulation to maintain sensory acceptance.

In relation to the reduction in/replacement of fat in cake products, three main classes of fat substitutes exist, namely, carbohydrate-based, protein-based and fat-based [22]. Carbohydrates are used as fat substitutes as they are similar to triglycerides in relation to physical and chemical structure [23] and they bind water and therefore contribute to texture and mouthfeel. In a study conducted using three commercial carbohydrate-based fat replacers derived from pectin, gums and oat bran, moisture content increased significantly in biscuits with a fat substitution [23]. Legumes such as beans are also used as fat replacers as beans are a source of complex carbohydrates roughly containing up to 60% total carbohydrates. Legumes are low glycaemic and high in protein and fibre and their consumption has been associated with a decreased risk of coronary heart disease [24]. In a study conducted on brownies, pureed black beans successfully replaced fat in relation to sensory acceptance at a level of 30% replacement [25]. Furthermore, this study reported that a 90% substitution of fat with black beans still produced an acceptable product. In a study conducted on oatmeal cookies, white beans were successful in replacing fat up to a level of 25% without significantly affecting sensory properties [26]. Cannellini beans have been used to successfully replace fat in brownies by up to 50% without affecting OA, texture and flavour properties [27].

For the above reasons, the present study utilised pureed butter beans (Pbb) for the sequential replacement of fat (0, 25, 50 and 75%) in reduced sugar sponge cake. Determining minimum levels of both sugar and fat needed to maintain sensory acceptance is imperative and could be a significant development in reducing the dietary intake of both fat and sugar.

Thus, the main aim of this work was, firstly, to determine the minimum levels of sucrose needed in sponge cake to maintain sensory acceptance similar to that of the control sucrose formulation. This was determined through sensory acceptance testing (SAT). Secondly, and following experimental results, the aim was to determine the minimum levels of fat needed to maintain sensory properties associated with liking and OA of reduced sugar sponge cake. This was determined by SAT and optimised descriptive profiling (ODP). Other characteristics such as texture, colour and compositional properties were studied for cakes during sequential FR of sucrose-replaced sponge cake.

2. Materials and Methods

Food ingredients used in this trial included free-range eggs (Upton brand, Ireland); caster sugar (99.9% sucrose, 0.3% moisture and 0.01% sodium, Tate & Lyle brand, UK); cream plain flour (82.7% carbohydrate, 2% of which were sugars, 11.7% protein, 3.4% fibre, 1.4% fat and 0.81% salt, Odlums brand, Ireland); Irish creamery butter (81% total fat, 65.4% of which were saturated and 15.1% moisture, Dunnes stores brand, Ireland); inulin (89% fibre and 8% sugar, Bioglan brand, Australia); Reb A, (Bulk Powders brand, Ireland); butter beans (16.8% carbohydrate, 1.4% of which were sugars, 0.6% fats 0.1% of which were saturated, 5.2% fibre, 0.01% sodium and 0.03% salt, Suma brand, UK); baking powder (3% sodium, Royal brand, US); and whole milk (4.7% carbohydrate, 4.7% of which were sugars, 3.5% fat, 2.3% of which were saturated, 3.4% protein and 0.1% salt, Dunnes stores brand, UK). All food products were obtained from a local supermarket and stored under refrigerated or cool, dry conditions prior to sample preparation.

2.1. Reb A Concentration Adjustment

Optimised descriptive profiling (ODP) was used to determine the concentration of Reb A needed to replace the sweetness concentration of sucrose, ensuring iso-sweetness. These intensity tests were carried out twice using 21 assessors. Concentration adjustments for Reb A were carried out according to the method of [28–31] using the same concentrations of stevia (0.06–0.16 g/L) and 24 g/L sucrose. In order to compare the means of the obtained date, one-way ANOVA was used. Tukey's post hoc test was used to adjust for multiple comparisons between treatment means using SPSS statistics 20 software (IBM, Armonk, NY, USA).

2.2. Sponge Cake Treatments

The formulation used for the preparation of the control sponge cake treatment was based on conversations had with local bakeries, cookbook recipes and associated websites. Three separate batches of sponge cake for all experimental treatments (8) were formulated and manufactured using recipes outlined in Table 1.

Table 1. Formulation of different sponge cake treatments.

	Samples	Sucrose	Inulin	Reb A	Butter	Pureed Butter Beans	Flour	Milk	Baking Powder	Eggs
						%				
Sucrose replacement	SC0/0	26.1	0	0	18.6	0	26.1	1.7	0.7	26.8
	SC25/0	19.55	6.49	0.02	18.6	0	26.1	1.7	0.7	26.8
	SC50/0	13.04	13.0	0.04	18.6	0	26.1	1.7	0.7	26.8
	SC75/0	6.51	19.49	0.06	18.6	0	26.1	1.7	0.7	26.8
Fat replacement	SC30/0	18.23	7.82	0.02	18.6	0	26.1	1.7	0.7	26.8
	SC30/25	18.23	7.82	0.02	13.96	4.7	26.1	1.7	0.7	26.8
	SC30/50	18.23	7.82	0.02	9.3	9.3	26.1	1.7	0.7	26.8
	SC30/75	18.23	7.82	0.02	4.7	13.96	26.1	1.7	0.7	26.8

SC; sponge cake, the first digit represents the sucrose replacement level and the second digit represents the fat replacement level of samples.

During the first phase of this trial, a control treatment with 0% SR was prepared. Sucrose was replaced by increments of 25% using a combination of inulin and Reb A (0, 25, 50 and 75%). Thus, 4 treatments were formulated during the first phase of this study and samples were identified as follows; SC0/0, SC25/0, SC50/0 and SC75/0. The most accepted sample was chosen from sensory acceptance testing and used for further analysis in determining the minimum levels of fat needed to maintain sensory properties associated with liking and OA of sugar-replaced sponge cake. Thus, four more formulations were prepared where fat was replaced sequentially by increments of 25% using pureed butter beans. The samples were identified as follows; SC30/0, SC30/25, SC30/50 and SC30/75, where the first digit denotes the SR replacement level and the second digit represents the FR level.

2.3. Sponge Cake Preparation

Butter was creamed in an electronic mixer (Kitchen Aid Professional mixer) and sugar was added gradually until soft and light in colour, at speed 2–3 for 4 min. Eggs (average and similar in size, without weighing) were added one at a time and beaten well between each addition at speed 2 for approximately 1 min, for each egg. Flour and baking powder were sifted in to the mixture and were mixed together at minimum speed for 2 min, before milk was added. For SR, inulin and stevia were added to the mixture during the creaming stage in partial replacement of sucrose. For FR, butter beans were firstly drained and then pureed in a Stephan mixer (UMC-5 Stephan u. Sohner & Co, Hameln, Germany) at 21 RPM for 5 min before being used in partial replacement of butter. Batter was poured into circular baking tins (9 inch) and cooked in a Zanussi convection oven (C. Batassi, Conegliano, Italy) for 30 min at 180 °C. After cooking, sponge cakes were left for 20 min in the tin. Then, cakes were placed on a rack to cool down to room temperature and were placed inside plastic storage containers before testing.

2.4. Sensory Analysis

2.4.1. Sensory Acceptance Testing (SAT)

SAT was carried out in the sensory science laboratory at University College Cork according to ISO 11136:2014 [32]. Using 25 untrained assessors ($n = 25$) who were all familiar with the products being tested, SAT took place over six separate sessions as three independent trials were carried out for both phases of this study (FR and SR). A three-digit random code was used to assign the samples ($20 \times 20 \times 20$ mm). Sensory sessions were carried out at room temperature under white light. Participants were instructed to use the water provided to cleanse their palates between tastings and used the following hedonic descriptors to rate their degree of liking of sponge cake samples: appearance, flavour, texture, colour and aroma liking. Assessors were asked to indicate their degree of liking for samples on a 9-point, numbers-only hedonic scale. Words were only used to anchor the scale at both ends with the term "extremely dislike" on the far-left end of the scale and the term "extremely like" on the far-right-hand side of the scale. Overall acceptability (OA) of samples was also determined using this scale. The 150 responses were collected for each sample ($25 + 25 + 25 \times 2$) in three independent trials for each phase, using 25 untrained assessors (samples were presented in duplicate). The statistical analysis method is shown in Section 2.7.

2.4.2. Optimised Descriptive Profiling (ODP)

Optimised descriptive profiling [30–32] was also carried out in the special sensory science laboratory of University College Cork. A separate panel of 21 assessors ($n = 21$) who were all regular consumers of sponge cake and who had all received prior training in descriptive analysis were trained and participated in this separate ODP test. In order to prevent the carry-over effect and first order, these assessors were served all samples randomly [33]. ODP only took place on samples during the second phase of this study after an accepted minimum level of sucrose content was determined by SAT. Thus, ODP sessions ran concurrently with SAT sessions during fat optimisation and therefore took place over three weeks. Sensory descriptors were selected from the panel discussion as the most appropriate and reflected the main variation in the samples profiled. The consensus list of intensity descriptors (Table 2) was measured on a 10 cm continuous line scale. The term "none" was used as the anchor point for the 0 cm end of the scale and the term "extreme" was used as the anchor point for the 10 cm end of the scale, unless stated otherwise in Table 2. The samples ($20 \times 20 \times 20$ mm) were served coded and presented simultaneously to assessors [34]. All samples were measured on the same scale for each intensity descriptor and presented in replicate. The statistical analysis method is shown in Section 2.7.

Table 2. Consensus list of sensory descriptors and definitions selected by panel and used in ranking descriptive analysis of sponge cake.

Attributes	Anchor Points on Scale	Definition
Springiness	None–extreme	Rate/speed by visual observation that the sponge cake returns to its original shape after the deforming force is removed
Appearance		
Crust darkness	Light–dark	Degree of colour darkness ranging from light brown to dark brown
Porous	None–extreme	Amount of bubbles and voids present in the inner mass of the sponge cake
Aroma		
Sweet aroma	None–extreme	Fundamental smell sensation of which sucrose is typical
Buttery aroma	None–extreme	Aromatics associated with butter produced from cow's milk
Caramel aroma	None–extreme	Odour produced when caramelising sugar without burning it
Texture		
Moisture	Dry–moist	Wet texture in mouth
Hardness	Soft–hard	The resistance of the cake to breaking upon pressure of the front teeth during biting
Crumbly	None–extreme	Easily broken up in the mouth into a lot of little pieces
Dense	Light–heavy	A heavy texture in mouth
Taste		
Sweet taste	None–extreme	Taste sensation associated with sucrose
Buttery flavour	None–extreme	Flavour sensation associated with butter; creamy mouthfeel and buttery aroma

2.5. Physicochemical Analysis

Physicochemical properties were determined for the following samples: SC0/0, SC30/0, SC30/25, SC30/50 and SC30/75.

2.5.1. Texture Profile Analysis (TPA)

Two sponge cake samples (30 × 30 mm) were obtained from the centre of each cake and used for texture analysis. Thus, results obtained for TPA on the Texture Analyser 16 TA-XT2I (Stable Micro Systems, Surrey, UK) represent a mean of 6 values (3 × 2 = 6). Sponge cake samples were sliced horizontally to a height of 30 mm. A double compression test for 50% was used for TPA with use of a 35 mm (in diameter) flat-ended cylindrical probe (P/35), at a speed of 1 mm/s with a 5 s time between the two cycles. This was carried out in accordance with the method of [35]. The test was carried out in triplicate for all treatments.

2.5.2. Colour

Two sponge cake samples (30 × 30 mm) from the centre of each cake were used for colour analysis. The CIE L*a*b* method was used for crust and crumb colour measurement. A Minolta CR-200B Chroma Meter (Minolta Camera Co. Ltd., Osaka, Japan) was used for L* (lightness), a* (green-red) and b* (blue-yellow) colour components measurement. Colour parameters were measured at two separate points directly from the top of each individual sponge cake sample. For this, the sponge cake samples were horizontally cut to remove the crust, and crumb colour was measured directly at two separate points. As two measurements for crust and crumb colour were taken for each individual sample and two samples were tested for each individual trial, of which there were three, crust and crumb colour values represent a mean of twelve measurements (2 × 2 × 3).

2.5.3. Moisture and Fat

Moisture and fat were detected according to Bostian et al. [36]. Two samples (30 × 30 mm) from each cake treatment were used for moisture and fat determination. As three independent trials were carried out, results obtained for moisture and fat represent a mean of 6 values (3 × 2 = 6). The Büchi Mixer B-400 (Büchi Labortechnik AG, Switzerland) was used in order to homogenise for compositional analysis. Moisture content was detected on the CEM SMART system and fat content was detected on the SMART Trac system (CEM GmbH, kamp-Lintfort, Germany). Two fibreglass pads were placed in the drying chamber of the CEM SMART system and their weight was tared. The pads were removed and the homogenised samples (2–4 g) were weighed accurately on the pads using separate weighing scales. Following this, one pad was placed over the sample, pressed together and placed back into the drying chamber to begin drying. The moisture (%) was displayed within a few minutes. A sheet of Smart Trac film was used to determine a percent of fats by wrapping the fibreglass pads with the sample. After wrapping, the samples were placed in a tube of the Smart Trac system and positioned in the Smart Trac NMR unit. After about 5 min, the percentage of fat was determined.

2.5.4. Protein

Results obtained from protein analysis represent a mean of 6 values (3 × 2 = 6). Protein content (%) was determined using the Kjeldahl method which was carried out according to [37]. Before testing commenced, the digestion block (Foss Tecator Digestor, Hillerød, Denmark) was pre-heated to 410 °C. Samples (0.5–2.0 g) were weighed accurately into digestion tubes. Two "kjeltabs" were added to each sample in the fume hood followed by 15 mL of sulphuric acid and 10 mL hydrogen peroxide. Two controls containing no sample were prepared in the same way. Tubes were placed in the heated digestion block and left there for roughly 30–40 min until they became colourless. At this point, the distillation unit was turned on and rinsed out by connecting a blank tube and receiver flask to the unit and pressing the steam button. Distilled water (50 mL) was added to each digested sample in the fume hood after the samples had cooled thoroughly. Samples were placed one by one into the distillation unit (Foss Kjeltec 2100, Hillerød, Denmark) along with a receiver flask containing 50 mL of 4% boric acid with an indicator. After the distillation process was complete, the contents of the receiver flask were titrated with 0.1 N hydrochloric acid until the green colour reverted back to the original red colour. Nitrogen content was converted to protein content using the factor 6.25.

2.5.5. Ash

Results obtained represent a mean of 6 values (3 × 2 = 6). A muffle furnace was used for ash content (%) detection. Homogenised samples (5 g) were weighed into small silica dishes and placed into the muffle furnace. The samples were heated up to 600 °C until only the inorganic material was left which was indicated by the light-white colour of the samples. The silica dishes containing the samples were then placed in a desiccator to cool and the dishes were weighed carefully. Ash (%) was calculated using the following equation:

$$\% \text{ Ash} = \text{weight of Ash} \times 100 \text{ weight of sample} \tag{1}$$

2.5.6. Carbohydrate

Carbohydrate (%) content was determined using the following calculation:

$$\% \text{ Carbohydrate} = 100 - (\text{Moisture\%} + \text{Fat\%} + \text{Protein\%} + \text{Ash\%}) \tag{2}$$

2.5.7. Dietary Fibre

Total dietary fibre was established according to the AOAC 985.29 method [38]. Sponge cake samples went through sequential enzymatic digestion using heat-stable α-amylase (95–100 °C, pH 6.0, 15 min), Subtilisin A (60 °C, pH 7.5, 30 min) and amyloglucosidase

(60 °C, pH 4.5, 30 min) to remove digestible carbohydrates and protein. To establish total dietary fibre content, the enzyme digests were precipitated with four volumes of 95% (*v/v*) ethanol, filtered and then washed with 78% (*v/v*) ethanol and acetone before being dried in an oven at 110 °C.

2.5.8. Total Sugars and Sucrose Content

Total sugars and sucrose content of samples was determined by ion chromatography, using a method developed internally by an independent accredited laboratory based in Ireland.

2.6. Sponge Cake Images

Photographs of the following sponge cake samples: SC0/0, SC30/0, SC30/25, SC30/50 and SC30/75, were taken one day after baking in the photograph room of the Food Science Building, University College Cork. Photographs were taken using a digital camera, Nikon D5300, equipped with a lens, Nikon DX AF-P NIKKON (18–55 mm, 1:3.5−5.6 G), Japan. Images were taken without flash with the following modes: macro, F6.3, 1/125, ISO 400. After images were taken, sharpness was increased to 100% and the saturation was adjusted for all images.

2.7. Statistical Analysis

Raw data obtained from sensory and physicochemical analysis were coded in Microsoft Excel. For sucrose optimisation, the significance of hedonic sensory properties in discriminating between the samples was analysed using ANOVA and Tukey's post hoc test (SPSS statistics 20 software (IBM, Armonk, NY, USA). The relationship between the set of samples (4) and the set of hedonic sensory variables was determined by partial least squares (PLS) regression using Unscrambler software (Unscrambler 10.3 CAMO software ASA, Trondheim, Norway). In the PLS regression, only hedonic sensory properties that discriminated significantly between samples were used. The X-matrix was defined as the different sample treatments. The Y-matrix contained the significant sensory variables of the design. For fat optimisation, the relationship between the set of sample treatments (X) and the set of sensory (hedonic and intensity) and physicochemical variables (Y) was examined by PLS regression. Again, only sensory and physicochemical properties that discriminated significantly between samples (4) were used. Both the sensory and physicochemical data were normalised during pre-processing by taking the logarithm to achieve uniform precision over the whole range of variation. Data were also standardised by dividing each variable (sensory and physicochemical) by its standard deviation. This process was necessary as the units of the studied variables were different. To achieve significant results for the relationships determined in quantitative PLS regression analysis, coefficients were analysed by jack-knifing which was based on custom cross-validation and stability plots [39]. Statistical significance for the relationships analysed by PLS was defined as $p < 0.05$–0.01 (significant), $p < 0.01$–0.001 (highly significant) and $p < 0.001$ (extremely significant).

TPA and proximate composition data were presented as a mean of six values ± standard deviation. Estimated fibre and sucrose contents were presented as a mean of three values ± standard deviation. Colour (crust and crumb) data were presented as a mean of twelve values ± standard deviation. One-way ANOVA was used to compare the means of the data obtained from physicochemical analysis. Tukey's post hoc test was used to adjust for multiple comparisons between treatment means using SPSS statistics 20 software (IBM, Armonk, NY, USA).

3. Results and Discussion

3.1. Reb A for Sucrose Replacement

The sweetness rankings of six different concentrations of Reb A and one standard solution of sucrose are presented in Table 3.

Table 3. Iso-sweetness of Rebaudioside A in aqueous solutions.

Sweetener	Concentration (g/L)	Mean Scores	Dilution Factor
Reb A	0.060	1.2 [a]	400
Reb A	0.069	5.4 [b]	350
Reb A	0.080	6.2 [c]	300
Reb A	0.096	6.3 [c]	250
Reb A	0.120	7.4 [d]	200
Reb A	0.160	9.4 [d]	150
Sucrose	24.0	5.6 [b]	n/a

[abcd] mean values (±standard deviation) in the same column bearing different superscripts are significantly different, $p < 0.05$. n/a: Only Reb A solution was deluted. Sucrose was used in solid state.

The Reb A solution containing 0.069 g/L did not obtain significantly different scores from the standard sucrose solution with regard to sweetness intensity. For this reason, a sucrose-to-Reb A ratio of 1:350 was chosen. This means that for samples containing 25% SR, which equates to a reduction of 58.3 g sucrose in the formulation, 0.17 g of Reb A was used to replace the sweetness (58.3/350). The same method was applied to samples with 50 and 75% SR. Inulin was added on a weight by weight basis.

3.2. Sensory Analysis

3.2.1. Sensory Acceptance of Sucrose-Replaced Cakes

The relationship between hedonic sensory variables (Y) and sponge cake samples prepared with increasing levels of SR (X) is visually represented by a partial least squares regression plot (PLSR) shown in (Figure 3).

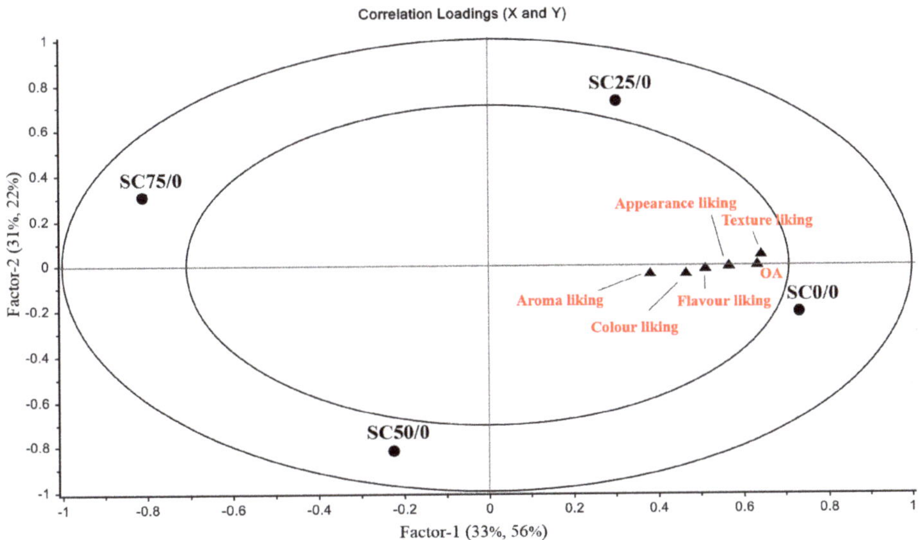

Figure 3. Partial least squares regression (PLSR) plot for the relationship between sponge cake samples with increasing levels of sucrose replacement: 0, 25, 50 and 75%, and hedonic sensory variables.

Most of the variation is shown in Factor-1, where 33% of the data in X explain 56% of the Y data. The SC0/0 sample which is situated in the outer circle of the lower right quadrant made a significant contribution to Factor-1. This sample was highly correlated with all liking parameters (aroma, appearance, colour, texture and flavour) and OA. The SC25/0 sample, which is positioned in the outer circle of the upper right quadrant, was also positively associated with all liking terms and OA. The SC75/0 sample, which is situated in the outer circle of the upper left quadrant, was negatively correlated with all liking parameters and OA. The SC/50 sample, which is positioned in the outer circle of the bottom left quadrant, made a significant contribution to Factor-2 and was slightly negatively correlated with liking terms and OA. It is evident from the plot that all liking parameters were correlated as shown by their proximity to each other, particularly so for texture liking and OA. To aid further understanding of the relationship between sensory terms and sponge cake samples, the significance of the estimated regression coefficients for the relationship between these two sets of variables can be seen in Table 4.

Table 4. Significance of estimated regression coefficients (ANOVA values) for the relationship of hedonic sensory parameters (Y) and sponge cake samples prepared with increasing levels of sucrose replacement (X).

	Hedonics					
	Aroma Liking	**Appearance Liking**	**Colour Liking**	**Texture Liking**	**Flavour Liking**	**Overall Acceptability**
SC0/0	0.57	0.68	0.56	0.57	0.68	0.56
SC25/0	0.55	0.84	0.56	0.57	0.76	0.66
SC50/0	0.57	0.84	0.57	−0.04 *	0.83	−0.03 *
SC75/0	−0.05	−0.03 *	−0.04 *	−0.01 **	−0.03 *	−0.02 *

Significance of regression coefficients * = $p \leq 0.05$, ** = $p \leq 0.01$, (−) indicates whether the relationship is negatively correlated.

Resembling results which are visually represented in the PLSR plot, the SC75/0 sample was significantly negatively associated with aroma, colour, flavour liking and OA ($p < 0.05$) and significantly negatively associated with texture liking ($p < 0.01$). Although it was evident from the PLS plot (Figure 3) that a 50% replacement of sucrose with inulin and Reb A was unsuccessful in terms of liking, the negative response was only significant for texture ($p < 0.01$) and subsequently OA ($p < 0.05$) (Table 4). In a study conducted on orange cakes, the addition of inulin negatively affected texture liking, whereas flavour liking and aroma liking were not affected [40]. In a study conducted by [17], muffins containing 30% SR with inulin and Reb A did not differ from the reference sample with regard to aromatics, browning, cracks on crust or buttery and sweet flavour. Texture liking was important to the acceptability of samples in this study; therefore, low texture liking scores were a good predictor for the rejection of samples. The liking of aroma, appearance, colour and flavour was not significantly affected by up to a level of 50% SR (Table 4). As samples containing 25% SR were so similar to the reference sample in this study and because a 30% replacement was achieved by [17] without significantly affecting important sensory attributes, an SR level of 30% was chosen for further experiments to determine the minimum levels of fat needed to maintain sensory properties associated with the liking and OA of samples.

3.2.2. Sensory and Physicochemical Properties of Reduced Sucrose Cakes with Increasing Levels of FR

The second part of this study involved the sequential replacement of fat in 30% sucrose-replaced sponge Cake with Pbb. The relationship between sensory and physicochemical variables (Y) and sucrose-replaced sponge cake prepared with 0, 25, 50 and 75% FR with Pbb (X) is visually represented by a PLSR plot in Figure 4.

The following hedonic sensory terms were left out of PLSR analysis because they did not significantly discriminate between samples: appearance, colour and texture liking. The following intensity sensory terms were omitted for the same reason: porous appearance, crumbly texture and dense texture. The following physical parameters and compositional properties were also left out of PLSR analysis for the same reason: hardness, gumminess and chewiness (N), springiness (mm) and cohesiveness, crumb a* and b* values and ash, carbohydrate, total sugars and sucrose content. Most of the variation in the plot is shown in Factor-1, where 33% of the X data explain 51% of the data in Y. The SC30/0 sample, which is shown in the outer circle of the lower right quadrant, makes a significant contribution to Factor-1. It is evident that the SC30/0 sample was positively associated with aroma and flavour liking and OA which can also be seen on the right-hand side of the plot. The following intensity sensory terms were correlated with liking parameters and OA and can be seen in close proximity with these hedonic parameters on the right-hand side of the plot: sweet, buttery and caramel aroma, sweet taste, butter flavour, crust and crumb darkness, moist texture, springiness and fat content (%). These intensity sensory terms can be considered as positive sensory variables associated with liking and OA. Actual fat content (%) was a big predictor of acceptability as it can be seen in very close proximity with aroma and flavour liking and OA. The SC30/75 sample, which is positioned in the outer circle of the lower left quadrant, was highly negatively correlated with liking parameters (aroma and flavour), OA and all sensory properties associated with liking and OA, situated on the right-hand side of the plot. This sample was correlated with intensity sensory terms and physicochemical variables situated on the left-hand side of the plot (perceived texture hardness, crust lightness, redness and yellowness and actual moisture, protein and fibre content (%)). These parameters were therefore negatively correlated with liking parameters and OA.

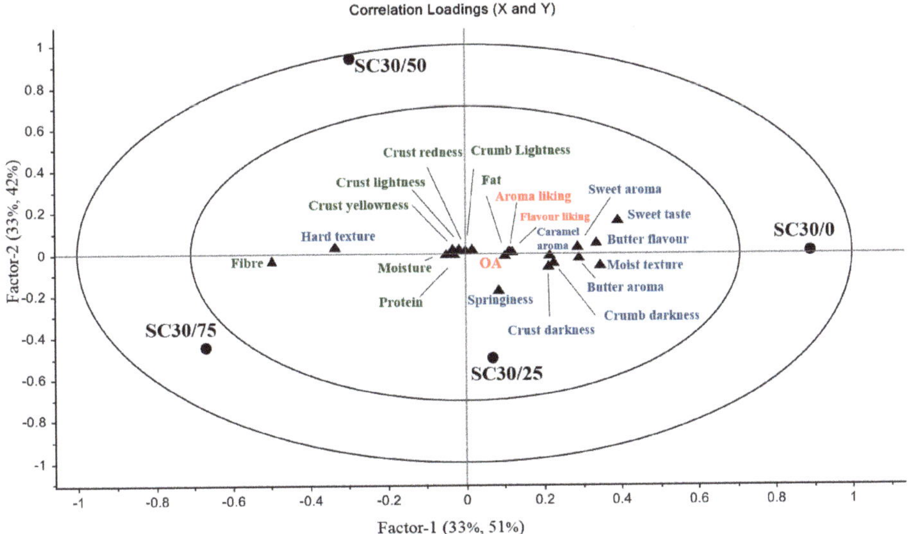

Figure 4. Partial least squares regression (PLSR) plot for the relationship between reduced sugar sponge cake (30%) prepared with increasing levels of fat replacement: 0, 25, 50 and 75%, sensory and physicochemical variables, hedonic and intensity sensory terms and physicochemical parameters.

The significance of the estimated regression coefficients for the relationship between samples and sensory and physicochemical parameters can be seen in Table S1 (Supplementary Materials). After custom cross-validation of PLSR analysis, the sample containing 75% replacement of fat with Pbb (SC30/75) was the only sample significantly negatively associated with aroma and flavour liking ($p < 0.05$) and significantly negatively associated with OA ($p < 0.01$). Pureed butter beans were therefore successful in terms of acceptance and liking in the replacement of fat in sucrose-replaced sponge cake up to a level of 50%. Similar results were reported by [27], who found that significant differences existed only when 75% of shortening was replaced with pureed cannellini beans in brownies.

The SC30/75 sample was significantly negatively associated with all intensity attributes associated with liking ($p < 0.05$), particularly sweet taste and butter flavour ($p < 0.01$) (Table S1, Supplementary Materials). A reduction in perceived butter flavour was expected at this level of butter replacement with Pbb and similar results on FR in biscuits have been reported by Laguna et al. [41]. The combined effect of fat and sugar on sensory acceptability has been demonstrated previously by Biguzzi et al. [42], who found that perceived sweetness intensity declined with fat reduction in biscuits. Therefore, it is not surprising that sweetness intensity was significantly affected at a level of 75% FR in this study. A reduction in aroma attributes such as butter, caramel and sweet is also not surprising. Butter aroma is stronger than the neutral aroma of butter beans, and a reduction in perceived butter aroma may have had a carry-over effect on caramel and sweet aroma perception, considering the synergistic relationship between sugar and butter in cake products. The decrease in perceived moist texture was also expected at this level of FR as fats have a lubricating effect and produce a sensation of moistness in the mouth [43]. Contradictory to sensory results obtained for moist texture perception, the SC30/75 sample was significantly positively ($p < 0.05$) associated with actual moisture content (%) (Table S1). The difference in results obtained for actual moisture content (%) and sensory results obtained for moist texture perception (Figure 3) highlights the importance of fat/butter content to the perception of moist texture, lubricity and, subsequently, acceptance. The moisture content of samples will be discussed in further detail in Section 3.4.

As mentioned, results obtained for texture profile analysis (TPA) did not discriminate between samples and so these parameters were omitted from the PLS plot. TPA results will be discussed in further detail in Section 3.3. Sensory results obtained for perceived sample hardness, however, did discriminate between samples and the SC30/75 sample was significantly associated with this intensity attribute ($p < 0.05$). Perceived hard texture was negatively correlated with positive sensory properties and liking parameters; however, it was not negatively correlated with texture liking as liking of texture did not discriminate between samples. As texture liking did not discriminate between samples, it can be said that perceived sample hardness was not a determinate for the lower acceptance scores obtained by the SC75/75 sample.

Crust lightness (L*), which is positioned on the left-hand side of the plot (Figure 3), was significantly positively associated with the SC30/75 sample ($p < 0.05$) (Table S1). This sample was significantly negatively associated with the sensory terms crust darkness and crumb darkness (Table S1), which were highly correlated with one another (Figure 3). Crust darkness and crumb darkness were intensity attributes most associated with samples containing low levels of FR (Figure 3). Confortiv et al. [20] also reported an increase in lightness of biscuit samples with FR using carbohydrate-based fat replacers. Samples containing 75% FR were also more yellow in colour than any other samples ($p < 0.05$) (Table S1). Instrumental results obtained for colour will be discussed in further detail in Section 3.3. Although perceived colour intensity and instrumental colour parameters discriminated between samples, as mentioned, liking of colour and appearance did not discriminate and, as a result, these liking parameters were omitted from the PLS plot. This finding suggests that in addition to textural parameters, colour parameters were not extremely important to the OA of samples. Contradictory to our findings on colour liking, in a study conducted on oatmeal cookies, samples containing higher levels of FR (50 and 75) with white beans were significantly different to samples containing 0 and 25% FR with regard to colour liking [27].

As flavour liking and aroma liking were the only hedonic parameters that significantly discriminated between samples, flavour (butter, sweet) and aroma (butter, sweet and caramel) intensity attributes were, therefore, determinants of OA. Although a dryer and harder texture was associated with samples containing 75% FR (Table S1, Supplementary Data), these textural parameters were not found to significantly affect the texture liking of samples. Samples containing this level of FR were perceived as lighter and more yellow in colour, which was supported by instrumental analysis.

However, colour or appearance liking did not significantly discriminate between sample treatments. These findings demonstrate that Pbb were successful in replacing fat in terms of colour and texture liking but were unsuccessful in terms of flavour and aroma liking up to a level of 75% replacement in 30% sucrose-replaced sponge cake. Hence, aroma liking and flavour liking were determinants of OA.

No significant differences were found in relation to any intensity sensory attributes or hedonic parameters up to a level of 50% FR in sucrose-replaced sponge cake. Thus, flavour liking and aroma liking and, subsequently, OA were unaffected by this level of FR.

3.3. Physical Properties and Images of Sponge Cake Samples

Mean values from physical analysis on the following samples of sponge cake are displayed in Table 5: SC0/0, SC30/0, SC30/25, SC30/50 and SC30/75. Cross-section images of sponge cake samples are depicted in Figure 5.

Table 5. Physical properties of sponge cake samples.

		SC0/0	SC30/0	SC30/25	SC30/50	SC30/75
TPA para-meters	Hardness (N)	8.2 ± 0.55 [a]	15.1 ± 0.14 [b]	13.2 ± 0.44 [b]	14.1 ± 0.76 [b]	14.2 ± 1.52 [b]
	Gumminess (N)	4.8 ± 0.66 [a]	6.0 ± 0.78 [b]	5.8 ± 0.33 [b]	5.9 ± 0.72 [b]	6.4 ± 0.60 [b]
	Chewiness (N)	8.3 ± 0.58 [a]	3.9 ± 0.59 [b]	3.9 ± 0.82 [b]	3.8 ± 0.75 [b]	3.7 ± 0.55 [b]
	Springiness (mm)	0.7 ± 0.12 [a]	0.6 ± 0.03 [a]	0.7 ± 0.04 [a]	0.6 ± 0.07 [a]	0.7 ± 0.05 [a]
	Cohesiveness	0.6 ± 0.71 [a]	0.4 ± 0.04 [a]	0.4 ± 0.03 [a]	0.4 ± 0.04 [a]	0.4 ± 0.08 [a]
Crust colour	Lightness (L*)	39.8 ± 0.65 [a]	40.5 ± 0.41 [a]	41.0 ± 0.77 [a]	45.5 ± 0.40 [b]	49.8 ± 0.32 [c]
	Redness (a*)	15.1 ± 0.25 [a]	15.8 ± 0.38 [a]	16.8 ± 0.88 [ab]	16.9 ± 0.60 [b]	17.5 ± 0.78 [b]
	Yellowness (b*)	24.1 ± 0.61 [a]	24.8 ± 0.88 [a]	25.2 ± 0.66 [ab]	29.8 ± 0.20 [b]	34.1 ± 0.40 [c]
Crumb colour	Lightness (L*)	69.2 ± 0.44 [a]	70.8 ± 0.50 [a]	71.6 ± 0.31 [ab]	72.3 ± 0.45 [b]	70.7 ± 0.51 [a]
	Redness (a*)	−3.1 ± 0.32 [a]	−2.8 ± 0.21 [a]	−2.7 ± 0.42 [a]	−2.8 ± 0.45 [a]	−2.7 ± 0.36 [a]
	Yellowness (b*)	31.0 ± 0.24 [a]	29.8 ± 0.60 [a]	27.7 ± 0.80 [a]	27.2 ± 0.30 [a]	28.0 ± 0.54 [a]

[abc] mean values (±standard deviation) in the same row bearing different superscripts are significantly different, $p < 0.05$.

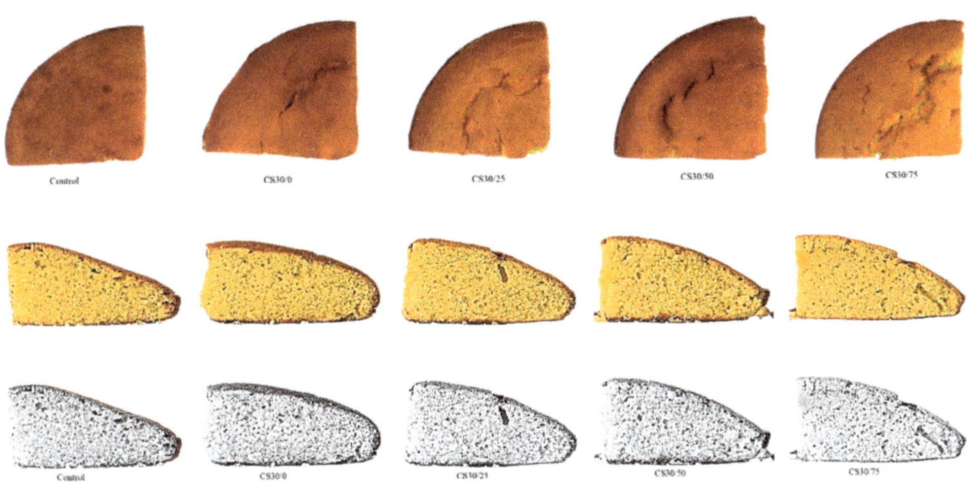

Figure 5. Top view (crust) and cross-section images of sponge cake fat reduced samples. From the left: control, S30/0, S30/25, S30/30, S30/50, S30/75.

A 30% SR in sponge cake significantly increased sample hardness ($p < 0.05$). This result was expected as sucrose plays a vital role in the tenderisation of baked goods, and similar results in relation to increased sample hardness with the addition of inulin have been reported by O' Brien et al. [44] and Volpini-Rapina et al. [40] in breadcrumbs and orange cakes, respectively. The chewiness (N) of samples also decreased significantly after a 30% SR. As previously mentioned, instrumental texture hardness values (N) did not significantly discriminate between samples with increasing levels of FR using Pbb (Table 5). As fats also play a vital role in the tenderisation of dough, an increase in crumb hardness was expected at higher levels of replacement. Souza et al. [45] reported an increase in the crumb firmness of pound cakes with only a 25% replacement of butter with green banana puree. However, as mentioned, instrumental TPA values did not correlate with sensory results and samples containing 75% FR were perceived as harder. This sample was also perceived as significantly dryer which may have had a carry-over effect on the perceived hardness of samples.

A 30% SR in sponge cake using inulin and Reb A did not significantly affect instrumental crust or crumb colour properties. This finding was supported by images depicted in Figure 5. Crust lightness, however, increased as FR increased in samples containing 30% SR in the range of (40.5 ± 0.41) for the SC30/0 sample and (49.8 ± 1.32) for the SC30/75 sample ($p < 0.05$) (Table 5). This finding was also visually supported by the images depicted in Figure 5. Crust yellowness (b* values) increased as FR increased in sucrose-replaced samples in the range of (15.8 ± 0.38) for the SC30/0 sample and (17.5 ± 0.78) for the SC30/75 sample ($p < 0.05$). Instrumental results obtained for crust colour therefore correlate with sensory results. The increase in crust yellowness observed with increasing levels of FR in our study could be attributed to the yellow colour associated with butter beans.

Crumb lightness values (b*) also discriminated between samples with increasing levels of FR; however, uneven trends were observed with the SC30/25 and the SC30/50 samples obtaining the highest values for crumb lightness ($p < 0.05$). This result did not correlate with sensory results as the SC30/75 sample was the sample most negatively correlated with crumb darkness, which means they were perceived as the lightest samples. A plausible reason for this is that sensory panellists failed to discriminate between crust and crumb colour, and this is supported by the high correlation between these two attributes, depicted in Figure 2. Images displayed in Figure 3 may help to explain the difficulty to distinguish between crust and crumb colour.

3.4. Proximate Composition of Sponge Cake Samples

As displayed in Table 6, a 30% SR in sponge cake did not significantly affect the moisture content (%) of samples. Moisture content, however, increased significantly as FR increased in 30% sucrose-replaced samples in the range of 22.1% for the SC30/0 sample to 32.4% for the SC30/75 sample ($p < 0.05$) (Table 6).

Table 6. Proximate composition (%) of sponge cake samples.

	SC0/0	SC30/0	SC30/25	SC30/50	SC30/75
Moisture	21.6 ± 0.55 [a]	22.1 ± 0.01 [a]	27.5 ± 0.03 [b]	28.6 ± 0.42 [b]	32.4 ± 0.06 [c]
Fat	12.9 ± 0.41 [a]	12.6 ± 0.74 [a]	10.1 ± 0.26 [b]	7.5 ± 0.35 [c]	4.5 ± 0.33 [d]
Protein	6.1 ± 0.47 [a]	6.0 ± 0.75 [a]	6.7 ± 1.40 [ab]	6.8 ± 0.51 [ab]	7.5 ± 0.33 [b]
Ash	1.0 ± 0.35 [a]	1.3 ± 0.06 [a]	1.6 ± 0.63 [a]	1.3 ± 0.18 [a]	1.4 ± 0.05 [a]
Carbohydrate	58.4 ± 0.28 [a]	55.0 ± 0.60 [b]	54.2 ± 0.25 [b]	55.5 ± 0.10 [b]	54.2 ± 0.55 [b]
Sucrose	28.9 ± 0.66 [a]	20.4 ± 0.55 [b]	21.0 ± 0.55 [b]	21.0 ± 0.54 [b]	21.1 ± 0.41 [b]
Total sugars	29.8 ± 0.52 [a]	20.8 ± 0.41 [b]	21.4 ± 0.33 [b]	21.5 ± 0.58 [b]	21.4 ± 0.65 [b]
Dietary fibre	1.3 ± 0.57 [a]	2.0 ± 0.71 [b]	2.4 ± 0.22 [bc]	2.8 ± 0.46 [c]	3.3 ± 0.22 [cd]
Energy (Kcal/100 g)	374	357	335	317	287

[abcd] mean values ($\pm$standard deviation) in the same row bearing different superscripts are significantly different, $p < 0.05$.

Butter beans contain a significantly higher percentage of moisture (75%) compared to the butter used in this trial (15%), which can explain the increase in moisture with the substitution of butter for Pbb. As mentioned, sensory results obtained for moist texture did not correlate with instrumental results, which highlights the importance of butter content to the perception of lubricity and moist texture. Fat content remained the same after a 30% SR as expected, but it decreased significantly with each level of FR in the range of 12.6% for the SC30/0 sample to 4.5% for the SC30/75 sample ($p < 0.05$). As a 50% FR was permitted in sucrose-replaced sponge cake samples in relation to sensory acceptance, this equated to a total fat reduction of 42%. According to the standards set by the Food Safety

Authority (2016) [46], this reduction in fat content permits the claim "reduced fat". A 30% SR did not affect the protein content of samples. Trends show that protein content increased with each level of FR; however, a significant increase was only observed for the sample containing 75% FR ($p < 0.05$). The increase in protein content can be attributed to the protein content of Pbb. Carbohydrate content decreased significantly after a 30% replacement of sucrose for inulin and Reb A, as expected ($p < 0.05$). Carbohydrate content remained constant in samples with increasing levels of FR at roughly 55%. Sucrose content and total sugars also decreased significantly after a 30% SR, as expected ($p < 0.05$). Similar to results obtained for carbohydrate content, sucrose content and total sugars remained constant with increasing levels of FR, as expected. Although carbohydrate content was high in all samples with increasing levels of FR (55%), less than half of the carbohydrate was in the form of sucrose and total sugars. As a 30% SR was achieved in terms of sensory acceptance, this equated to a 28% reduction in total sugars. Dietary fibre increased significantly with the substitution of 30% sucrose for inulin and Reb A ($p < 0.05$), which can be attributed to the fibre present in inulin. Trends show that fibre content increased with each level of FR but only significantly so after a 50% FR with Pbb ($p < 0.05$). The optimised sample (SC30/50) contained 2.8 g/100 g of dietary fibre which was a total increase of 115% from the original sample, and a total energy reduction of 15% was achieved.

4. Conclusions

Texture liking and, subsequently, OA were negatively affected by a 50% SR using a combination of inulin and Reb A. However, other important hedonic attributes such as flavour liking and aroma liking were not negatively affected by this level of replacement. This demonstrates the value of these replacers in relation to maintaining flavour properties associated with liking. Further work to improve texture, such as the addition of emulsifiers, for example, polysorbates, or the addition of hydrocolloids, is necessary to permit higher levels of sucrose replacement using this combination of replacers.

Pureed butter beans were very successful in the replacement of fat by up to a level of 50% replacement. Flavour and aroma intensity attributes and, subsequently, liking and OA were negatively affected by a 75% FR. Perhaps the addition of flavourings to compensate for the reduction in flavour lost with butter would permit higher levels of fat replacement using pureed butter beans.

The present study achieved a simultaneous 50% FR in 30% sucrose-replaced sponge cake, using natural substitutes and without negatively affecting OA. Sponge cake samples contained 42% less fat and 28% less sucrose than the original sample. Thus, the criteria to meet the standard for a "reduced fat" claim were achieved. More work is necessary to permit further sucrose and fat replacement using these ingredients and without negatively affecting important quality parameters, in order to meet the criteria for reduced/low-sugar and low-fat health claims. Optimised samples contained 115% more dietary fibre than the original sample.

Supplementary Materials: The following are available online at https://www.mdpi.com/2304-815 8/10/2/254/s1, Table S1: Significance of estimated regression coefficients (ANOVA values) for the relationship of sensory & physicochemical parameters (Y) and reduced sugar Sponge cake samples prepared with increasing levels of fat replacement (X).

Author Contributions: A.A.T.: data curation, writing—review and editing, visualisation, resources, supervision. A.M.R.: conceptualisation, methodology, validation, formal analysis, investigation, resources, writing—original draft, visualisation, project administration. K.N.K.: project administration, funding acquisition. E.G.: project administration, funding acquisition. M.G.O.: validation, resources, supervision, writing—review and editing, project administration, funding acquisition. J.P.K.: resources, supervision, writing—review and editing, project administration, funding acquisition. All authors have read and agreed to the published version of the manuscript.

Funding: This research received funds from the Food Institutional Research Measure (FIRM), grant number 14F 812, administered by the Department of Agriculture, Food and the Marine.

Institutional Review Board Statement: Not applicable.

Informed Consent Statement: Not applicable.

Conflicts of Interest: The authors declare no conflict of interest.

References

1. Blundell, J.E.; Macdiarmid, J.I. Passive overconsumption. Fat intake and short-term energy balance. *Ann. N. Y. Acad. Sci.* **1997**, *827*, 392–407. [CrossRef] [PubMed]
2. World Health Organization. Guidelines. 2011. Available online: https://www.diabetes.ie/are-you-at-risk-free-diabetes-test/get-sugar-smart/ (accessed on 14 December 2020).
3. National Taskforce on Obesity. Obesity the Policy Challenges—The Report of the National Taskforce on Obesity. Available online: http://www.hse.ie/eng/health/child/healthyeating/taskforceonobesity.pdf (accessed on 5 May 2019).
4. Irish Universities Nutrition Alliance. National Adult Nutrition Survey (NANS) Summary Report. Available online: http://www.iuna.net/wp-content/uploads/2010/12/National-Adult-Nutrition-Survey-Summary-Report-March-2011.pdf (accessed on 1 April 2019).
5. Azaïs-braesco, V.; Sluik, D.; Maillot, M.; Kok, F.; Moreno, L.A. A review of total & added sugar intakes and dietary sources in Europe. *Nutr. J.* **2017**, *16*, 1–15. [CrossRef]
6. Bennion, E.B.; Bamford, G.S.T. Sugars. In *The Technology of Cake Making*, 6th ed.; Bent, A.J., Ed.; Springer Science + Business Media Dordrecht: Bristol, UK, 1997; pp. 84–99. Available online: http://10.1007/978-1-4757-6690-5 (accessed on 22 May 2019).
7. Wilderjans, E.; Luyts, A.; Brijs, K.; Delcour, J.A. Ingredient functionality in batter type cake making. *Trends Food Sci. Technol.* **2013**, *30*, 6–15. [CrossRef]
8. Ureta, M.M.; Olivera, D.F.; Salvadori, V.O. Baking of Sponge Cake: Experimental Characterization and Mathematical Modelling. *Food Bioprocess Technol.* **2016**, *9*, 664–674. [CrossRef]
9. Pyler, E.J.; Gorton, L.A. *Baking Science & Technology*, 14th ed.; Sosland Publishing Company: Kansas City, MI, USA, 2017.
10. O'Sullivan, M.G. A Handbook for Sensory and Consumer-Driven New Product Development. Technologies for the Food and Beverage Industry. In *Food Science, Technology and Nutrition*; Woodhead Publishing: Cambridge, UK, 2017; p. 370.
11. Struck, S.; Jaros, D.; Brennan, C.S.; Rohm, H. Sugar replacement in sweetened bakery goods. *Int. J. Food Sci. Tech.* **2014**, *49*, 1963–1976. [CrossRef]
12. Shoaib, M.; Shehzad, A.; Omar, M.; Rakha, A.; Raza, H.; Sharif, H.R.; Shakeel, A.; Ansari, A.; Niazi, S. Inulin: Properties, health benefits and food applications. *Carbohydr. Polym.* **2016**, *147*, 444–454. [CrossRef]
13. Roberfroid, M. Inulin-Type Fructans: Functional Food Ingredients. *J. Nutr.* **2005**, *137*, 2493–2502. [CrossRef]
14. Stephen, A.M.; Champ, M.M.; Cloran, S.J.; Fleith, M.; van Lieshout, L.; Mejborn, H.; Burley, V.J. Dietary fibre in Europe: Current state of knowledge on definitions, sources, recommendations, intakes and relationships to health. *Nutr. Res. Rev.* **2017**, *30*, 149–190. [CrossRef]
15. Meyer, D.; Bayarri, S.; Tárrega, A.; Costell, E. Inulin as texture modifier in dairy products. *Food Hydrocoll.* **2011**, *25*, 1881–1890. [CrossRef]
16. Rodríguez-García, J.; Salvador, A.; Hernando, I. Replacing Fat and Sugar with Inulin in Cakes: Bubble Size Distribution, Physical and Sensory Properties. *Food Bioprocess Technol.* **2014**, *7*, 964–974. [CrossRef]
17. Struck, S.; Gundel, L.; Zahn, S.; Rohm, H. Fiber enriched reduced sugar muffins made from ISO-viscous batters. *LWT Food Sci. Technol.* **2016**, *65*, 32–38. [CrossRef]
18. Kinghorn, A.D.; Wu, C.D.; Soejarto, D.D.; Compadre, C.M. Stevioside. *Alternative Sweeteners*, 3rd ed.; Nabors, L.O., Ed.; Dekker Marcel, Inc.: New York, NY, USA, 2001; pp. 167–183.
19. Jorge, K. Soft drinks. Chemical Composition. In *Encyclopedia of Food Sciences and Nutrition*, 2nd ed.; Elsevier Ltd.: London, UK, 2003; pp. 5346–5352. [CrossRef]
20. Izawa, K.; Amino, Y.; Kohmura, M.; Ueda, Y.; Kuroda, M. 4.16-Human–Environment Interactions—Taste. In *Comprehensive Natural Products II*; Liu, H.-W., Mander, L., Eds.; Elsevier: Amsterdam, The Netherlands, 2010; pp. 631–671.
21. Li, Y.; Li, Y.; Wang, Y.; Chen, L.; Yan, M.; Chen, K.; Xu, L.; Ouyang, P. Production of Rebaudioside A from Stevioside Catalyzed by the Engineered *Saccharomyces cerevisiae*. *Appl. Biochem. Biotechnol.* **2016**, *178*, 1586–1598. [CrossRef] [PubMed]
22. Marcus, J. *Lipids Basics: Fats and oils in Foods and Health: Healthy Lipid Choices, Roles and Applications in Nutrition, Culinary Nutrition: The Science and Practice of Healthy Cooking*; Elsevier: Amsterdam, The Netherlands, 2014; pp. 231–277.
23. Confortiv, F.D.; Charles, S.A.; Duncan, S.E. Evaluation of a carbohydrate-based Fat replacer in a fat reduced baking powder biscuit. *J. Food Qual.* **1997**, *20*, 247–256. [CrossRef]
24. Bazzano, L.A.; Ogden, H.J. Legume Consumption and Risk of Coronary Heart Disease in US Men and Women. *Arch. Intern. Med.* **2001**, *161*, 2573–2578. [CrossRef] [PubMed]
25. Uruakpa, F.O.; Fleisher, A.M. Sensory and Nutritional Attributes of Black Bean Brownies. *Am. J. Food Sci. Nutr.* **2016**, *3*, 27–36. Available online: https://pdfs.semanticscholar.org/3549/3d339792903d0a076577861e654496a5d064.pdf (accessed on 5 May 2019).
26. Rankin, L.L.; Bingham, M. Acceptability of oatmeal chocolate chip cookies prepared using pureed white beans as a fat ingredient substitute. *J. Am. Diet. Assoc.* **2000**, *100*, 831–833. [CrossRef]

27. Szafranski, M.; Whittington, J.A.; Bessinger, C. Pureed cannellini beans can be substituted for shortening in brownies. *J. Am. Diet. Assoc.* **2005**, *105*, 1295–1298. [CrossRef] [PubMed]
28. dos Silva Navarro, R.D.C.; Minim, V.P.R.; da Silva, A.N.; Gonçalves, A.C.A.; Carneiro, J.D.D.S.; Gomide, A.I.; Lucia, S.M.D.; Minim, L.A. Validation of Optimized Descriptive Profile (ODP) technique: Accuracy, precision and robustness. *Food Res. Int.* **2014**, *66*, 445–453. [CrossRef]
29. dos Santos Navarro, R.D.C.; Minim, V.P.R.; Simiqueli, A.A.; da Silva Moraes, L.E.; Gomide, A.I.; Minim, L.A. Optimized descriptive profile: A rapid methodology for sensory description. *Food Qual. Prefer.* **2012**, *24*, 190–200. [CrossRef]
30. dos Santos Navarro, R.D.C.; Minim, V.P.R.; Carneiro, J.D.D.S.; Nascimento, M.; Della Lucia, S.M.; Minim, L.A. Quantitative sensory description using the optimized descriptive profile: Comparison with conventional and alternative methods for evaluation of chocolate. *Food Qual. Prefer.* **2013**, *30*, 169–179. [CrossRef]
31. dos Santos Navarro, R.D.C.; Minim, V.P.R.; da Silva, A.N.; Peternelli, L.A.; Minim, L.A. Optimized descriptive profile: How many judges are necessary? *Food Qual. Prefer.* **2014**, *36*, 3–11. [CrossRef]
32. International Organisation for Standardisation. 11136:2014 Sensory Analysis-Methodology-General Guidance for Conducting Hedonic Tests with Consumers in a Controlled Area. Available online: https://www.iso.org/home.html (accessed on 15 May 2017).
33. MacFie, H.J.; Bratchell, N.; Greenhoff, K.; Vallis, L.V. Designs to balance the effect of order of presentation and first-order carry-over effects in hall tests. *J. Sens. Stud.* **1989**, *4*, 129–148. [CrossRef]
34. Stone, H.; Bleibaum, R.; Thomas, H.A. Test strategy and design of experiments. In *Sensory Evaluation Practices*, 4th ed.; Elsevier Academic Press: Cambridge, MA, USA, 2012; pp. 117–157.
35. Janjarasskul, T.; Tananuwong, K.; Kongpensook, V.; Tantratian, S.; Kokpol, S. Shelf life extension of sponge cake by active packaging as an alternative to direct addition of chemical preservatives. *LWT Food Sci. Technol.* **2005**, *72*, 166–174. [CrossRef]
36. Bostian, M.L.; Fish, D.L.; Webb, N.B.; Arey, J.J. Automated methods for determination of fat and moisture in meat and poultry products: Collaborative study. *J. Assoc. Off. Anal. Chem.* **1985**, *68*, 876–880. [CrossRef]
37. Jeddou, K.B.; Bouaziz, F.; Zouari-Ellouzi, S.; Chaari, F.; Ellouz-Chaabouni, S.; Ellouz-Ghorbel, R.; Nouri-Ellouz, O. Improvement of texture and sensory properties of cakes by addition of potato peel powder with high level of dietary fibre and protein. *Food Chem.* **2017**, *217*, 668–677. [CrossRef] [PubMed]
38. AOAC International. *Official Methods of Analysis of AOAC International*, 18th ed.; AOAC International: Gaithersburg, MD, USA, 2007.
39. Martens, H.; Martens, M. Modified Jack-knife estimation of parameter uncertainty in bilinear modelling by partial least squares regression (PLSR). *Food Qual. Prefer.* **2000**, *11*, 5–16. [CrossRef]
40. Volpini-Rapina, L.F.; Sokei, F.R.; Conti-Silva, A.C. Sensory profile and preference mapping of orange cakes with addition of prebiotics inulin and oligofructose. *LWT Food Sci. Technol.* **2012**, *48*, 37–42. [CrossRef]
41. Laguna, L.; Primo-Martín, C.; Varela, P.; Salvador, A.; Sanz, T. HPMC and inulin as fat replacers in biscuits: Sensory and instrumental evaluation. *LWT Food Sci. Technol.* **2014**, *56*, 494–501. [CrossRef]
42. Biguzzi, C.; Schlich, P.; Christine, L. The impact of sugar and fat reduction on perception and liking of biscuits. *Food Qual. Prefer.* **2014**, *35*, 41–47. [CrossRef]
43. Rios, R.V.; Pessanhai, M.D.F.; Almeida, P.F.D.; Viana, C.L.; Lannesi, S.C.D.S. Application of fats in some food products. *Food Sci. Technol.* **2014**, *34*, 3–15. [CrossRef]
44. O'Brien, C.M.; Mueller, A.; Scannell, A.G.M.; Arendt, E.K. Evaluation of the effects of fat replacers on the quality of wheat bread. *J. Food Eng.* **2013**, *56*, 265–267. [CrossRef]
45. Souza, N.C.O.; Oliveira, L.D.L.; Alencar, E.R.; Moreira, G.P.; Santos Leandro, E.; Ginani, V.C.; Zandonadi, R.P. Textural, physical and sensory impacts of the use of green banana puree to replace fat in reduced sugar pound cakes. *LWT Food Sci. Technol.* **2018**, *89*, 617–623. [CrossRef]
46. Food Safety Authority Ireland. *Information on Nutrition and Health Claims*; FSAI: Dublin, Ireland, 2006. Available online: https://www.fsai.ie/science_and_health/nutrition_and_health_claims.html (accessed on 15 December 2020).

MDPI
St. Alban-Anlage 66
4052 Basel
Switzerland
Tel. +41 61 683 77 34
Fax +41 61 302 89 18
www.mdpi.com

Foods Editorial Office
E-mail: foods@mdpi.com
www.mdpi.com/journal/foods